Communicatio Innovation

Rethinking Agricultural Extension

Third Edition

Cees Leeuwis

with contributions from

Anne van den Ban

Life Sci.

630.715
B22i
2004

D0902654

Blackwell
Science

CTA

First Dutch edition © Boom-Pers 1974, 1985
First English edition © A. W. van den Ban & H. S. Hawkins 1988
Second edition © 1996 by Blackwell Science Ltd
Third edition © 2004 by Blackwell Science Ltd
a Blackwell Publishing company

Editorial offices:
Blackwell Science Ltd, 9600 Garsington Road, Oxford OX4 2DQ, UK
 Tel: +44 (0) 1865 776868
Iowa State Press, a Blackwell Publishing Company, 2121 State Avenue, Ames,
Iowa 50014-8300, USA
 Tel: +1 515 292 0140
Blackwell Science Asia Pty Ltd, 550 Swanston Street, Carlton, Victoria 3053, Australia
 Tel: +61 (0)3 8359 1011

The right of the Authors to be identified as the Authors of this Work has been asserted in accordance
with the Copyright, Designs and Patents Act 1988.

All rights reserved. No part of this publication may be reproduced, stored in a retrieval
system, or transmitted, in any form or by any means, electronic, mechanical, photocopying,
recording or otherwise, except as permitted by the UK Copyright, Designs and Patents
Act 1988, without the prior permission of the publisher.

First published in Dutch by Boom-Pers 1974, 1985
Modified English edition co-published by Longman Scientific & Technical 1988 and John Wiley &
Sons, Inc. under the title *Agricultural Extension*
Second edition published by Blackwell Science 1996
Reprinted 1996, 1998, 1999
Third retitled edition published by Blackwell Publishing Ltd 2004

Library of Congress Cataloging-in-Publication Data
Leeuwis, Cees.
  Communication for rural innovation : rethinking agricultural extension/Cees Leeuwis,
  with Anne van den Ban.
    p. cm.
  Includes bibliographical references and index.
  ISBN 0–632–05249–X (softcover: alk. paper)
  1. Agricultural extension work. I. Leeuwis, Cees. II. Ban, A. W. van den.
  III. Technical Centre for Agricultural and Rural Cooperation (Ede, Netherlands)
  IV. Title.

S544.L35 2003
630′.71′5—dc22
2003058372

ISBN 0–632–05249–X

A catalogue record for this title is available from the British Library

Set in 10/12.5 pt Times
by Graphicraft Limited, Hong Kong
Printed and bound in India
by Replika Press Pvt. Ltd, Kundli

The publisher's policy is to use permanent paper from mills that operate a sustainable forestry policy,
and which has been manufactured from pulp processed using acid-free and elementary chlorine-free
practices. Furthermore, the publisher ensures that the text paper and cover board used have met
acceptable environmental accreditation standards.

For further information on Blackwell Publishing, visit our website:
www.blackwellpublishing.com

Contents

Preface

This book provides a follow-up for Van den Ban and Hawkins' classic *Agricultural Extension* (1988, 1996), of which some 35 000 copies have been printed in 10 languages[1]. It does so in several ways.

First, the book attemps to catch up with recent thinking about the relationship between communication and change. The origins of Van den Ban and Hawkins' book can be traced back to the 1970s, which was the period in which the first (Dutch) edition of the book was compiled (Van den Ban, 1974). Since then, the practice and theory of extension and development communication have changed fundamentally. Although many efforts were made to incorporate new ideas into later editions, we feel that it is now time for a totally new book as we can no longer do justice to the changes in extension thinking by merely adding to or adapting a pre-existing text. In this new book we have maintained and adapted those insights and conceptual models which are still of value today, but at the same time we have incorporated a variety of new ideas, angles and modes of thinking, some of which derive from disciplines that did not feature much in extension discourses of the past. The product of our efforts, we hope, is a book that is ready for the 21st century, and will help to shape and inspire new forms of communicative intervention.

Secondly, the new book provides a follow-up in that it aims at a slightly different audience from the book *Agricultural Extension*. The original book was primarily aimed at practitioners in classical agricultural extension organisations. However, since the 1980s, the landscape of organisations that apply communicative strategies to foster change and development in agriculture and resource management has become much more varied. In this context, we want this new volume to offer inspiration to communication professionals who would never think of themselves as 'extensionists'. Moreover, since the 1980s, a large number of practical handbooks have been published on extension, development communication, participation, etc. (e.g. Blackburn, 1994; Pretty et al., 1995; Swanson et al., 1997). We do not want to repeat what is already widely available. Thus, in this book we tend to discuss methods and methodological issues in the context of wider conceptual debates. We pay relatively more attention to novel (e.g. internet-based) methods and to new ideas regarding the management of interactive processes. In conclusion, this book is aimed in particular at those who function in the higher echelons of public, private and non-governmental organisations that use communication in order to facilitate change in agriculture and resource management. Here we think, for example, of process facilitators, communication division staff, knowledge managers, training officers, consultants, policy makers, change managers and – last but not least – extension (and research!) managers or

---

[1] Of the earlier Dutch and German versions, an additional 30 000 copies were printed. Some translations have been edited instead of being translated literally (e.g. the French book by Van den Ban et al., 1994).

specialists at district, province and national level. At the same time, the book can be used as an advanced introduction into issues of communicative intervention for BSc or MSc students. Indeed, at our own university the book is used on the international MSc programmes Management of Agro-ecological Knowledge and Social Change (MAKS) and International Development Studies.

Finally, this book originates from the Communication and Innovation Studies group at Wageningen University, which was founded originally by Van den Ban in 1964 as the Department of Extension Education. Thus, the book fits a particular tradition of thinking about communication and change, a tradition that was started by Professor Van den Ban and later continued by his successors Niels Röling, Cees Van Woerkum and their academic staff. Both successors published introductions to communication and innovation studies (Röling, 1988; Röling et al., 1994; Van Woerkum & Van Meegeren, 1999; Van Woerkum et al., 1999), mostly in Dutch. Therefore, it was felt that it was high time for a new English language overview of our field of study. Clearly, the present book has benefited greatly from the insights and ideas of a range of scholars that work or have worked in and around the Communication and Innovation Studies group, and from the contribution of H.S. Hawkins to the previous book. In that sense, it is very much a collective achievement.

Leeuwis is greatly indebted to Van den Ban for several reasons. Apart from the numerous useful intellectual inputs, Van den Ban also provided the opportunity to write a follow-up book in the first place, and gave Leeuwis the space to make it to a large degree 'his own thing'.

The CTA

The Technical Centre for Agricultural and Rural Cooperation (CTA) was established in 1983 under the Lomé Convention between the ACP (African, Caribbean and Pacific) Group of States and the European Union Member States. Since 2000, it has operated within the framework of the ACP-EC Cotonou Agreement.

CTA's tasks are to develop and provide services that improve access to information for agricultural and rural development, and to strengthen the capacity of ACP countries to produce, acquire, exchange and utilise information in this area. CTA's programmes are designed to: provide a wide range of information products and services and enhance awareness of relevant information sources; promote the integrated use of appropriate communication channels and intensify contacts and information exchange (particularly intra-ACP); and develop ACP capacity to generate and manage agricultural information and to formulate ICM strategies, including those relevant to science and technology. CTA's work incorporates new developments in methodologies and cross-cutting issues such as gender and social capital.

CTA
Postbus 380
6700 AJ Wageningen
The Netherlands

PART 1
Rethinking extension

In the first four chapters of this book we set out to put into context the concept and societal role of what was previously labelled 'agricultural extension'. In Chapter 1 we outline the main challenges that agriculture is facing today and the implications this has for communicative intervention practice. This is followed by a discussion of the changing ideas regarding agricultural extension at the conceptual level (Chapter 2). We explain the evolution of the concept of 'extension' into the notion of 'communication for innovation'. The political and ethical dimensions of communication for innovation are discussed in Chapter 3, while two basic approaches to communicative intervention (the instrumental and the interactive approach) are discussed Chapter 4. In the subsequent chapters of this book we will further explore the details and implications of changing views on agricultural extension.

1 Introduction

As the problems and challenges faced by agricultural sectors change over time, we will have to adapt our ideas about the role and meaning of 'agricultural extension'. In this introductory chapter we outline some of the challenges that agricultural extension is facing, and point to the need to reinvent agricultural extension as a professional practice. The final section of this chapter provides a more detailed overview of the set-up and contents of the book.

1.1 Challenges for agricultural extension practice

The challenges to agricultural extension in the early 21st century derive, on the one hand, from the challenges that farmers and agriculture face in view of their ever-changing social and natural environment, and, on the other hand, from the changes that emerge within extension organisations themselves in connection with, for example, new funding arrangements, developments in extension theory, and the emergence of new computer-based communication technologies. Different people may have different ideas about what is a challenge for agricultural sectors and/or extension. Thus, the challenges we present are neither complete nor an absolute truth; they are open for debate. Moreover, challenges can often be associated with threats. Our use of the word 'challenge' is deliberate, because we feel it is often more productive to deal with problems and threats in a pro-active way, rather than to run away from them or go into a defensive mode.

1.1.1 Challenges for farmers and agriculture at large

Some of the challenges that face today's agriculture have been with us for a long time, while others have arisen more recently. We will briefly discuss them more or less in order of age.

Food production, food security and intensification

Although the overall world food situation has improved, there are still 800 million people who are chronically undernourished (Zijp, 1998). Improving food security is a challenge which is not simply about producing more food, as many of the causes of food insecurity relate to insufficient *access* to available food, insufficient economic development outside agriculture, bad governance, detrimental trade relations, debt crisis, inadequate functioning of agricultural institutions, etc. (see Koning et al., 2002). Nevertheless, sufficient food production remains an important condition for alleviating food insecurity. Moreover, the demand for food is likely to increase significantly in the near future, as the world population is still growing, and also since higher incomes in many countries result in greater food consumption. Much of the increased food production will have to be realised on land that is already under cultivation, as the availability of new land suitable for agricultural production is limited. Similarly, the scope for expanding irrigated agriculture is constrained due to increased competition for water.

This means that intensification will have to be realised in diverse and risk-prone rain-fed areas, for which available Green Revolution technologies have proved to be largely ill-suited in technical and/or social-organisational terms (Chambers & Gildyal, 1985; Lipton, 1989; Reijntjes et al., 1999). This failure of Green Revolution technologies indicates that agriculture may have to look for routes of 'intensification' other than through the intensive use of external inputs (chemical fertilisers, high yielding varieties, pesticides, machinery, etc.) in mono-cropping systems (Reijntjes et al., 1999). We may, for example, look for forms of 'intensification' that are labour intensive and make use of more complex cropping systems, based on locally adapted knowledge (Van der Ploeg, 1999). It is important, however, to realise that we have learned from the past that no generally applicable agricultural development model exists. What is important is that agricultural systems are flexibly adapted to their environment, and this does not coincide with dogmatic views of what agriculture should look like. In any case, increasing agricultural production through the development and stimulation of technical and/or organisational innovations remains an important concern for agriculture and agricultural extension.

Poverty alleviation, income generation and future prospects

According to estimations by the World Bank (World Bank, 1997; Zijp, 1998) there are some 1000 million economically active people worldwide whose livelihood depends at least in part on subsistence and/or commercial farming. The majority of these have incomes of less than one US dollar a day. For the improvement of their livelihood these people depend directly or indirectly on agricultural development – directly, in the sense that agricultural development may allow them to have a higher income from farming, and indirectly since agricultural development is widely regarded as an important trigger and condition for non-agricultural economic growth (IFAD, 2001; Koning et al., 2002). Contributing to agricultural development, therefore, remains an important challenge. From the perspective of poverty alleviation too, farming that relies on high external input does not seem to be the most feasible development model for many of the rural poor, as it is notoriously difficult for them to acquire necessary inputs.

A problem with many forms of agricultural development is that they usually imply that the same amount of produce can be produced with fewer people, which means that levels of employment in agriculture tend to come under pressure (see also Chapter 3 and Van den Ban, 2002). Where no alternatives exist outside agriculture, greater prosperity for some may mean increased poverty for others. Where prosperity in cities is growing and access to markets in other countries improving (see below), this effect may be ameliorated by possibilities for small farmers to venture into new high value-added and labour intensive agricultural products such as fruit, vegetables, flowers and processed food. However, such products are often more risky than staple food crops, and frequently require specialised knowledge and skills. Moreover, marketing channels are usually not readily available, while international competition can be fierce. Nevertheless, it can be worth exploring and supporting such options, not least to maintain labour and income opportunities in agriculture.

At the same time, it is perhaps significant to note that many of the rural poor see little future in agriculture (Farrington et al., 2002), and would like to see their children get a good education and not become farmers. In some regions agriculture is increasingly looked upon as a 'last resort' activity (e.g. Khamis, 1998), i.e. as something one does if everything else fails. With this cultural outlook on agriculture it will be difficult to meet any of the challenges mentioned in this chapter because few may be willing to make a real investment, and many capable people may prefer to leave the sector. Thus, an additional challenge for agricultural sectors may in some cases be to improve its own credibility and image as a promising and valuable economic sector.

Sustainability, ecosystems and natural resource management

Across the world, agriculture has been criticised sharply for its detrimental effects on the natural environment and the world ecology at large. Soil degradation, erosion, water pollution, excessive use of chemicals, waste of water, decreasing ground water tables, destruction of natural habitats for wildlife, and limited animal welfare are just a few of the concerns raised by environmentalists, ecologists, nature conservationists and the public at large. This had led to a call for agriculture to become less exploitative and more 'sustainable', which means that agriculture will have to be carried out to make the best use of available natural resources and inputs, and regenerate conditions for future production (e.g. soil fertility, resilience of the ecosystem, water availability). There are different schools of thought on the precise technical, social, economic and ethical criteria and characteristics that should be used to assess and describe the 'sustainability' of agriculture. For some, sustainable agriculture means agriculture with low external input, while others argue that this kind of agriculture is unsustainable since it requires a large increase in cultivated area, and that the use of fertilisers and high-tech machinery can also be sustainable. Various labels have been coined, such as integrated agriculture, ecological agriculture, organic agriculture, biological agriculture, permaculture, precision agriculture, etc. Regardless of one's convictions, sustainable agriculture and natural resource management represent important challenges for primary agriculture, agro-industries and service institutions.

As several authors have pointed out (e.g. Berkes & Folke, 1998; Röling & Wagemakers, 1998) 'sustainability' cannot just be looked at in biophysical or ecological terms, because the state of 'hard systems' depends crucially on interactions between multiple human beings (i.e. on the 'soft system'). The hydrological state in a water catchment area, for example, cannot be understood properly in hydrological terms only, i.e. without taking into account the practices of various water users. This is because hydrological processes and their outcomes are shaped and influenced by the way farmers irrigate their land, use stream banks, make wells, plough their land, manage contours, choose crops, etc. This in turn depends partly on wider social–organisational circumstances, such as water laws and regulations, the way markets for different agricultural products are organised, population pressure, the functioning of agricultural service institutions, etc. Thus, when one wishes to improve, from a sustainability point of view, the management of water in a catchment area, one will essentially have to foster new agreements, modes of co-ordination and/or forms of organisation among farmers themselves, and between farmers and other societal stakeholders, including other water users (e.g. industries, urban communities).

The example on water catchment management indicates that the management of natural resources often transcends community and regional boundaries. It is argued by many that some of the environmental problems faced by the world (e.g. climate change, water shortage) can only be solved if co-ordination is achieved on a trans-national or even global level. And to further complicate the matter, it is sub-optimal – at least from an ecological perspective – to manage the use of different natural resources (e.g. water, biodiversity, energy, etc.) separately, because ecological cycles are closely intertwined. Even if local or regional stakeholders in agriculture are often not in the position to foster co-ordination at such a scale and level of complexity, global ecological issues may well have local and regional implications for farmers and others if one accepts the motto 'think globally, act locally'.

At the local level, then, an additional challenge is posed by the experience that sustainable agriculture requires different types of agricultural knowledge from that previously developed by research institutes and disseminated by extension organisations (Röling & Van de Fliert, 1994; Röling & Jiggins, 1998). When we limit ourselves to 'agro-technical' knowledge, three issues arise. Although there may be disagreement over the precise meaning of the term 'sustainable', it is self-evident that such types of agriculture require farmers to manage and co-ordinate ecological processes and cycles carefully. In crop-protection, for example, it is no longer sufficient merely to apply a number of preventive sprayings according to a standard recipe. Instead, a balance must be maintained between pests and their natural predators, and keeping the ecosystems in which the latter exist. The management of this kind of balance requires that farmers have a good insight into complex ecological processes and interconnections, and at the same time that they can anticipate the inherent unpredictability of such systems (Holling, 1985, 1995).

A second feature that seems to be important is that, especially with low external input, sustainable agrarian practices will probably need to be more varied than conventional practices. The crop rotations of biological farmers, for example, involve a greater number of crops, and a certain amount of integration with stock grazing would seem an obvious step. This relative 'de-specialisation' means that farmers need to be conversant with a broad spectrum of knowledge. Lastly, ecological processes and situations are by nature locally specific since important differences can exist within individual regions or even individual fields. Awareness of the local situation is therefore essential. In short, the nature of the requisite knowledge could be described as *complex*, *diverse* and *local*. Much of this knowledge is not readily available and needs to be developed and/or adapted 'on the spot' with close co-operation between farmers, researchers and extensionists.

In summary, it can be argued that, if agricultural branches are to become more sustainable, farmers and other stakeholders will – more than in the past – have to take into account and link inherently complex knowledge regarding both *global* and *local* processes and circumstances. The emergence of new practices and forms of co-ordination depends in essence on joint learning and negotiation between stakeholders (Daniels & Walker, 1996). As discussed in section 1.1.2, this may require different forms of extension practice from the modes of operation we have seen in the past.

Globalisation and market liberalisation

Due to huge changes in communication and transport technologies, the exchange of goods, people and ideas has become much easier and more widespread than before. Even the most remote rural areas often have numerous direct or indirect connections with the wider world economy. Moreover, under the influence of World Trade Organisation (WTO) agreements, and World Bank and International Monetary Fund (IMF) policies, as well as national policies, this world economy becomes increasingly organised according to the principles of the 'free market'. Many economists regard the free market as the most efficient means to allocate scarce resources. And even where – according to neo-classical economic theory itself – the conditions for such a market to operate effectively are not provided (e.g. perfect competition and perfect information), we witness attempts being made to create a free market and/or to create appropriate conditions for it. Although one can legitimately question the effectiveness, morality, political implications and cultural connotations of the current free market ideology, we cannot ignore the consequences for agriculture, especially non-subsistence. The emerging world market provides both constraints and opportunities for agriculture. The gradual removal of trade barriers and agricultural protection systems may allow producers in, for example, Africa to venture into new products (e.g. flowers, labour-intensive crops) that can be exported to industrialised countries, but it may also effectively wipe out agricultural branches (e.g. maize or milk production for local markets) where products can be imported more cheaply.

In connection with this, it is worrying that huge differences exist in labour productivity between industrialised and non-industrialised countries, and that these differences are increasing rather than decreasing. According to the World Bank (2000), many industrialised countries have a labour productivity in agriculture that is 50 to 100 times higher than non-industrialised countries. Of course, there are also enormous differences in costs (e.g. in terms of land, equipment, inputs, etc.), and the quality of the World Bank data may well be contested, but the threat is real that important agricultural products may increasingly be produced more cheaply in industrialised countries. In addition, these countries spend over 70 times as much on income support for their own farmers than on development assistance (IFAD, 2001). This restricts the opportunities for non-industrialised countries to export their agricultural products. In this regard, market liberalisation is rhetoric, not reality. Where and when market liberalisation progresses, regions will have to increasingly adapt their market-oriented agricultural systems according to their competitive potential vis-à-vis other regions of the world. This implies that there is an increased need to use and collect information on opportunities and consumer demands elsewhere in the world, and on the developments that take place in competing regions. If, alternatively, regions wish to escape from the pressures of the world market, they will have to deliberately maintain and/or establish protection and/or non-market arrangements (e.g. in the form of contracts, family or tribal networks, joint ventures, etc.) between agricultural producers, processing companies and consumers. This too provides important challenges, as it tends to run counter to the dominant economic regime. In any case, as farmers are often relatively weakly represented in debates on world market arrangements, a final challenge here may be to stimulate and strengthen new forms of farmer organisation at various levels (local, national and international) in order to have a greater leverage vis-à-vis other market parties.

Multi-functional agriculture

In connection with the societal debates on environmental issues and sustainability, it has been realised that agriculture has, or can have, many more functions than producing food and non-food plant or animal products. Farmers may or may not 'produce' clean air, a beautiful landscape, biodiversity, attractive space for recreation, clean water, a healthy soil, animal welfare. In other words, there can be many things that farmers 'produce' for which they are not directly rewarded in financial terms. Of course, it is in the interest of farmers themselves to maintain a healthy soil and clean water, and several governments have introduced laws and licensing systems to prevent environmental degradation. Thus, some of these 'products' can be regarded as something that farmers need to deliver 'free of charge'. However, when farmers are functioning in a liberalised world market, it cannot be taken for granted that they preserve the landscape, maintain recreational spaces and improve animal welfare if their immediate competitors elsewhere in the world are not required to take such often production-limiting measures as well. Mainly in rich industrial societies where citizens would like farmers to maintain such landscape, recreational and/or natural values, we see the introduction of new arrangements (e.g. a nature conservation contract, certified value-added marketing chains for 'nature friendly' food) through which farmers can be rewarded financially for the provision of non-agricultural functions. Even apart from an ecological merit, such reward systems for 'multi-functional agriculture' are in some countries rapidly becoming an economic prerequisite for the survival of the agricultural sector.

The Netherlands, for example, is a small and densely populated country, in which space, nature and land are extremely scarce. Although agriculture, in the narrow sense, is technically advanced and highly productive, it is increasingly becoming non-viable since the costs for acquiring or even inheriting land and production rights are much higher than in nearby surrounding countries (e.g. Eastern Europe), while the same is true for variable costs such as labour. Thus, in order to make agriculture survive, government bodies and farmers are looking for new value-added products, including non-agricultural ones such as recreational services, nature conservation and even agro-health care services. Hence, developing suitable arrangements for multi-functional agriculture is a challenge that an increasing number of regions in the world will have to face in view of ecological and/or economic pressures. This challenge includes the need for the agricultural sector to establish effective communication and co-operation with other actors in society, such as one-issue action groups and non-agricultural sectors. As we have learned in the Netherlands this is not always easy, as some of these parties have come to look upon each other as 'enemies' with competing interests in a 'struggle' over land-use (e.g. Aarts & Van Woerkum, 1999).

Agrarian reform

In different parts of the world we witness different types of agrarian reform. In many industrialised countries farm sizes have steadily increased while the numbers of farms has dropped significantly. This trend was facilitated by technological developments and agro-economic policies, and seems to continue in view of market liberalisation and efforts to reduce over-production. In the former communist countries of Eastern Europe very large co-operative and state farms are being divided into smaller landholdings, with former workers becoming farm managers. In parts of Southern Africa too large commercial farms are being redistributed into smaller farms for people with insufficient land, or new ownership arrangements are being forged between large scale commercial farmers and former employees. Each of these situations has its own history, and produces specific problems. In the Netherlands, for example, many retrenching farmers have emotional problems in giving up farming, while in both Russia and Southern Africa it is difficult to establish adequate agricultural infrastructures for redistributed farms. Moreover, Russian officials complain that farmers find it difficult to take up the culture of enterpreneurship, while in Southern Africa it often proves difficult to overcome animosity between different racial communities. Although the challenges posed by policy-induced changes in the agrarian structure vary across regions, it is important that they are tackled.

Food safety and chain management

Increasingly, urban consumers of food products are concerned about the safety of the food they consume. The shops and markets in our globalising economy can be full of vegetables, processed food and meat that were produced in far-away places. Similarly, the feed and fodder on which animals were raised before being slaughtered can originate from across the world. In recent decades we have witnessed several food scares, when more and less serious problems emerged with food. Cattle in Europe were given feed compounds that contained bone material from diseased sheep, and developed a dangerous disease called BSE which may be transferred to human beings when they eat certain parts of infected animals. Similarly, contaminated oils and fats were fed to chickens or added to olive oil, which caused health hazards. Other horror stories revolve around illegal use of growth hormones for meat production, and residues of pesticides and other toxic components in vegetables and milk products. Similarly, many consumers, rightly or wrongly, worry about the consequences of consuming food that has been prepared on the basis of genetically modified organisms. In view of such experiences, a significant number of consumers have lost trust in food production chains. Basically, they fear that anonymous primary producers, food processors, animal feed industries, etc. may be more concerned with earning money than with the health of consumers (and/or other values they care about, such as animal welfare, the environment, etc.). Consumer organisations and large retailers call for better guarantees and transparency in food production chains. This is often put into effect by 'integrated chain management'. This basically means that all major steps and transactions in the food production chain are monitored and made traceable, so that if, for example, a contamination problem is discovered, its origin can be analysed, as well as who is responsible, and where other contaminated food products or raw materials may be found. The establishment of effective control systems is far from easy, and will still take considerable time. However, even today many large retailers and food processing companies are only willing to buy primary agricultural products when farmers can give guarantees concerning the production methods and inputs used. Meeting such increasingly stringent criteria is a challenge for farmers and the agricultural sector as a whole.

Knowledge intensity, knowledge society and commoditisation of knowledge

Many of the challenges mentioned above can only be tackled if the agricultural sector develops and uses more sophisticated and better adapted knowledge and information (e.g. on localised agro-ecological processes, market developments, risks, etc.). This derives in part from the nature of the required innovations (intensification in fragile rain-fed areas, sustainability, keeping in tune with the world market, etc.), but also from a more general development in society whereby knowledge becomes an increasingly significant economic production and growth factor (World Bank, 1998; FAO & World Bank, 2000; Little et al., 2002). The basic idea is that the competitive advantage of companies and sectors increasingly depends on the quality and timely use of the knowledge and ideas of those who work in it, rather than on, for example, the relative advantages with regard to labour costs. The Dutch glasshouse horticultural sector is an example. When compared to, for example, Southern Europe or sub-tropical areas, the sector is characterised by very high energy costs (heating systems), high investments (glasshouses), and high labour costs. Despite these disadvantages the sector has so far remained competitive internationally, mainly because growers invest a lot in the generation, mutual exchange and application of new insights.

According to the World Bank (1998), knowledge production in society is accelerating, while at the same time the accessibility of such knowledge tends to improve in view of rapid developments in information and communication technologies (i.e. the internet), at least for those who are 'connected' and have sufficient resources. This latter reservation on the issue of access is an important one, especially since, in view of its economic importance and potential, applied knowledge is more and more regarded as a marketable product for which a price needs to be paid. This is a development which does not only affect farmers (who increasingly have to pay for extension services), but all parties in the knowledge network. Free exchange of knowledge and information between (and even within) fundamental research, applied research, extension and farmers becomes less and less self-evident. For agricultural sectors across the world it is a challenge to keep in touch with, contribute to, and/or catch up with the rapid developments in knowledge, science and technology. It has to be achieved in circumstances where access to relevant knowledge becomes easier from a technological point of view, but perhaps more difficult in financial terms.

1.1.2 Reinventing extension

When challenges change, the organisations which are supposed to support farmers in dealing with them will have to change as well. Besides, there are internal challenges that extension organisations will have to meet if they wish to play a role in the future. Taken together, it can be argued that agricultural extension will have to be reinvented as a professional practice; that is, it will have to significantly adapt its mission, rationale, mode of operation, management and organisational structure (see Chapter 16). This will have to be accompanied by conceptual changes regarding agricultural extension; these are discussed in Chapter 2 and subsequent chapters. Below, we indicate briefly some of the practical changes that may be required; they too will be discussed in more detail in other parts of the book.

Dealing with collective issues

In the past, extension and extension theory have focused on supporting individual farm management, and on the promotion of farm-level innovations. However, when we look at the challenges of today, many of these transcend the level of individual farms or farm households. Issues like the management of collective natural resources, chain management, collective input supply and marketing, organisation building, multi-functional agriculture and venturing into new markets typically require new forms of *co-ordinated action* and *co-operation* among farmers, and between farmers and other stakeholders. In addition, we have learned from the past that the successful application of most farm-level innovations is also often dependent on factors that transcend the farm level (e.g. input supply, marketing, community support, transport, processing). In other words, many innovations have been mistakenly looked at as being individual in nature. The conclusion that most of the innovations needed in present day agriculture have collective dimensions (i.e. they require new forms of interaction, organisation and agreement between multiple actors) has important implications for extension practice and extension theory. In the past, much of the thinking about extension has, for example, revolved around individual decision-making and adoption processes (see Chapter 8). Clearly, a greater emphasis on collective processes would require that we pay more attention to issues like dealing with diverging interests, different actor perspectives, and conflicts, and hence shift our attention to processes like conflict resolution, organisation building, social learning and negotiation. This shift in emphasis requires us to rethink what extension is all about, and what type of people and organisations we need for it.

Co-designing rather than disseminating innovations

The tendency among extension organisations to promote indiscriminately badly adapted and pre-defined innovations, many of which were developed by researchers with little understanding of farmers' problems and priorities, has been documented and criticised widely (Van der Ploeg, 1987; Röling, 1988; Leeuwis, 1989; Van Veldhuizen et al., 1997). In view of the challenges presented in the previous section, the idea of selling pre-defined packages becomes perhaps even less appropriate.

No simple and ready-made solutions exist for the intensification of rain-fed agriculture under highly diverse social and natural conditions. Similarly, sustainable agriculture typically requires relatively complex solutions that are carefully adapted to *local* agro-ecological and social conditions, and hence must be tailor-made. Moreover, no blueprints exist for the types of collective innovations described in the previous paragraph; such innovations can only grow and emerge out of the interactions between various stakeholders.

All this implies that extension needs to play a more active role in *processes of innovation design and adaptation*. Such a role may, for example, entail the shaping, organisation and facilitation of innovation processes (i.e. process management), and/ or the making of 'translations' between farmers' and external researchers' views and concerns. Of course, whenever promising locally adapted innovations have emerged, extension staff may consider how these may get a wider application without having to repeat the innovation process from point zero. However, experience has shown that locally developed innovations and knowledge cannot be transferred through conventional transfer of technology approaches (see Röling & Van de Fliert, 1994; Van Schoubroeck, 1999). Rather, the 'scaling-up' of tailor-made innovations to different contexts and people will always have to include elements of redesign, encompassing new processes of learning and negotiation, and hence should not be looked at merely as 'dissemination'. Clearly, playing a role in innovation design and process management towards innovation would mark a break-away from traditional forms of extension.

Matching the technical and social dimensions of an innovation

In order to contribute to innovation processes, it is important that extension organisations have a clear idea of what exactly constitutes an 'innovation' and what kind of process is needed to arrive at it. Many scientists regard an 'innovation' as a new technical product or procedure that is created in a research facility; in line with this, innovation processes are primarily associated with 'doing research'. However, it is well known that many of the new ideas, products and processes developed in laboratories and the like never reach the stage of being applied in everyday practice (Little et al., 2002). For purposes of extension, therefore, we need a more pragmatic conception of an 'innovation'. Following Roep (2000), we propose to define an innovation more pragmatically, in terms of its successful application. From this perspective, then, an innovation needs to be understood as a 'novel working whole' (Roep refers to a 'reordered working whole'). In other words, it may be 'a new way of doing things' or even 'doing new things', but it can only be considered an innovation *if it actually works* in everyday practice. Looking at an innovation in this way helps us to understand that an innovation is not only composed of novel technical devices or procedures, but also of new or adapted human practices, including the conditions for such practices to happen.

A good example is provided by Van Schoubroeck (1999) who describes how scientists and farmers discovered that the Chinese citrus fly in Bhutan could in principle be combated by means of splashing poisonous bait into mandarin tree canopies, rather than spraying huge amounts of pesticide. However, the adequate functioning of this technique required that communities developed a range of social–organisational arrangements. These included the design and application of monitoring routines to determine the timing of bait splashings, procedural arrangements for the collective preparation of bait and the sharing of costs, and organisational arrangements to ensure that all community fields were treated at the same time. Strikingly, different communities of mandarin growers had varying capabilities to organise themselves to this end, and hence required different technical devices to combat the Chinese citrus fly. This example shows that we can only speak of a complete innovation if there exists an appropriate mix and balance between new *technical* devices and novel *social–organisational* arrangements. Thus, innovations have a technical and a social dimension, and contributing to innovation means that one needs to work on both dimensions simultaneously. Furthermore, the example indicates once more that innovations have a collective dimension in that they require co-ordinated action between different actors. Hence, apart from 'doing (joint) research', working on innovation has a lot to do with *creating support networks* and negotiating new arrangements between various stakeholders. As indicated earlier, this may require new tasks, skills and activities by extension organisations.

Catering for diverse farming and livelihood strategies

For a long time economists as well as extensionists have assumed, implicitly or explicitly, that agricultural development is something that progresses in one particular direction (e.g. towards high input, high output, high-tech farming). The idea was that given certain conditions there is basically *one* optimal way of managing a farm. Much used categorisations of farmers such as 'vanguard farms', 'followers', 'early adopters', 'late adopters' and 'laggards' (Rogers, 1983) reflect this idea, namely that everybody is (or should be) moving in the same direction, even if some may do so more quickly than others. In recent years, many studies have indicated that this idea is flawed. It transpires that farms that are initially characterised by comparable layouts and household composition, and which operate under very similar conditions, can still develop along different, economically viable paths (Bolhuis & Van der Ploeg, 1985; Leeuwis, 1989, 1993; Van der Ploeg, 1990). Key factors in explaining such different patterns of farm development are the diverse strategies and aspirations that farmers may have regarding their social and natural environment, as well as variations in the way they organise their livelihoods and in the role agriculture plays with respect to non-agricultural activities[1]. Some farmers may, for example, prefer to organise their farms relatively autonomously (i.e. independent from input markets), while others do not mind buying in external inputs. Similarly, some enterpreneurial farmers like to operate on a large scale resulting in bulk production, while others capitalise on their craftsmanship and engage in smaller scale, quality production.

When implemented properly, different strategies may yield positive results. In the past, this kind of diversity has often not been properly valued by extensionists, who regularly preferred one particular model of farm development (Leeuwis, 1989; Roep

---

[1] Many studies on livelihood strategies show that a substantial part of the income generated by rural households stems from non-agricultural sources (Hebinck & Ruben, 1998; Ellis, 2000; IFAD, 2001).

et al., 1991). A study in Ireland, for example, showed that extensionists regarded farmers who did not develop along the lines proposed by the extension organisation as backward and stagnant, while closer investigation reveiled that these 'laggard' farmers had adopted similar numbers of innovations – albeit different ones – when compared with those who followed extension advice (Leeuwis, 1989). Rather than being 'less innovative' or 'stagnant' they showed a *different* dynamism, which was not recognised (and perhaps deliberately ignored) by extensionists. More than in the past, extension organisations will have to anticipate diversity among farmers, which means that they have to be able to give different advice to different people, and treat diversity as a resource rather than as a burden.

Managing complexity, conflict and unpredictability

In the past, extension and other development oriented organisations have often looked at change and innovation as something that could and should be *planned* (e.g. Havelock, 1973). The idea was – and for many still is – that it makes sense to formulate goals in advance, and that it is possible to then organise a rational process that eventually results in achieving the desired outcomes. Implicit in such conceptions is the assumption that people and developments in society can in principle be predicted and steered, if only there is adequate knowledge about the causes and effects of societal problems. In line with this, a similar trust existed in the predictability and controllability of technical and natural processes. Typically, many styles and methods for project-planning reflect this kind of control-oriented thinking. However, in recent decades we have learned that human beings often act in unexpected ways and that interactions between people have a dynamic of their own which cannot be predicted. Moreover, human actions and interventions may well have unintended consequences, and one often has to deal with unanticipated developments outside interventions.

In a similar vein, technical and agro-ecological processes do not always 'behave' in expected ways, and at the point where the social and the technical meet, many unforeseen developments may take place (Holling, 1995). Not surprisingly then, many projects have never realised their original objectives. It has been shown that projects were hampered by the fact that fixed objectives, activities and budget allocations were formulated in advance, which made it difficult to incorporate later developments, insights and priorities (Leeuwis, 1993). Moreover, it is striking that when one analyses the history of positive developments and innovations, one often finds that these have not resulted from formal planning. Rather, 'unplannable' phenomena like accidental discoveries, coincidence, informal networking, creativity, enthusiasm and 'personal chemistry' played a major role. In all, we have come to think of change and innovation as inherently messy, chaotic, complex and unpredictable; in other words, as quite incompatible with the idea of planning. Often, part of this messiness is connected with tensions and conflicts between people that tend to emerge whenever meaningful changes are considered; after all, there are always vested interests and values at stake in such a process. The challenge for extension and other development organisations, then, is to organise their interventions in a much more adaptive and flexible way, so that learning experiences and emergent developments can be incorporated in ongoing activities. This may require new forms of monitoring, evaluation and securing accountability in connection with development efforts.

Becoming learning organisations

From the perspective of organisation theory (e.g. Mintzberg, 1979), the survival of organisations depends eventually on whether or not they can adapt to changing circumstances. When we look around, we see that our society and agro-ecological environment is changing continuously, and according to some the pace of change becomes more and more rapid in view of the development of 'knowledge society'. This would imply that, in order to survive, organisations have to change and adapt more or less permanently, and in a way that is consistent with changing characteristics and 'demands' of the environment. If organisations do not reflect critically on their mission, services, products, culture, procedures, etc. on a regular basis, they may well become dysfunctional and go bankrupt or be abolished.

In order to adapt, it has been argued that organisations must become 'learning organisations' (Senge, 1993; Van den Ban, 1997; Easterby-Smith et al., 1999). This essentially means that, within and between hierarchical levels in the organisation, the members of the organisation need to share both positive and negative experiences (i.e. successes, mistakes and problems) and learn from them. This sounds quite simple and straightforward, but it is not. In practice, it appears that organisations often choose to ignore and avoid threatening developments in the environment, that there is little institutional space for critical thinking, and that problems and mistakes are hidden away from other people or organisational levels. Often, rewards systems in organisations do not encourage employees to be critical and open about their failures, while it may well be that such failures carry the seeds for future successes. Moreover, especially in formal and hierarchical bureaucracies – which extension organisations frequently are – communication often takes place from top to bottom rather than the other way round. Consequently, the higher levels in the organisation may have very little knowledge of the real activities, problems and concerns of their frontline workers, which considerably reduces the chance that they take appropriate management decisions (Wagemans, 1987). Given the challenges ahead, and the continuous changes that extension organisations face, it will be imperative for many extension organisations to improve their capacity and mechanisms for learning.

Being brokers in an era of participation

Particularly when funded by donors or government agencies, extension workers and change agents often find themselves in a broker position; that is, they are placed in the difficult position of having to marry, and/or mediate between, different interests. On the one hand, they are paid by the government or a donor, which typically is interested in stimulating a particular type of development, change or innovation (e.g. increasing cash crop production or strengthening the position of women in agriculture), and they have to somehow show to such funding agencies that they are doing a good job. On the other hand, they have to work and maintain credibility with their immediate clients (e.g. farmers), who may have totally different priorities from funding agencies and hence expect support from extension workers in altogether different areas. The extension agents, then, are squeezed in the middle. On certain occasions (e.g. when reporting), they will have to frame or translate the things they do into the discourse, terminology and criteria of those who fund them, whereas in other contexts they have to link in to the language and aspirations of farmers. This 'juggling with discourses' (Hilhorst, 2000) requires considerable creativity and skill. One might expect that this tension could be somewhat ameliorated by the fact that many donors and governments nowadays advocate decentralisation and participatory approaches, which would in theory grant primacy to the priorities and criteria of clients. However, in practice funding agencies often limit the boundaries within which people can participate (Craig & Porter, 1997; Zuñiga Valerin, 1998; Amankwah, 2000; Pijnenburg, 2003); that is, they still have fairly explicit ideas as to what should be the outcomes of the participatory process.

Here we touch on a general 'participation paradox'; on the one hand participatory approaches start from the idea that people are capable, knowledgeable and active, while on the other hand participatory projects are, at least to a degree, outside interventions which build on the assumption that 'something specific is missing', which in turn comes close to 'people cannot do it themselves'. In addition, funding agencies and clients may have rather different ideas concerning the meaning of 'participation' itself. To donors and funding agencies 'participation' may mean 'empowerment' or 'handing over responsibility to people', while citizens may interpret it as 'getting paid for work done' or as 'government laziness and avoidance of responsibility'. In any case, it is clear that – despite all the rhetoric and good intentions – participatory trajectories are often far from smooth, and quite often produce disappointing results for those that initiate or participate in them (Eyben & Ladbury, 1995; Leeuwis, 1995; Mosse, 1995; Wagemans & Boerma, 1998; Brown et al., 2002; Pijnenburg, 2003). Thus, there is still a need to further clarify what exactly participation means in an intervention context, what the role of extensionists can be in participatory processes, and what institutional and funding arrangements may be helpful in ameliorating some of the tensions that practitioners face.

Coping with dwindling resources

Many public extension services across the world face dwindling resources. The reasons for this differ. In some cases, governments are more or less forced to cut budgets in view of structural adjustment policies or economic crisis. Elsewhere, extension organisations have not been able to show convincingly to governments and/or donors that they deliver value for money. And in industrialised countries especially, governments feel that farmers should and can pay for extension services themselves (Van den Ban, 2000). A typical response to limited resources is inertia, and blaming the government. Understandable as this may be, such responses do not solve anything. Therefore, one of the challenges that extension organisations face is to devise innovative ways both of working with limited resources and of accessing new sources of income. Some basic modes of doing this are well known, including various cost recovery strategies, co-operation with non-governmental organisations (NGOs), supporting farmer to farmer extension, and/or total privatisation of the extension organisation. However, there is still scope for inventing more creative solutions. Moreover, developing and implementing strategies as mentioned above is often far from smooth, and requires new forms of organisational management and new skills and attitudes on the side of extension staff, farmers and government officials (see Chapters 16 and 18).

Changing professional identities

In view of the developments, conceptual changes and challenges indicated in this section, it becomes pertinent to ask who we mean when talking about extension agents, and whether we should maintain the term 'extension' at all. We can see that conventional public extension organisations are in decline, and that there is reason to rethink seriously their missions, activities, skills and organisational forms. Already, we see that some of the tasks and functions that used to be performed by public extension services are partly taken over by private extension services, commercial companies and non-govermental organisations, or by decentralised governmental bodies that are no longer part of a centrally managed public service. And if extension organisations change their roles, they may well provide services that are also provided by others such as NGOs, local governments and research institutes. Phrased differently, there are many things that the authors of this book would consider to be extension activities, that are carried out by people who would never think of themselves as extension agents. Rather, they may think of themselves as a 'development worker', a marketing employee, an external communication manager, a trainer, a mediator, a public relations officer, a process facilitator, an organisation development consultant, etc. What these people have in common is that their work mainly centres around the deliberate use of communication to stimulate change. In relation to these (and other, see Chapter 2) developments in the professional environment, some authors in the field of extension studies have abandoned the notion of 'extension worker' altogether; rather they refer to 'communication specialist' or 'communication worker' (e.g. Van Woerkum et al., 1999). Another term used at times is 'change agent' (Havelock, 1973; Buchanan, 1992). In this book we will use these terms interchangeably. In any case, we can conclude that the professional identity and organisational environment of the 'extension agent' are changing. This book, then, aims at the new style communication worker, be it a former extension agent or not.

1.1.3   In conclusion: a new societal function for extension

The challenges outlined above have many implications. Among others, we have to adapt our view of what extension is (see Chapter 2), and also our ideas of *why* it is important. In the early days of extension, the latter question was usually answered with reference to the need to increase food production and to encourage economic development. To this end, extension was mainly seen as a function that *fostered knowledge and technology transfer* between farmers and researchers, or among farmers themselves. As we have outlined above, we have now come to realise that these tasks are not as simple as perhaps assumed earlier. Improving food production and fostering economic development is *not* just a matter of individuals receiving messages and adopting the right technologies, but has much more to do with altering inter-dependencies and co-ordination between various actors. In addition, we recognise new challenges, problems and developments – some of which operate at a larger scale than before (e.g. ecological degradation, globalisation, knowledge society) – that further complicate matters. Hence, the issues we are dealing with are not just agricultural or relating to land-use only; they are concerned more broadly with *rural resource management*. Resources in this context include not only water, land, biological processes and biophysical inputs, but also human relations, forms of organisation, economic and legal institutions, knowledge or skills (see also Uphoff, 2000).

A key point here is that we have come to realise that the 'real' and perceived problems we face as human beings are, at least partly, *man-made*; that is, they emerge from the way in which human beings interact with each other and with their natural environment, which in turn is guided by man-made rules, organisational forms, modes of interpretation, etc. Moreover, problems emerge from interactions between different sets of people, at different levels of society, in multiple localities and at varying points in time. Essentially, this means that effective, societal problem solving involves the bringing about of new patterns of co-ordination between different sets of people and their natural environment. As Röling (2002) has argued, it is an illusion to think that such patterns can be brought about through the technological solutions and market mechanisms that tend to be proposed with reference to the biophysical and economic sciences. First and foremost, new forms of human co-ordination require novel, and at least partly shared, modes of thinking as well as agreements between societal stakeholders on how to organise things differently. In any case, communication will have to be an important element in deliberate efforts to arrive at these. Hence, we propose that extension organisations should take on a new role, and aspire *to manage communication in processes that are somehow aimed to bring about new patterns of co-ordination*. We will think of such processes mainly in terms of network building, learning and negotiation.

1.2   Objectives and outline of this book

We have outlined a number of challenges for agricultural extension in the first decade of the 21st century. We feel that it is neither useful nor possible to present detailed recipes in this book for how to deal with these. Rather, we hope to offer modes of thinking and building blocks that inspire those who lead extension organisations which aim at catalysing change, to renew their organisations and improve professional practice. In order to help staff or students to process the contents of this book and translate them to their own situation, each chapter concludes with some questions for discussion that can be used in small discussion groups. Below, we present an outline of the specific issues that are dealt with in the various parts and chapters of the book.

Part 1: Rethinking extension

The introductory part of this book (Part 1) outlines first the changing context and challenges for communication and innovation in the domain of agriculture and rural resource management (Chapter 1). It then proceeds in Chapter 2 with a discussion of the changing ideas regarding 'extension' at the conceptual level. The chapter indicates why we nowadays prefer to speak about 'communication for innovation' rather than 'extension', and presents some of the major strategies and functions of communicative intervention as well as additional practices and terminology that are connected with the changing idea of extension. Subsequently, Chapter 3 addresses some of the ethical and political implications of communication for innovation. Finally, two basic approaches to communicative intervention – the instrumental and the interactive approach – are introduced in Chapter 4.

Part 2: The relationships between human practice, knowledge and communication

Part 2 deals with the relationships between human practice, knowledge and communication, and is divided into three chapters. Chapter 5 introduces a conceptual model for understanding why people do what they do, or do not do, at a given point in time, and how this may be influenced by their social and biophysical environment. Several important implications for the theory and practice of communication for innovation are discussed. Among others, it will be shown that knowledge and perception play a vital role in shaping human practice. This, then, leads to an elaborate theoretical introduction to knowledge and perception in Chapter 6. Attention is paid to the ways in which knowledge may be constructed socially, and to the significance of diverging perceptions across different social actors. Moreover, the chapter discusses how knowledge inherently implies ignorance, and touches on the differences and similarities between scientific and non-scientific knowledge. Last but not least, the complex interrelations between knowledge and power are brought to the forefront. The final chapter in Part 2 deals with human communication as a process that may contribute significantly to changing social actors' knowledge and perception, and hence to innovation. In doing so, Chapter 7 introduces several characteristics and peculiarities of human communication, as well as different conceptual modes of thinking about it. Subsequently, various practical problems with regard to communicative intervention are identified.

Part 3: Innovation as a process of network building, social learning and negotiation

The central themes in Part 3 of the book are innovation and the processes through which innovations come about, including network building, learning and negotiation. Chapter 8 explains how and why, in our field of study, the initial focus on the adoption and diffusion of innovations has moved towards understanding processes of innovation design. It is proposed that the key role for communication in innovation processes is to enhance network building through social learning and negotiation. Therefore, Chapter 9 presents key insights on human learning. This includes a discussion of some of the key forms and mechanisms of learning, as well as an overview of possible stimulants, obstacles and pre-conditions for learning. The significance of understanding and looking at processes of negotiation is underlined in Chapter 10. It is argued that change and innovation hardly ever occur without tension and conflict, and that communication professionals need to anticipate such dynamics in order to be effective. The chapter introduces various forms of negotiation, as well as key pre-conditions that need to be met for productive negotiations towards innovation. Important implications of adopting a learning and negotiation perspective for intervening in innovation processes are discussed in Chapter 11, which explores the relations and balance between top–down and participatory forms of communicative intervention in innovation processes. In connection with this, the chapter further clarifies the possible roles and contributions of 'outsiders' (including communication specialists and researchers) and 'insiders' in innovation processes, and specifies the different areas of knowledge that need to be mobilised in processes of change and innovation.

Part 4: Media, methods and process management

The more practical aspects of communication for innovation are dealt with in Part 4, which elaborates on methodological issues from different entry points. In Chapter 12, the entry point is formed by the basic forms and media for communication. Hence, the focus is on the potential of mass media, interpersonal communication and new hybrid (internet or computer-based) media. This is followed by a discussion of concrete and usually multi-media methods in Chapter 13. It is proposed that different methods need to be combined and woven together through time in a contextual, creative and emergent fashion, and that standard recipes (e.g. in the form of popular methodologies) are unlikely to be of much use in most problem situations. This principle leads to an unconventional grouping of methods according to their main communicative and/or learning function (i.e. awareness raising, active exploration, information provision and training). A range of specific methods is introduced, with the emphasis on more recently developed exploratory methods for multi-stakeholder situations, including computer-based methods. Chapter 14 deals with the management and facilitation of interactive process management. It starts with a number of critical reflections on conventional thinking about participation. In particular, it is argued that the idea that maximum participation is always possible and desirable is both practically and conceptually misleading. Subsequently, important process management tasks are identified, and in relation to each task several possibly relevant guidelines are discussed. Although it is emphasised in Chapter 14 that processes cannot be planned and prepared in detail beforehand, it is argued that the preparation of individual activities remains an important area of attention. This is further elaborated in Chapter 15.

Part 5: Organisational and interorganisational issues

Part 5 of the book focuses on specific (inter)organisational aspects of communicative intervention. Chapter 16 focuses on organisational issues, and starts with an overview of different modes of thinking about organisations and their management, resulting in a number of key points of attention for managers of communicative intervention organisations. From here, attention is drawn to the need to formulate new missions for conventional extension organisations, without throwing overboard everything that such organisations stood for in the past. Several organisational implications of embracing a novel and wider mission are discussed, and specific attention is paid to the challenge of building learning organisations. Several practical suggestions to this end are formulated. The chapter ends with a discussion of different forms of decision-oriented organisational research that may be relevant in the context of communicative intervention.

Issues regarding interorganisational co-operation are the focal point in the following three chapters. Chapter 17 introduces knowledge systems thinking, which proposes to look at communicative intervention in the context of a wider set of institutions and/ or actors in a knowledge network. In connection with this we also pay attention to the systems methodology RAAKS and to networking strategies. Subsequently, Chapter 18 discusses some important changes in agricultural knowledge networks that are associated with privatisation and commercialisation of knowledge. Attention is paid to the conceptual and political underpinning of privatisation policies, to diverse forms of privatisation, and to the changing dynamics of innovation processes in emerging 'knowledge markets'. With regard to the latter, it is argued that, besides positive changes, there are a number of serious risks with regard to the innovative capacity of knowledge networks under privatised conditions. Part 5 of the book closes, (in Chapter 19) with a discussion of co-operation across diverse epistemic and disciplinary communities. At this point, special attention is given to the differences between the social and the natural sciences, and to different conceptions of cross-disciplinary co-operation. Moreover, existing obstacles to co-operation between scientists from different disciplines and between scientists and societal stakeholders are identified.

Part 6: Epilogue

The final part of the book is specifically addressed to those who are interested in doing MSc or PhD research in Communication and Innovation Studies. Several themes for conceptual research are suggested in Chapter 20 under the overall heading of studying communication (and not only communicative intervention) in the context of socio-technical design processes. Comparative process ethnography is proposed as a useful methodological approach that can be fruitfully combined with different strands of network analysis.

Questions for discussion

(1) Which of the challenges mentioned for agriculture are relevant to your situation and which not? What additional challenges do you see?

(2) Which of the challenges mentioned for agricultural extension are relevant to your situation and which not? What additional challenges do you see?

(3) Choose a challenge that you find relevant. How are different (public and/ or private) organisations dealing with this challenge? Which organisation seems most effective, and why?

2 From extension to communication for innovation

In Chapter 1 we outlined some of the challenges that extension is facing, and argued that agricultural extension as a professional practice is changing, or will have to change, considerably. In this chapter we focus more on the conceptual implications of such changes. In doing so, we first give a historical overview of how extension has been seen in the past, and then propose a novel definition. Subsequently, we touch on some general issues such as different types of communication services and functions.

We conclude by touching on some concepts that are closely affiliated to the notion of extension, including knowledge systems, extension education, extension research and extension science. We discuss their original meaning, and propose some adapted terminology in view of our changed definition of extension.

2.1 Historical roots and evolving conceptions of extension

The meaning of the term 'extension' has evolved over time, and has different connotations in different countries. In this section we touch on such different conceptions.

2.1.1 Origins, early meanings and international terminology

Throughout history, and across the world, there have existed patterns of agricultural knowledge exchange, with some people (e.g. religious leaders, traders, elders, etc.) often playing special 'advisory' roles in this respect. According to Jones and Garforth (1997), more or less institutionalised forms of agricultural extension existed already in ancient Mesopotamia, Egypt, Greece and Phoenicia. The term 'extension' itself is more recent; it orginates from academia, and its common use was first recorded in Britain in the 1840s, in the context of 'university extension' or 'extension of the university'. By the 1880s the work was being referred to as the 'extension movement'. In this movement the university extended its work beyond the campus. In a similar vein, the term 'extension education' has been used in the USA since the early 1900s to indicate that the target group for university teaching should not be restricted to students on campus but should be extended to people living anywhere in the state. Here extension is seen as a form of adult education in which the teachers are staff members of the university.

Most English-speaking countries now use the American term 'extension'. In other languages different words exist to describe similar phenomena. The Dutch use the word *voorlichting*, which means 'lighting the pathway ahead to help people find their way'. Indonesia follows the Dutch example and speaks of lighting the way ahead with a torch (*penyuluhan*), whereas in Malaysia, where a very similar language is spoken, the English and American word for extension translates as *perkembangan*. The British and the Germans talk of *advisory work* or *Beratung*, which has connotations of an expert giving advice but leaving the final responsibility for selecting the way forward with the client. The Germans also use the word *Aufklärung* (enlightenment) in health education to highlight the importance of learning the values underlying good health, and to emphasise the need for arriving at more clarity on where to go. They also speak of *Erziehung* (education), as in the USA where it is stressed that the goal of extension is to teach people to solve problems themselves. The Austrians speak of *Förderung* (furthering) meaning something like 'stimulating one to go in a desirable direction', which again is rather similar to the Korean term for 'rural guidance'. Finally, the French speak of *vulgarisation*, which stresses the need to simplify the message for the common man, while the Spanish sometimes use the word *capacitacion*, which indicates the intention to improve people's skills, although normally it is used to mean 'training'.

2.1.2   Evolving definitions

Enlightenment definitions of extension

Initial meanings of the term 'extension' – as well as international equivalents of the term – have been influenced significantly by 'enlightenment thinking'. Although different nuances exist, the basic thrust is that 'the common folk' are to a degree 'living in the dark', and that there is a need for well-educated people to 'shed some light' on their situation by means of educational activities. This reflects that the early conceptions of extension were somewhat paternalistic in nature; that is, the relationship between the extensionist and their clients was essentially looked at as being similar to the teacher/student or parent/child relationship, placing the extension agent in an 'expert' and 'sending' position and their audience in a 'receiving' and 'listening' role. In line with this tradition, many definitions of agricultural extension emphasise its *educational* dimensions:

> 'Extension is a service or system which assists farm people, through educational procedures, in improving farming methods and techniques, increasing production efficiency and income, bettering their levels of living, and lifting social and educational standards.' (Maunder, 1973:3)

> 'Extension is an ongoing process of getting useful information to people (the communicative dimension) and then assisting those people to acquire the necessary knowledge, skills and attitudes to utilise effectively this information and technology (the educational dimension).' (Swanson & Claar, 1984:1)

It must be noted that each definition is a product of its time. When 'enlightenment' conceptions of extension were formulated there was still a firm belief in the potential and blessings of science as an engine for modernisation and development, and there was a genuine concern that everybody should be able to pick the fruits of science. The belief then was that by adopting science-based innovations, and by grounding their practices and decisions in rational scientific insight and procedures, farmers and agriculture would benefit almost automatically. In view of the experiences of the last decades, however, science has nowadays become much more contested and the belief in science as a neutral and objective engine to progress has eroded significantly (Knorr-Cetina, 1981a; Callon et al., 1986; Van der Ploeg, 1987; Beck, 1992; see also Chapter 6). Although science has contributed significantly to agricultural change and production increases in high potential areas, its impact in other regions has remained much more limited. Moreover, science-based agriculture in high potential areas was accompanied by a number of serious problems related to, among other topics, the environment and health (see Chapter 1). Furthermore, even in high potential areas scientists regularly produced innovations and recommendations that were of limited use to many farmers. It was realised that successful innovation required as much input from farmers themselves as from scientists (see Chapter 8).

In line with 'enlightenment' thinking, there was great concern in the 1950 to 1970 period with the 'adoption and diffusion' of science-based innovations (see Chapter 8). Extension scientists developed an interest in so-called adoption *decisions*. In the context of diffusion, it was also recognised that farmers could gain a lot from each other's knowledge and experience (regarding new technologies, among other topics) when solving agricultural problems (Van den Ban, 1963). Inspired by such interests and insights, the emphasis in definitions of extension shifted slightly from 'education' to supporting *decision making* and/or *problem solving*:

> 'Agricultural extension: Assistance to farmers to help them to identify and analyse their production problems and to become aware of the opportunities for improvement.' (Adams, 1982:xi)

> 'Extension is a deliberate and systematic attempt – by means of the transfer of knowledge and insight – to help and/or develop someone in such a way that the person is able take decisions in a specific situation with a maximum level of independence, consciousness, and conformity with his own interest and well-being.' (Van Gent & Katus, 1980:9, translated by the authors)

> 'Extension involves the conscious use of communication of information to help people form sound opinions and make good decisions.' (Van den Ban, 1974; Van den Ban & Hawkins, 1996:9)

The last definition is the one which was used in the predecessor of this book (Van den Ban & Hawkins, 1988, 1996). Like most definitions presented so far, it still carries the idea that extension is mainly about 'help' in the interest of the farmer.

Intervention definitions of extension

The definitions mentioned so far are in essence *normative* definitions, in that they indicate what the authors feel extension *should be* and/or *should do*. In other words, they 'prescribe' what the authors *would like* extension to look like ideally, e.g. as a practice that is experienced as 'help' and 'assistance' and leads to 'good decisions' and 'development'. Alternatively, one could also try to define extension more descriptively in terms of what people who call themselves extensionists *actually do*, which frequently might not correspond with normative definitions (see also Röling

& Kuiper, 1994). When taking a closer look at what extensionists do in practice, one might, for example, discover in some cases that their work has little to do with 'help' but rather with imposing technologies and/or enhancing state control over farmers (e.g. Ferguson, 1990). Along these lines, it was recognised during the 1980s that extension could not just be regarded as 'help' and 'being in the interest of the recipient'. It was realised that extension is in many ways also an *intervention* that is undertaken and/or paid for by a party who wants to influence people in a particular manner, in line with certain policy objectives. Thus, it was realised that there was often a tension between the interest of the extension organisation (and/or its funding agency) and the interest of recipients such as farmers. Government extension services could, for example, aim at increasing the production of export crops, while farmers would be more interested in other issues or crops. In this more descriptive conception of extension, there needed at least to be a partial *overlap* or *link* (see Röling, 1988) between the interests of clients and extension organisations, otherwise people would obviously not be willing to change (unless they were forced/persuaded to by other means than just extension messages). In line with such views new definitions of extension emerged:

> 'Extension is helping behaviour consisting of – or preceding – the transfer of information, usually with the explicit intention of changing mentality and behaviour in a direction that has been formulated in a wider policy context' (Van Woerkum, 1982:39, translated by the authors)

> 'Extension is a professional communication intervention deployed by an institution to induce change in a voluntary behaviour with a presumed public or collective utility' (Röling, 1988:49)

The phrase added by Röling on 'presumed public or collective utility' is important, because it was used to distinguish extension from other forms of communication intervention such as:

- *Commercial advertising*, where the goal is to sell products in the interest of a limited group (salesmen, shareholders).
- *Political propaganda*, where the goal is to influence people's ideological beliefs and/or perceptions of reality in order for some to gain or maintain power.
- *Public relations*, where the goal is to manage one's own reputation or public image.

At the same time, this phrase exemplifies that these definitions still contain normative elements. After all, it is more or less implicit in Röling's definition that extensionists should not be involved in, for example, trade, advertising or political propaganda, and if they are this cannot be regarded as 'extension'. As Röling and Kuiper (1994) point out, it is impossible to avoid normative elements in a definition of extension if one's purpose is not only to study extension as a societal phenomenon, but also to inform extension practitioners on how they can do better. From a purely descriptive point of view, the definition of extension would be something like:

> 'Extension is everything that people who think of themselves as extensionists do as part of their professional practice.'

A book written on the basis of such a definition of extension could reveal very interesting activities and phenomena, but as soon as one wants to draw lessons for a wider audience one needs to assume certain criteria as to what it is, and is not, that extension aspires to, and how.

Extension as communication for innovation

The two 'intervention' definitions of extension still start largely from the premise that extension derives from a semi-state institution that is concerned with the public interest or public policy. As indicated in Chapter 1, this situation is rapidly changing in view of the emergence of private and NGO-based extension and communicative intervention. In addition, we need various changes in the definition of extension if we are to take the challenges formulated in Chapter 1 seriously. As elaborated in Chapter 1, these include a need to:

- Shift away from a focus on *individual* behaviour change which has characterised most of the definitions so far, and incorporate the idea that extension is about fostering new patterns of co-ordination.
- Move away from the idea that extension works mainly on the basis of *pre-defined* directions, policies and innovations, and emphasise its *generative* dimensions.
- Indicate that changes usually have a dual (material-technical and social-organisational) component.
- Transcend the idea that extension is mainly concerned with decision-making, and emphasise the importance of *social learning* and *negotiation* in extension processes.
- Define extension as a *two-way* or *multiple-way process*, in which several parties can be expected to contribute relevant insights, and which may have action implications for all parties (not only farmers, but also researchers, extensionists, policy makers, agricultural industries, etc.) involved in the process.

In view of such significant needs for redefinition (see also Sulaiman & Hall, 2002), some senior authors in the field of extension have chosen to completely *abandon* the notion of 'extension' altogether (e.g. Röling & Wagemakers, 1998; Van Woerkum et al., 1999; Ison & Russell, 2000). They feel that the word 'extension' has misleading connotations, and that it is practically impossible to stretch the meaning of the concept as necessary. In line with this, Van Woerkum and Röling no longer use the concept in many of their writings, and they have in their university renamed the field of Extension Science as Communication and Innovation Studies. Similarly, Ison and Russell (2000) speak of 'second-order research and development'. In many ways we agree with such proposals to move away from the term 'extension'. The main reason why we still use the term is that this book is aimed not just at a small group of academics, but also at a wider group of practitioners in training and management positions of which many still identify strongly with the term 'extension'. As outlined in Chapter 1, however, this group is likely to erode, while alternative audiences who do similar work are likely to expand. Hence, we have chosen to start with the term 'extension' in this introductory part of the book, and emphasise the need to change our conception of it. In view of the above, we propose to define extension as:

'a series of professional communicative interventions amid related interactions that is meant, among others, to develop and/or induce novel patterns of co-ordination and adjustment between people, technical devices and natural phenomena, in a direction that supposedly helps to resolve problematic situations, which may be defined differently by different actors involved.'

Or in a more condensed form:

'a series of embedded communicative interventions that are meant, among others, to develop and/or induce innovations which supposedly help to resolve (usually multi-actor) problematic situations.'

Let us look more closely at some of the ingredients of this, mainly descriptive, definition:

(1) The definition maintains that extension is a *professional* activity, practised by people who are somehow paid and/or rewarded for it. We do not call everyday communicative interactions, for example, between farmers, 'extension', even if they contribute to innovation.

(2) Extension is regarded as an *intervention*, as it is usually subsidised or paid for by external agencies (donors, governments, private companies) whose aspirations for doing so are not the same as those of the supposed beneficiaries. Nevertheless, extension can only be effective if there is sufficient overlap or compatibility between the aspirations of change agents and clients.

(3) Extension draws heavily on *communication* as a strategy for furthering aspirations. Communication is the process through which people exchange meanings (e.g. through the use of information). Thus, extension is an activity that is geared towards bringing about *cognitive* changes, used as a trigger for other forms of change (e.g. human practices, growth of crops, water availability, regulations). At the same time, the emphasis on 'communication' marks a shift away from a focus on *education* to a focus on *learning*.

(4) Extension is a *process* involving a *series* of communicative interventions and interactions. It is not a once-only event. People respond to communicative interventions, and such interventions have consequences, which usually bring about other communicative interventions.

(5) Extension takes place *amid other interactions*, which indicates that there are many other interactions going on between people that do not involve extension and/or change agents, but which are still very relevant to the process. Farmers in a village, for example, interact a lot with each other, with other service providers and with community and/or religious leaders, and this is bound to have an impact on innovation processes.

(6) Although communication workers are usually interested, albeit with different degrees of intensity, in bringing about change and innovation of some kind, we cannot explain the dynamics of the process by just looking at such intentions. Whenever people interact, multiple goals and intentions play a role. Change agents too may have *other aspirations*, some very mundane, that impinge on the way they go about their work; these may include pleasing their boss, aquiring social status, enhancing control over farmers, reserving time for side-line activities, visiting home regularly, etc.

(7)   The statement that extension aims to '*develop and/or induce*' innovation emphasises that we cannot simply look at extension as 'dissemination of innovations'. Frequently, extension activities are, or need to be, geared towards
designing new innovations. And even if extension activities aim at the 'diffusion' of existing innovation packages, this can often not be effective without
including elements of 'redesign'. The term 'to induce' is chosen here to capture
this mixture of dissemination and adaptation. The definition does not further
specify what *kind* of processes are involved in 'developing' and 'inducing', thus
leaving space for all sorts of social processes, including social learning, network
building, decision-making, negotiation and human capacity building.

(8)   The '*innovations*' that extension seeks to contribute to are regarded as '*novel
patterns of co-ordination and adjustment between people, technical devices and
natural phenomena*'. The latter phrase is used to convey that effective innovations – especially in the field of agriculture and resource management – include
a balanced mixture of social, technical and natural elements and processes.

(9)   Extension activities are usually legitimised by referring to the need for solving
a *problematic situation*. Whether or not this problematic situation is resolved,
and to what extent, is of course something that remains to be seen as the
process unfolds. Hence, the use of the term '*supposedly*' in the definition.

(10)   The term '*supposedly*' is used to point to a different issue as well. Although in
an extension process solutions and innovations are often *presented* as contributing to problem solving, this does not mean that they are promoted by
extensionists or others solely or mainly for this purpose. In an extension process, change agents may have various aspirations (see also point 6). Thus they
may, for example, induce integrated pest management innovations mainly in
order to improve their own experience and job opportunities.

(11)   Finally, the definition mentions '*multi-actor problematic situations*' (rather than
of problem situations) in order to indicate that the solving of such situations
usually depends on the activities of several interconnected actors, who may in
fact have different views of what the problem is, and what criteria the solution
should meet. Even in situations where an individual farmer raises a seemingly
individual problem, there are usually more people involved (e.g. other household members, family, labourers, contractors), who are part of the problem or
its solution. For a male farmer, the cost of pest infestation may be a problem
because it reduces cash income available for socialising in a bar, while his wife
may regard it as a problem because it prevents her from buying school uniforms.
Thus, the availability of male and female labour for labour intensive pest
management strategies may depend on an agreement on the distribution of the
extra cost incurred. Similarly, the feasibility of adopting a disease resistant
crop variety – which also happens to be early ripening – may depend on the
willingness of others to provide labour at an earlier time in the season.

As can be seen from these discussions, we have tried to arrive at a mainly *descriptive* definition of extension. This is because one cannot hope to contribute to extension without describing what it entails in practice. At the same time, however, it is
impossible to make practical contributions without a vision of how it can be done

better. Thus, points 3 and 7 are more normative in nature as they indicate what we feel extension *should* do, even if we know that change agents often also use non-communicative strategies to promote change (which contradicts with point 3), and still regard and organise their work largely as 'dissemination' (which is at odds with point 7). We are aware that the descriptive ingredients of the definition in particular may raise additional normative issues for the reader, for example on whether or not we can accept that change agents go against the interests of certain clients, have hidden agendas, personal goals, etc. We will discuss such issues under the topic of 'ethics and politics' in Chapter 3.

Terminology from this point onwards

In the preceding sections we have described how the concept of 'extension' has evolved historically, and emphasised the need for a novel definition. Essentially, we intend to look at extension as 'communication for innovation'. From here on we use the latter term whenever possible, or use the term communicative intervention. Similarly, we minimise the use of the terms 'extensionist' and 'extension worker', and   following Van Woerkum et al., 1999 – write of communication specialists, communication workers or change agents instead.

2.2   Different types of communication services and strategies

In practice, communication for innovation can take many forms, not just in terms of the methods and techniques used (see Chapter 13), but also with regard to the *wider intervention purpose*, which again relates closely to the *assumed nature of the problematic situation*. Depending on the situation, the problem may, for example, be regarded as 'a lack of adequate technology', 'conflict over collective resources', 'lack of organisational capacity' or as 'an individual farm-management problem'. Clearly, the practice of communication for innovation (and the theories on which this is based) will have to differ accordingly. In Table 2.1 we have summarised several types of communicative intervention, which we will call different *communication services* (as a shorthand for 'communication for innovation' services), since they essentially define different kinds of '*products*' that can be 'delivered' by communication workers. At the same time, however, they can be seen as different *communication strategies* because they refer also to the *way in which* communicative intervention is supposed to contribute to societal problem solving. Depending on one's analysis of a problem, one may decide that providing a specific type of *service* is an appropriate *strategy* towards improving the situation.

The first two services in Table 2.1 we group together under the term 'farm management communication'. This involves modes of communicative intervention that are particularly geared towards supporting 'individual' farm households in identifying, interpreting and solving problems on their specific farms. Even if supporting horizontal knowledge exchange clearly involves working with farmer groups (see section 13.3.2), the focus in both types of communicative intervention is on dealing with 'individual' farmers' problem situations ('individual' in quotes because different household members are often involved) which do not require collective action. That

Table 2.1 Different communication for innovation services/strategies and their characteristics.

Strategy/service	Intervention goal	Role of communication worker	Role of 'client(s)'	Key process(es) involved	Basis of legitimation
Focus on 'individual' change/farm management communication					
Advisory communication	• problem solving • enhancing problem solving ability	• consultant • counsellor	• active problem owner	• problem solving • counselling	• active demand
Supporting horizontal knowledge exchange	• knowledge exchange • diffusion of innovations	• source of experience • facilitator	• active learners/sources of experience	• learning • networking • problem solving	• active demand • public interest • limited resources
Focus on collective change/co-ordinated action					
Generating (policy and/or technological) innovations	• building coherent innovations	• facilitator • resource person • supporting vertical knowledge exchange	• active participants	• problem solving • social learning • network building • negotiation	• societal problem solving • ensuring progress • qualities of interactive mode of working
Conflict management	• managing pre-existing conflict	• mediator • facilitator	• stakeholder participant	• negotiation • social learning	• wish to remove obstacles to progress
Supporting organisation development and capacity building	• strengthening the position of a group or organisation	• organiser • trainer • facilitator	• active participants	• social learning • negotiation	• 'political' sympathy with a group
Focus can be individual or collective change					
Persuasive transfer of (policy and/or technological) innovations	• realisation of given policy objectives • pre-defined behaviour change	• social engineer	• 'unexpecting' receiver (initially)	• adoption • acceptance	• (democratic) policy decision • preceding interactive process

Table 2.2 General communication functions which can be relevant within different communication services and strategies.

Function	Intervention sub-goal	Role of communication worker	Role of 'client(s)'
Raising awareness and consclousness of pre-defined issues	• encouraging people to define a situation as problematic • mobilising interest	• providing (confrontational) feedback • raising questions	• unexpecting receiver or relatively passive participant
Exploring views and issues	• identifying relevant views and issues	• stimulating people to talk • active listening • active learning	• source of information • active participant/ learner
Information provision	• making information accessible to those who search for it	• translating and structuring information	• active learner
Training	• transferring and/or fostering particular knowledge, skills and abilities	• educator/trainer	• student

is, although farm households can assist each other in managing such issues by means of horizontal knowledge exchange, farm management communication focuses on problems for which the locus of control and responsibility lies with individual farm households, which can take action independently. In contrast, there are several other communication services which inherently require forms of co-ordination which transcend the household level (see Table 2.1).

Apart from these different communication services/strategies, there are also some general *communication functions* which may be relevant within *each* of the strategies described in Table 2.1. A function like 'information provision' (see Table 2.2), for example, can at some point be relevant to all strategies mentioned in Table 2.1. This implies that even if there are differences with regard to eventual intervention goals, and even if operational methods are likely to be different, there can also be considerable overlap regarding sub-goals and methods (for more details on these general functions see Chapter 13).

Together, these services/strategies and functions give an overview of the types of things that communication workers do, and for what purpose. As we will elaborate in Chapter 4, all these services and functions can be performed in different ways, depending among other factors on whether one starts from an 'instrumental' or 'interactive' mindset.

2.2.1   Basic rationale of different communication services and strategies

Communication strategies differ not only in terms of their intervention purpose, but also with regard to the preferred role division between communication workers and clients. Similarly, each distinct strategy requires a different emphasis to the key processes that change agents may usefully support during the interaction. Finally, the grounds on which such services/strategies are, or can be, deemed socially acceptable, desirable and/or legitimate can diverge. As we will see in Chapter 16, it is important that organisations for communicative intervention have a clear idea of the types of services they wish to provide, as it has important implications for the training of staff, recruitment policy, organisational management, etc. Below we will outline the basic rationale behind the different communication strategies indicated in Table 2.1. A more elaborate discussion of underlying theories follows in Parts 2 and 3 of the book, while methodological issues are addressed in Part 4.

Advisory communication

Advisory communication happens when farmers ask communication workers to share their ideas on how to deal with a particular management problem. These problems can be immediate and operational (e.g. 'how to fight the disease I discovered yesterday'), or have a longer time-scale (e.g. 'what crops can I grow best next year'; 'should I continue farming in the long run') (see also section 5.1.3). In helping farmers to deal with such problems, communication workers may not only provide relevant substantive knowledge, but can also offer guidance on the process of problem solving, or can enhance the clients' own problem-solving ability. It can be important to help farmers become more aware of what their goals and aspirations are in the first place (Zuurbier, 1984), so that they can define more clearly what is problematic and what is not (see also section 12.2.2).

In principle, the initiative for advisory communication lies largely with the farmer. Of course, communication workers can 'advertise' that they are able to help solve particular types of problems, but it is essentially up to the farmer whether to use such services. It is this active or expected demand by clients that is often used to legitimise the provision of this kind of communication service. In advisory communication, the communication worker's role is basically that of a consultant or counsellor, depending on whether the emphasis is on providing knowledge or process guidance. For the adequate provision of these kinds of services, it is particularly important that communication workers have, or have access to, relevant kinds of expertise, and that they have adequate skills to elicit the needs and expectations of farmers, as well as the capacity to adjust to these (for further details see sections 12.2.3 and 13.3.1).

Supporting horizontal knowledge exchange

Individual farmers usually have much expertise – based on experience, on-farm experimentation and/or training – which could be relevant to other farmers. Farmers are aware of this and as a result there are often informal means of farmer-to-farmer (i.e. horizontal) exchange of knowledge and information. Typically, markets, work parties, funerals, bars, celebrations, community meetings and church services provide opportunities for farmers to talk about agriculture, while observation of other farmers' practices is also an important mechanism for horizontal exchange. If needed, communication workers can stimulate or help to improve farmer-to-farmer exchange in various ways. They can, for example, organise meetings or festivities that are conducive to this kind of exchange, induce the formation of study groups, support existing groups and networks with training and logistics, develop more systematic modes of farm comparison, correct uneven exchange of knowledge within communities, communicate experiences from other communities, organise excursions, etc. (see also section 13.3.2). The role of the communication worker here is not that of a consultant or expert, but rather of a *facilitator*; that is, of someone who brings people together (networking) and acts as a catalyst for, and/or directs, learning and exchange processes, either in general or around a specific problem. Sometimes farmers actively demand these kinds of services, while in other cases governments support farmer-to-farmer exchange for the benefit of the public (e.g. more rapid diffusion of innovations). In addition, public extension organisations in particular often use farmer-to-farmer exchange to make efficient use of increasingly limited resources, i.e. to reach a relatively high number of farmers with limited inputs and/or to stimulate knowledge exchange in the absence of professional communication workers.

Generating policy and/or technological innovations

As indicated in the previous chapter, there is an increased need for communication workers to organise processes through which new innovations are designed, rather than to 'sell' pre-defined packages to farmers. 'Innovations' here are 'novel working wholes' (see Chapter 1; Roep, 2000) that involve a variety of practices and multiple actors. Often innovations have technological components, but some are more 'policy-oriented' such as novel market arrangements, new government regulations and/or alternative forms of organisation. The main purpose of this type of communication service, then, is to arrive at appropriate and coherent innovations in the face of certain challenges and/or problems. Due to the collective nature of innovations, this communication service usually requires the bringing together of various stakeholders in group sessions and/or semi-permanent 'platforms' (Röling, 1994a). Here a wide range of activities can take place, including joint experimentation and exploration, aimed at generating new knowledge, insight and mutual understanding (see section 13.5). In addition, forging effective links and knowledge exchange between such platforms and various knowledge institutions (e.g. applied research, universities, etc.) can be an important stimulant to innovation (see also section 11.2.3 and Chapter 17). Again, the key function for communication workers here is to facilitate the process, and it is important to work towards a balance between new *technical* devices and novel *social–organisational* arrangements. Thus, besides learning-oriented activities such as experimentation and exploration, sufficient attention should be paid to the creation of support networks and the negotiation of new arrangements between various stakeholders (see Chapter 1). As we will discuss in Chapter 10, this often means that communication workers have to deal with tensions and conflicts that emerge during the innovation process. Investments in these kinds of innovation processes are often made because of specific societal problems and/or the desire to foster progress in areas where this is thought to be lacking. Moreover, this type of communicative intervention is inherently interactive (at least to some extent) and is frequently legitimised with reference to specific qualities attributed to an interactive mode of working (see Chapter 4).

Conflict management

In some situations, serious tensions and conflicts among stakeholders form the start-ing point for communicative interventions, rather than the intervention emerging during an interactive process (see above)[1]. In many communities or regions conflicts exist around the distribution and use of collective resources (e.g. water, arable land, grazing land, fish, etc.; see also our discussion on social dilemmas in section 5.2.2). Such conflicts often have cultural, ethnic, moral and/or political dimensions too (see section 10.2.3). In some cases conflicts are productive in the sense that innovative solutions arise from the pressures and competition that accompany conflict. All too frequently, however, conflicts have negative consequences (e.g. natural resource degradation) and/or hinder progress and innovation; that is, in some cases it can be a long time before conflicts are resolved and/or become productive. Communication workers are often confronted with conflicts that affect their work and they can even become entangled in them. From the literature on conflict resolution (e.g. Pruitt & Carnevale, 1993), however, we know that the involvement of relative outsiders – in the form of mediators, facilitators or referees (e.g. judges) – may help to partly resolve conflict and/or to make conflicts productive. Thus, rather than becoming a party in the conflict, communication workers may at times be able to play a positive role in conflict management. Depending on the situation, this can be either by adopting a mediating or facilitating role, or by encouraging the handling of the conflict by others who are in a better position – in terms of status, skills and authority – to contribute positively. As Röling (1994a, b) suggests, such efforts may take place on a 'platform' where different stakeholders are brought together to overlook the situ-ation and learn and negotiate towards more productive outcomes (i.e. co-ordinated action). Although conflict management has not been a traditional extension service or strategy, we feel that 'new style' extension organisations may have to become better equipped for it. This is because innovation, conflict and intervention are closely intertwined (see Chapter 10), which essentially means that conflict manage-ment is something that change agents cannot run away from. Dealing with tension and conflict requires insights and skills that, in our experience, are not yet widely available in public or private organisations that apply communicative intervention.

Supporting organisation development and capacity building

In many cases innovation involves and/or depends on the adequate functioning of farmer and community organisations or groups (see also Chapter 5), such as irriga-tion management committees, credit groups, marketing co-operatives, commodity groups, study groups, etc. We will see in Chapter 10 that for purposes of conflict

---

[1]   Note that in some cases such conflicts are in part the result of previous interventions by others or one's own organisation.

resolution it can be important too that weaker parties become better organised and improve their ability to make claims. Thus, an important role for communication workers can be to contribute to organisation development and human capacity building, so as to strengthen a particular group's capacity to innovate, help themselves and/or make claims. The role of change agents here can range from initiating organisation development, contributing to organisational activities and processes, providing training in organisational skills, facilitating processes of organisation change, etc. Such activities are often inspired by 'political' sympathy with particular, often disadvantaged, groups. The term 'political' here does not refer to political parties or movements, but rather to the fact that 'strengthening a group' means almost automatically to improve their 'power position' with regard to others. In Chapter 3 we will discuss in more detail the relationships between communication for innovation and 'politics'.

Persuasive transfer of policy and/or technological innovations

The most widespread form of communicative intervention is to persuade farmers or other target groups to adopt specific technological packages and/or to accept certain ideas or policies. The main intervention goal here is to help realise specific policy objectives (e.g. increase export earnings) by the stimulation of pre-defined behaviour changes (e.g. the adoption of cash crops and/or new varieties). Typically, such efforts have been in the form of comprehensive extension campaigns, which in their eventual form and method partly resembled what we have called 'advisory communication' and 'horizontal knowledge exchange'. However, whenever external[2] persuasive concerns enter an interaction between communication workers and farmers, we would prefer to call it *'persuasive transfer'* rather than 'advisory communication' or 'horizontal exchange' – even if the form may be the same – because it means that a different intervention goal and operational logic enters the scene. As part of this logic, the required role of the communication worker in persuasive transfer is much more that of a social engineer who tries to manipulate strategically the farmers' behaviour, rather than that of a consultant or facilitator. Similarly, the role of the client is different in persuasive forms of communicative intervention. Usually people do not ask to be persuaded in a specific direction, so farmers are more at the receiving end than the demanding end. Although persuasive transfer has become increasingly unpopular in discussions of communicative intervention (see Chapter 4), persuasive transfer of innovations still exists widely. Often this form of intervention is based on local or national policy decisions (e.g. to increase cotton production, or reduce the use of pesticides), or an earlier interactive process in which stakeholders agreed on the promotion of certain behaviour changes.

In this section we have tried to unravel different types of communication services and strategies, distinguished mainly on the basis of their underlying intervention

---

[2] By 'external' we mean persuasive interests that derive from donors or governments who play a role in the 'back of the mind' of the communication worker. This is in contrast to a situation where a change agent presents advice persuasively to emphasise that he really ('internally') believes that it is in the best interest of the – perhaps even paying – farmer to solve a problem in a particular way.

goal and not on the basis of their method. In practice, several intervention goals can play a role within particular activities, in which case the distinction is more analytical than practical. In other cases these types of services can be associated with specific activities. In any case, the distinction is important in that it may help communication workers and their organisations to think about what their mission and mandate is or should be.

2.3   Agricultural knowledge systems and other extension-related concepts

The term 'agricultural extension' refers not only to a professional practice, but also to an area of study which has generated knowledge and insight and can be studied in agricultural colleges and universities. In this section we clarify several terms used in connection with this, and also propose alternative terms in view of our wish to move away from the concept of 'extension'.

Agricultural knowledge systems

Conventional extension organisations have always been looked upon as playing a role among other institutions, functions and actors who are active in the area of agricultural knowledge, such as universities, strategic research, stations for applied research, farmers, agri-business, agricultural magazines, agricultural schools and colleges, etc. This collection of actors is often referred to as the Agricultural Knowledge and Information System (AKIS) (Röling, 1989; Engel, 1995; FAO & World Bank, 2000). For a long time the role of extension and communicative intervention was looked on as transferring and disseminating ready-made knowledge from research to farmers, or from 'early adopters' to other farmers. This is often referred to as the 'Transfer of Technology' model of extension (Chambers et al., 1989), which fits in with a linear model of innovation (see Chapter 8). As shown earlier in this chapter, we now look at the role of communicative intervention in a much broader way. The emphasis is much more on the facilitation of network building, social learning and conflict management among a variety of actors with a view to arriving at new innovations (see Chapters 8, 9 and 10). Thus, communication workers are seen as interacting with a wider set of actors than the knowledge institutions. Nevertheless, it remains relevant to look at issues like knowledge exchange and links between clients, extensionists, research and other parties in the agricultural knowledge system; not least since other knowledge institutions may well have an influence on whether or not communication workers can effectively play their newly envisaged roles. Communication workers might, for example, aspire to engage in interactive technology design, but find out that research institutes are unable or unwilling to co-operate and co-ordinate activities to that end. Thus, when talking about agricultural knowledge systems, one is immediately confronted with issues of interinstitutional co-operation and associated problems.

In order to understand the functioning and potential of communicative intervention, it remains vital to look at it in the context of other actors in the knowledge system. In Chapter 17 we will elaborate on the various actors in agricultural knowledge systems, the processes that occur between them, and the feasibility of system management.

Extension Science/Communication and Innovation Studies

In agricultural universities, groups have emerged that study the phenomenon of 'agricultural extension', as described by its evolving definitions (see section 2.1). Röling (1988) has called this academic tradition 'extension science', and new names are being invented to describe this field of study. In the Netherlands we now speak about 'Communication and Innovation Studies'. Scholars in this field systematically investigate communication for innovation processes and experiences, and connect their conclusions with more abstract and general concepts and theories. In the early days extension science was predominantly an applied science in that most of the questions and conclusions were aimed at informing communication workers how to do a better job. Thus, many theories were formulated on, for example, how to use media effectively, how to develop effective communication plans, how to manage agricultural knowledge systems. More recently, studies have appeared which are more oriented towards describing and interpreting what happens around communication for innovation processes, and which do not start from a wish to arrive at practical, prescriptive theories and recommendations. One can, for example, analyse how communication workers cope with the contradictory pressures from farmers and the government, without wishing to inform them on how to do so better. But such studies can usually be used by others to derive valuable practical lessons. Typically, Communication and Innovation Studies borrows insights from, and sometimes adds insights to, several other social science disciplines. Originally, these were mainly Communication Science, Social Psychology, Adult Education and Rural Sociology. More recently, many more disciplines have offered inspiration to our field of study, including the Sociology of Science and Technology, Management Science, Systems Theory, Political Science and Anthropology.

Although Communication and Innovation Studies is a *social* science field, it has also attracted interest from natural scientists, not least because it often focuses on those interactions between people that concern their agro-ecological environment. This meeting of social and natural scientists' views is partly what makes our field of study so interesting, but it also generates tensions since social scientists and natural scientists often have very different ideas about the role and potential of scientific knowledge (see Chapters 6 an 19). Apart from this, social scientists and natural scientists tend to face rather different methodological challenges, and thus tend to work in very different ways (for more details see section 19.1). Thus, when natural scientists start to get involved in Communication and Innovation Studies (for example by reading this book) it often takes them a while to adjust.

Extension training/communication for innovation training

Conventionally, the term 'extension training' referred to the process through which extension staff became equipped to do their job. This kind of training has also been referred to as 'extension education'. However, in view of the strong educational connotations in early definitions of extension (see section 2.1), this latter term has also been used synonymously with extension practice itself (e.g. Supe, 1983). In any case, we now prefer the term 'communication for innovation training'. Such training provides change agents at different levels in organisations (management, field workers, etc.) with insights and experiences for taking strategic and operational decisions in communicative intervention. It may cover technical, methodical and/or management issues, and it can take place in various ways; for example through formal courses, fixed or flexible curricula, practicals, supervision, distance education, workshops, and organisation development trajectories (see Chapter 16).

Ideally, the findings from Communication and Innovation Studies offer inspiration to those who perform communication for innovation training. Thus, we hope that trainers can pick out elements of this book, connect these with other experiences, and translate the resulting mix of insights into training modules for communication workers.

Extension research/communication for innovation research

We can distinguish two types of extension, or communication for innovation, research. First, as an integral part of communication for innovation activities, change agents regularly need to engage in investigation and research, such as situation analysis, exploration, literature research, on-farm research, pre-testing, monitoring and evaluation. Typically, such research is 'decision oriented' in that it helps communication workers and others to make decisions about the nature and content of their future activities. We refer to this kind of activity as '*decision-oriented research*', which is part of an intervention process. In order to yield useable results, this kind of research cannot be a detached activity carried out by an isolated investigator. Rather, considerable interaction with prospective clients is needed to make sure that the decisions taken are in line with their needs and requirements (see also section 16.5).

Earlier we touched on a second type of research, which is usually carried out by scholars in Communication and Innovation Studies (or Extension Science; see above). Although in some cases their research may overlap with decision-oriented research, academics usually have an additional interest, which is to conceptualise and theorise about the findings. In other cases, this kind of '*conceptual research*' has little connection with decision-oriented research. It may, for example, involve social psychological experiments in a laboratory setting, or may be oriented merely towards observing the communicative intervention arena, with no intention to directly inform decision-making by communication professionals. As we have discussed under the heading of extension science, conceptual research is still often applied research since it frequently seeks to develop or test theories that have practical or even prescriptive implications for communication workers. Thus, the main difference between decision-oriented and conceptual research lies in the level of abstraction and the intention of the researcher, and not so much in its applicability. As Kurt Lewin stated: 'There is nothing so practical as a good theory'. Given its theoretical aspirations, however, conceptual research (see also Chapter 20) will often have to meet different (i.e. academic) standards, in terms of preparation, methodology and analysis, from those of decision-oriented research.

Questions for discussion

(1) Choose a communicative intervention programme or organisation that you are familiar with. Which definition of extension is used (implicitly or explicitly) by this programme/organisation? How appropriate is the definition used in that specific context?

(2) Do you prefer a normative or a descriptive definition of extension/communication for innovation? Why?

(3) Choose a communicative intervention programme or organisation that you are familiar with. Which communication for innovation strategies and functions do you recognise in their activities? Are these strategies and functions the most appropriate for addressing the problems that the programme/organisation tries to work on?

3 The ethics and politics of communication for innovation

When describing different communication services/strategies in the previous chapter, we have touched briefly on how they are usually legitimised in practice. However, we can approach the question of legitimacy from a more normative and ethical perspective as well. Is it legitimate and ethical for governments and others to use such forms of communicative intervention? Why? For what goals? And under what conditions? We distinguish four dimensions in discussing such issues:

- the political implications of communication for innovation;
- the acceptability of government communicative interventions;
- the acceptability of non-governmental communicative interventions;
- professional standards.

3.1 The political implications of communication for innovation

As we have emphasised in our definition, communication for innovation must be regarded as an *intervention*; it is a service that is usually provided or subsidised by certain parties (governments, NGOs, commercial firms) to achieve ends. Even if outcomes are not always pre-specified in great detail, goals are still there at a more abstract level (e.g. to improve farm management, to encourage innovation, to empower people to help themselves, etc.). Thus, the goal-oriented nature of communication for innovation is equally true for persuasive *and* more participatory forms of communicative intervention, for example, an interactive process that is induced with the *purpose* of securing more sustainable resource management. All this implies that communication for innovation is *not* a *neutral* activity, as pursuing certain goals is always to the benefit of some and to the detriment of others.

An example: keeping Cochrane's 'agricultural treadmill' going

An enlightening example of this is Cochrane's 'agricultural treadmill' (see also Leeuwis et al., 1998; Röling et al., 1998)[1]. Cochrane (1958) analysed the economic processes which propel technological change in a group of firms which produce the same commodity while no individual firm can affect the price, so that all try to produce as much as possible against the going price. Adopting an efficiency enhancing innovation gives the first adopters a windfall profit because they can produce more efficiently

---

[1] The description of Cochrane's treadmill derives largely from the two papers mentioned.

than everyone else against a price which is still determined by the old technology. But as soon as others also adopt the innovation, the total amount of product increases, which causes prices to drop. In the case of farmers, this implies that those who do not use an innovation see their incomes drop. They are now more or less forced to adopt the technology if they want to continue to grow the same crop and make a profit. However, as markets become quickly saturated, it is more or less inevitable that some farmers are squeezed out because less land and fewer farmers are needed to grow the same amount of produce. In the Netherlands, for example, this kind of mechanism has led to a massive exodus from farming over the last decades, and currently some 3% of farmers stop farming every year because they or their successors are no longer able or willing to stay on the 'treadmill'. While farmers are often unhappy about such a 'rat-race', governments tend to like it because 'treadmill' processes tend to be accompanied by: (a) lower prices of agricultural products for consumers; (b) increased competitiveness of agriculture in comparison with other countries; and (c) release of labour for non-agricultural work. The point to note here is that *all types* of agricultural communication services (and not just the persuasive ones; see Table 2.1) can in principle keep the treadmill going and/or make it pick up speed. In fact, in many countries this contribution provides at least part of the rationale for funding agricultural extension (and also agricultural research). Not surprisingly, agricultural extension is regularly regarded as a rather suspect and biased activity by those farmers and scientists who, on moral or political grounds, oppose 'treadmill' dynamics and/or the types of innovations (e.g. environmentally unsustainable ones) that are produced in the process (Van der Ploeg, 1987; Röling & Groot, 1999).

The above example shows that communicative intervention may be to the detriment of some (e.g. farmers who are squeezed out), and that the ends to which communication services are used – and/or the types of messages that accompany these – can be controversial and contested. Outside agriculture such a controversy arises, for example, around the prevention of AIDS. Some churches strongly oppose the idea of encouraging the use of condoms as an AIDS prevention strategy. Thus, health educators who give advice to youngsters about the use of condoms are regarded by religious leaders as spreading and stimulating morally unacceptable forms of behaviour, while others greet their services as a blessing.

We can conclude that communicative intervention always has moral and political implications, and that it can be used to further conflicting political ends and/or moral convictions. Instead of 'greasing' the dominant treadmill, for example, communicative intervention can be used to improve opportunities to escape from it by supporting the formation of alternative agricultural chains and networks, ruled by *other* criteria for efficiency (e.g. ecological sustainability instead of monetary cost per product unit), and going along with a different innovation direction, including alternative reward mechanisms. Having said that, it is important to realise that those who have most access to resources are often in a better position to use and direct communicative intervention than less powerful groups. In that sense the fate of communicative intervention – like other services such as education and infrastructure – is that it is more likely to be supportive of relatively powerful interests. A consoling thought here may be that both priority ends and dominant groups can sometimes change quite rapidly; what is marginal today may be dominant in the future.

3.2   The acceptability of government communicative intervention

We have seen that communicative intervention is always about *influencing*, and sometimes even persuading, people and processes towards specific or abstract ends. Government subsidised communicative intervention means that a government tries to influence the way in which its citizens think and behave. One can wonder whether this is, in principle, an acceptable practice. An argument against such practices is that, from the viewpoint of democratic principles, it is citizens who should be influencing the government, rather than the other way round. Moreover, one could argue that there is a risk that communicative intervention is abused for purposes of political propaganda (see section 2.1). However, there are a few arguments in favour of government communicative intervention as well:

- In most political systems it is formally the government's job to develop and implement policies in the public interest, which equals trying to be effective in fostering change. Fostering change is impossible without communication.
- Sometimes governments are confronted with problems that need to be tackled for the benefit of future generations (who cannot speak for themselves yet), and which can only be resolved by policy measures (e.g. restrictions, tax increases) that are inherently unpopular with the current generation. In such cases communication with citizens may be needed to help improve policy support.
- Communicative interventions may not only allow governments to influence citizens, but can also allow citizens to influence governments. As we will see in Chapter 4, policies are unlikely to be effective unless sufficient effort has been made during the policy design process to interact with and 'listen' to those citizens on whom the effectiveness of policies depends.
- One could argue that 'influencing others' is more or less synonymous with 'human (inter)action', and that it would be ridiculous or even harmful to the public interest if it were unacceptable for governments to do what ordinary people and non-governmental organisations do in everyday practice.
- In order for citizens to evaluate, anticipate and make optimal use of government policies and services, it is essential that governments are transparent. In order for governments to be transparent they have an obligation to inform citizens, explain the rationale behind policies, indicate how people can meet requirements, etc. This can only be done by communicative intervention.
- Finally, it must be realised that not to use communicative intervention can in many situations also be regarded as an intervention that may have serious consequences. In other words, there may be situations in which it is simply unacceptable for governments *not* to use communication services. For many people it would, for example, be unacceptable if governments made no effort to communicate about AIDS prevention, or if governments withheld information on how to combat important plant-diseases, restore soil fertility, preserve the environment, etc.

In total, we would say that it is acceptable for governments to try and influence citizens through communicative intervention provided that: (a) the government is legitimate and can be regarded as having a mandate from its citizens; (b) there is sufficient consensus among constituents (e.g. as represented in a parliament) that the intervention goals strived at are indeed in the public interest; (c) the communication activities do not violate (inter)national laws and/or basic human rights; and (d) that communicative intervention is carried out in accordance with professional standards (see section 3.4). Let us look a bit closer at some of these conditions.

In relation to the first condition, there are different procedures through which governments can acquire 'a mandate'. Even if the authors value general democratic principles, we do not believe that there is one ideal model (e.g. that of Western multi-party democracy) for organising legitimate government. In any case we feel one can no longer speak of 'a mandate' when a government loses public support and can only remain in power by repression and/or fraud. Even under such unfortunate conditions there may be forms of communicative intervention which in themselves are not problematic because they are mainly in the interest of the recipient (e.g. advisory communication on, say, pest management). However, the chances that communication services and communication workers are abused under such circumstances are quite high.

The criterion of 'sufficient consensus' implies that governments should not use communicative intervention in connection with intervention goals on which a society or community is deeply divided. If, for example, a society is deeply divided about the acceptability of genetic-engineering, it would in our view be inappropriate for a government to embark on persuasive communication to stimulate the adoption of genetically engineered seed varieties. This does not necessarily mean that nothing can be done in communicative terms. The government may use communicative intervention here as an attempt to bring deeply divided societal stakeholders closer together, for example by organising an interactive process with the aim of designing a mutually acceptable policy on genetic engineering.

Finally, the criterion of 'sufficient consensus' may have to be overruled when this consensus is against (inter)national laws, or when the protection of basic human rights (e.g. of minority groups) is at stake. In the Netherlands, for example, there are religious minorities who object to preventive inoculation against contagious diseases. Even if the large majority of citizens would, in the public interest, like to make inoculation compulsory for everyone, the government refrains from doing so in order not to violate the integrity and right to self-determination of these minorities. Thus, special policies, *and* accompanying communication services, are provided to preserve these human rights. It is perhaps interesting to note here that no exception is made for the farm animals that belong to farmers from these religious minorities; these have to be compulsorily inoculated against certain contagious animal diseases.

3.3   The acceptability of non-governmental communicative intervention

In most countries, non-governmental profit or non-profit organisations are also engaged in communicative intervention activities. In some cases, the term 'extension' is used to mask activities that could be better described as commercial advertising, political propaganda or public relations management (see section 2.1.1). Many non-governmental organisations, however, do have a genuine wish to contribute to resolving what they see as societal problems. Here we can think of 'one-issue' interest groups (which focus solely on, for example, combating environmental degradation, racism or discrimination of women), local development associations (e.g. trying to revive a local economy), or private companies (e.g. a biotechnology firm aspiring to be responsive to public demand). This collection of non-governmental organisations is often referred to as 'civil society'. In connection with pluralistic and democratic societies especially, it is argued that the relative influence and power of civil society in relation to the state has increased during the last decades, meaning that civil society groups, and especially large business corporations, play a more significant role in shaping society than before. This implies that it becomes even more relevant to ask ourselves whether or not conditions and standards exist for the acceptable use of communicative intervention by civil society organisations, and how these differ from those mentioned in connection with the state.

Based on wider democratic political philosophies and principles, civil society organisations are granted many liberties (e.g. freedom of speech, freedom of organisation, etc.) in pluralistic societies, including – although not explicitly mentioned – the right to use forms of communicative intervention. The basic idea behind the latter seems to be that it is important for citizens to take their own responsibility in contributing to identifying and solving societal problems. Civil society organisations can make use of their civil liberties as long as they do not violate basic human rights or civil laws. The latter may regulate the use of information and communication, and stipulate, for example, that it is illegal to knowingly mislead people when giving advice. In this context, a key difference between governmental and non-governmental organisations is that the latter need not be legitimised by the public. The communicative intervention activities of a farmers union, for example, may need to be legitimised by its membership but not by members of an environmental movement and/or other members of society.

Under non-pluralistic and/or non-democratic forms of government, civil society organisations are often much more constrained in their activities, especially in communicative intervention. At the same time, one could argue that in cases where governments cannot be seen as having a mandate from their citizens (see section 3.2), non-governmental organisations have a moral right (and perhaps even a legal right under international law) to use communicative means in order to mobilise people in solving the countries' problems, if necessary in violation of national communication regulations that are in breach of international (e.g. United Nations) standards and laws.

In conclusion, our view would be that civil society organisations have an important role to play in society, and in fact create a lot of 'policy'. From that perspective, the use of communicative intervention by such organisations is important for reasons similar to those mentioned in connection with government. Such forms of communicative intervention are acceptable, provided that: (a) they emerge in accordance with the internal regulations of the organising institution; (b) they do not violate basic human rights or the laws of a legitimate government; and (c) that communicative intervention is carried out in accordance with professional standards (see section 3.4). It may well be that the eventual pattern of communicative intervention to emerge from civil society interventions can be regarded as unbalanced. Large commercial interests, for example, may be seen as dominating communication for innovation. However, in pluralistic societies citizens can be seen as having their own responsibility in taking initiatives to correct such imbalances. In the 1970s and 1980s, for example, citizens in many parts of the world formed environmental organisations which engaged in communicative campaigns. This was to correct what they saw as the misleading picture that governments and private interests painted of the environmental consequences of economic growth, and to stimulate environmentally friendly behaviour. At the same time, governments can support efforts to correct imbalances. In many countries, for example, idealistic non-profit organisations – including environmental movements, societies for nature conservation and development organisations – are subsidised by the government to organise activities in the sphere of communicative intervention.

3.4   Professional standards

We have already referred to the need to uphold certain professional standards in communicative intervention. However, a problem here is that no (inter)nationally accepted 'code of conduct' exists in this respect. Thus, it is in many ways the responsibility of individual communication workers and their organisations to decide on what is and is not politically and morally acceptable in a given situation. Taking this responsibility requires that communication workers engage regularly in *critical (self-)reflection*. Some issues that they may need to consider are:

- *Wider government legitimacy*: In line with the issues raised above, it is important for communication workers to reflect on whether it is right to work for a particular government. Where a government is not legitimate, they will need to estimate whether it is possible to 'keep their hands clean' and/or to contribute to change from 'within'.

- *Intervention goals and impact*: More generally, it is important to reflect on whether one can agree with the intervention goals strived for by the organisation one works for or is considering working for. One may also consider whether or not the, sometimes unintended, consequences of communicative intervention for different groups (e.g. men, women, small farmers, large farmers, etc.) are morally and politically acceptable. Who the organisation is working for and with is also relevant. It will usually be impossible to find an organisation that completely matches one's personal views and values regarding intervention, so one will have to strike a balance, also considering the possibilities of contributing to change from within. A problem here is that neither organisational intervention goals nor societal consequences are always transparent, and may change over time (as is the case with one's own goals and values). Thus, it may be important to investigate these.

- *Truthfulness*: It would be nice to say that communication workers should always 'speak the truth'. However, there is often considerable dispute, even among scientists, about what exactly is the truth, and even on whether or not 'the truth' exists (see Chapter 6). At present, for example, there is still considerable debate on whether or not man-induced climate change is real, and if so, whether or not reduction of carbon dioxide emissions is the most appropriate answer. Even apart from this, scientists usually speak of 'likelihoods' and not 'certainties', and find it notoriously difficult to predict the future. Thus, communication workers usually have to deal with many uncertainties, especially with regard to future developments. At the same time, communicating all the nuances, uncertainties and pros and cons to clients may be a practical impossibility, even apart from the fact that clients often prefer, and even pay for, a fairly clear message. Perhaps the best advice we can give to communication workers is: (a) to make sure they keep informed and search actively for arguments from different sources; (b) to be honest when they do not know about something; and (c) to give as balanced an account of different views as possible, and not withold any arguments and uncertainties if these can be expected to affect the interest of clients, and/or when these are explicitly asked for.

- *Respectfulness*: In general, and even more so when dealing with people in change processes, it is important to respect people for what they are and consider them as valuable and capable human beings. Being respectful can include many aspects such as: being open and honest about one's intentions, making a real effort, taking people seriously, being willing to see things from other people's perspective, accepting when people say 'no', and readiness to withdraw when people do not appreciate one's services.

- *Clear separations between roles*: Besides communicative intervention there are other activities that communication workers can, or have to, be involved in. In some cases, for example, it is part of their job to provide inputs to farmers, collect statistical data, or to enforce and/or control certain government regulations. Also, they may be asked by their employers to get involved in political propaganda or commercial advertising. In our view, some of these functions are *incompatible* with effective communication for innovation, and hence need to be avoided whenever possible. These include political propaganda, commercial advertising and policing functions, which tend to seriously hamper one's credibility and trustworthiness as a change agent. More generally, it is important to clearly separate – e.g. in time, space and/or announced identity – the different roles and activities one needs to perform. This makes it clearer for clients to know who they are talking to: the communication worker, the salesman, the middleman, the government representative, etc. It is not uncommon for the same person to take on different identities in a relationship; in different contexts a farmer may also be 'colleague', 'friend', 'family member', 'expert' or 'political ally' to another person. People are used to dealing with such complexity, but it may help to try and avoid confusion by creating different contexts for different roles.

3.5   Dilemmas regarding ethics and politics: an example

It is relatively easy for us to write about the moral and political acceptability of communicative intervention, but in practice it is not always easy to decide what is the right thing to do, and when. Moreover, upholding moral standards may involve considerable sacrifices. Let us give a real-life example. In autumn 2000 there was a controversy about land in Zimbabwe in which extension played a role. As parliamentary elections were approaching in 2000, the ZANU-PF president of the country launched a special emergency policy to redistribute land owned until then by white commercial farmers, and give it to people with insufficient land. Preceding and during the period that this policy was announced, people who identified themselves as war veterans occupied a number of designated farms, which at times was accompanied by violence. The legality of both the land-redistribution policy and the occupations has been contested in Zimbabwean courts, and the highest judges of the country ordered the occupants to leave, and ruled that farms could not be confiscated without paying compensation. However, the Zimbabwean government and police did not enforce the ruling of the judges, and the land-redistribution programme continued without compensation. Meanwhile, for the first time in post-colonial Zimbabwean history, an opposition party was successful during the elections. It won a considerable number of seats in parliament, and went on to challenge a number of seats in court, accusing ZANU-PF of fraud and vote rigging in some 38 districts. Independent opinion polls indicated that the majority of Zimbabweans supported the principle of land-redistribution, but – like the judges and the opposition party – did not agree with the way in which it was administered by the government.

In this complex context, laden also with post-colonial and post-liberation war sensitivities, the extension organisation Agritex – which comes under the Ministry of Lands, Agriculture and Water Development – was ordered to make sure that the 'fast-track' land-redistribution programme ran smoothly. This included identifying farms for re-distribution, pegging land, selecting farmers, settling people, giving advice, getting them started, etc. Although most if not all Agritex workers are official members of ZANU-PF, most of those that one of the authors spoke with expressed their unhappiness about the whole situation. They felt the speedy land-redistribution policy led to chaos, destruction of the agricultural and wider economy, and was highly questionable from a legal point of view. Despite such widespread feelings in the organisation, the large majority of Agritex staff did, albeit grudgingly, follow orders. As they explained, they were likely to lose their much needed jobs if they did not, and they were afraid that they could face even more severe sanctions owing to what would be seen as their 'disloyalty' to ZANU-PF. In a similar vein, the author administered an organisation-development course for Agritex staff, agreed on long before, in order to strengthen the organisation in providing services to farmers. In this particular context, it was clear that many people felt the government of Zimbabwe had lost its mandate and, even apart from that, had used and even abused the extension organisation in order to further a highly controversial policy.

Despite all this, the story illustrates that it is can be very difficult to take responsibility and act in a morally sound way. Extension staff were not prepared to run serious risks and make sacrifices, and neither was the author (even if the risks he was facing were far less serious). The latter did not want to disappoint his colleagues in Zimbabwe (who after all suffered from the situation as well), and did not want to run the risk of being accused of neo-colonial western arrogance. Nevertheless, taking a firm stance is sometimes needed, because if nobody does, nothing ever changes. In order to avoid or spread risks, it may be wise to act collectively rather than individually. Moreover, choosing an appropriate time and place is important (but difficult to determine in advance) if one wants to be effective.

Questions for discussion

(1) Choose a communicative intervention programme or organisation that you are familiar with. What key guidelines would you like to include in a 'code of conduct' for communication workers in this programme/organisation?

(2) What would you advise communication workers to do when they are faced with complex ethical and political dilemmas?

4 The role of communicative intervention in policy planning: instrumental and interactive approaches

When discussing different communication strategies in sections 2.2. and 2.3, we mentioned the existence of *persuasive* and more *interactive* modes of communicative intervention. Historically, more interactive forms have been advocated in response to – or better, as a critique of – persuasive forms. Underlying these different modes of communicative intervention are fundamentally different theoretical ideas of *steering* and *directing* change and innovation. Thus, the division between persuasive and interactive modes of intervention is strongly connected with changing ideas about *planning*. Below we present two episodes in the evolution of thought on this, with an emphasis on the corresponding ideas concerning the role of communicative intervention in change. Because of the substantially different ideas, assumptions and attitudes that characterise the two views on planning and the role of communication in it, we will refer in this section to *instrumental* versus *interactive approaches* to communicative intervention. The term 'instrumental' is closely associated with 'persuasive', but not fully identical as some forms of communication are not in themselves 'persuasive' but are still part of a wider 'persuasive' policy (e.g. one can simply inform people about the existence of a new law, without making an attempt to persuade people to act in accordance with it). In speaking of *approaches*, we transcend the earlier idea of just speaking about different *strategies* (see sections 2.2 and 2.3). We are in fact dealing with two fundamentally different philosophies and mindsets regarding communication and change. This also implies that, in principle, the strategies (Table 2.1) and functions (Table 2.2) identified in section 2.2 can be performed in different ways, depending on the mindset of those involved.

4.1 Top–down planning and instrumental communication

4.1.1 'Blueprint' planning and problem solving

During roughly the 1950 to 1990 period, it was quite common in development circles to think of change and innovation as something that could be *planned* (e.g. Havelock, 1973). It was thought useful to define in advance clear goals and outcomes for the future (in connection with the problems of that moment), and possible to organise rationally a series of steps that would eventually lead to the desired outcomes. Many ideas and procedures regarding project planning reflected, and still do, this type of thinking. Models for project planning typically include a number of steps such as:

(1)   problem definition and problem analysis;
(2)   setting eventual and project goals;
(3)   further diagnosis of the causes of problems/problem analysis;
(4)   identifying alternative solutions;
(5)   comparing and evaluating alternative solutions in relation to goals/criteria for goal achievement;
(6)   choosing between alternative options and solutions;
(7)   developing an action plan to realise solutions;
(8)   implementing the plan;
(9)   monitoring and evaluation of goal achievement;
(10)   adapting the action plan.

The above steps are heavily inspired by rational decision-making theory. The idea is that if one follows these steps rationally and thoroughly (i.e. on the basis of adequate knowledge and information), one can achieve the desired outcomes in a given time (i.e. the duration of a project). Projects, then, are perceived as relatively short inputs that mostly come from outside, and serve to achieve pre-specified ends (see Figure 4.1, and also Long & Van der Ploeg, 1989 for a thorough analysis and critique).

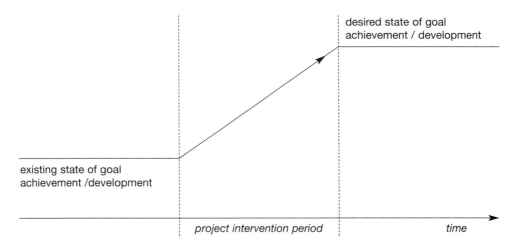

Figure 4.1   The impact of a project conceptualised in blueprint planning.

4.1.2   The instrumental model of communicative intervention

In line with this somewhat 'mechanical' view of social change and innovation (i.e. the idea that one could 'design' future society in a rational, organised and predictable way, in much the same way as one would design a machine), the role of communication was looked at in an 'instrumental' way.

The 'instrumental model' of communicative intervention is characterised by two important and interrelated features. First, instrumental forms of communicative intervention take place *after* the goals and corresponding policies and/or innovations have been defined by outside agencies. The prime idea is to persuade as many people as possible to *accept* a given policy (as developed by policy-makers), or to persuade as many people as possible to *adopt* a given innovation (as developed by scientists). Thus, this type of persuasive communication is more or less an '*end of pipe*' phenomenon (i.e. it takes place only at the end of the policy pipeline); it only becomes important after the policy and/or technology design process has been completed (Aarts & Van Woerkum, 1999).

A second feature of the instrumental model of communicative intervention is that communication is used deliberately as a policy instrument (in conjunction with other instruments) in order to steer and direct human behaviour, which is thought of as being largely predictable. This feature is best illustrated by the 'sorting scheme' introduced by Van Woerkum (1990a; see also Van Woerkum et al., 1999) (see Figure 4.2).

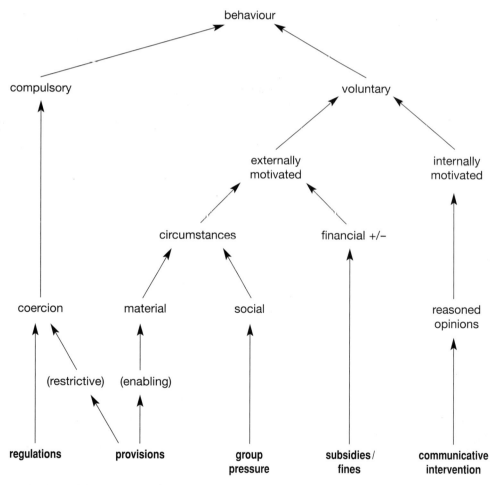

Figure 4.2 The relationship between communicative intervention and other policy instruments aimed at stimulating behavioural change, as conceptualised by Van Woerkum (1990a). Policy instruments in bold.

In this scheme a rather strict distinction is made between 'voluntary' and 'non-voluntary' (or 'compulsory') behaviour[1]. As Van Woerkum's scheme (1990a:268) points out, 'compulsory' behaviour can arise from coercion that derives from the sanctioning of laws and regulations or constraints caused by restrictive provisions. For example, car drivers may be coerced into reducing their speed by adequate enforcement of speed restrictions through intensive policing and high fines, and by provisions such as speed-reducing road barriers. Voluntary behaviour, then, can be either 'externally' or 'internally' motivated. 'Externally motivated' voluntary behaviour originates from material and social circumstances or financial impulses (subsidies/fines/taxes) brought into being by corresponding policy instruments. Farmers can, for example, be stimulated to hand in used tractor oil or other chemical wastes by installing collection facilities in garages and agricultural shops, by organising farm-to-farm collections, or by giving financial rewards for each returned item. 'Internally motivated' voluntary behaviour is seen as arising from reasoned opinions (e.g. a conviction that reducing speed or handing in used oil is a sensible thing to do), that can be influenced by communicative intervention.

Van Woerkum stresses that it is important to recognise that communication is not a particularly 'strong' policy instrument, and that a careful balance needs to be worked out between communicative intervention and other policy instruments. In some cases, communication could play a dominant role in the intervention, for example in cases where the visibility of the desired behaviour is limited (so that the behaviour is hard to control, e.g. in cases of sexual behaviour), when sanctioning is very expensive, or when a quick response is needed in the absence of a suitable legal framework for sanctioning. Here, in the absence of more 'forceful' policy instruments, communication is used in a directly persuasive mode, to convince people about, for example, the importance of safe sex. In other situations, communicative intervention could be subordinate to other policy instruments and limit itself to merely communicating the existence of regulations, sanctions, provisions, subsidies, taxes, fines, etc. Here communication is not directly persuasive, even if it is part of a wider persuasive policy campaign. In such a campaign, non-communicative instruments at the same time communicate for themselves in that their very existence can be seen as a 'message' from those that introduce them. In more ambiguous cases, communication and other policy instruments could play a more complementary role in efforts to change people's behaviour (Van Woerkum, 1990a:269–70).

As seen from the above, the 'sorting scheme' can be used in two interrelated ways: (a) for the analysis of current behaviour in connection with existing policy mixes; and (b) for the design of alternative policy mixes. If, for example, the current situation is that farmers do *not* return chemical wastes, the model can be used to analyse which policy instruments are the limiting factor(s). Are people internally motivated to return chemical wastes? If not, what externally motivating instruments are in place? Are sufficient provisions available? Is the financial reward system stimulating? In relation to this type of use, the scheme can be combined with the model for

---

[1]   Although a gradual distinction between more and less 'enforced' forms of human behaviour is useful, we argue in Chapter 5 that all human behaviour is subject to social influences and pressures, which implies that the term 'voluntary' is somewhat misleading.

explaining farming practices that is developed in Chapter 5, as there are some close interconnections between the variables that explain practices, and the policy instruments that may be used to influence them effectively (see section 5.3.2).

4.1.3   In conclusion

In all, the instrumental model of communicative intervention has much in common with top–down/blueprint planning, as it starts from essentially similar assumptions, such as:

* the idea that human behaviour/society can be manipulated in a predictable way, if only one has an adequate understanding of current behaviour/society;
* the conviction that it is possible to gain an adequate understanding of the 'causes' (or 'determinants') of human behaviour, which are in turn the 'causes' of certain problems in society.

4.2   Process management and interactive communication

4.2.1   Process management towards innovation

From the 1980s onwards the ideas about the feasibility of planning change and innovation have altered quite dramatically. A number of experiences and findings have contributed to this:

* A great many changes and innovations that were planned never materialised, and original project or policy goals were hardly ever realised. Moreover, intervention efforts often appeared to have unintended and unanticipated social and agro-ecological consequences. Hence, human behaviour and society (and at times also the agro-ecological environment) proved to be far less predictable and controllable than expected. In connection with this, top–down planning often appeared to be an *obstacle* to change and innovations; while new developments and insights called for the adaptation of goals and plans, project planning procedures often did not allow this (e.g. Leeuwis, 1993, 1995).

* Although on paper (e.g. in project documents and evaluation reports) projects looked straightforward and rationally organised, the everyday practice of projects was much more chaotic. Project staff had to deal with ever-changing circumstances and unanticipated dynamics, including internal conflicts, tensions with surrounding institutions, and antagonisms between societal stakeholders (Crehan & Von Oppen, 1988; Long & Van der Ploeg, 1989). Thus, formal project documents appeared to be highly artificial reflections of project practice, needed merely to satisfy the bureaucratic needs of donors and government institutions.

* The idea that one could generate unequivocal knowledge and understanding of causes and effects in human behaviour and the functioning of society at large has eroded rapidly since the 1980s. It became increasingly clear that multiple realities (see Chapter 6) exist not only among societal stakeholders, but also among scientists belonging to different schools of thought. This considerably hindered the aspiration of developing 'objective' knowledge (see section 6.5) on

which rational planning could be based. Moreover, it was increasingly realised that human action was less determined by external factors and conditions than assumed. Instead, it became increasingly clear that human beings are active agents (rather than externally determined 'zombies'), that respond to external circumstances in an active, creative, flexible and contextual way (Long, 1989; Long & Long, 1992). In fact, the whole idea of 'the person' as an independent entity with relatively stable characteristics has been called into question, as individuals appear to be highly adaptive to different circumstances, and tend to have different identities and opinions in different interaction settings. In line with this, Long argues that 'individuals' are part of, or even constituted by, a wider web of relationships that impinge on their practices and actions, and hence he speaks of the *social actor* rather than the individual (Long, 1990:9).

● In view of such complexities, instrumental forms of communicative intervention frequently suffered from typical practical problems and tensions:

   (a) Instrumental campaigns tended to be based on the perceptions and purposes of an intervening party, in relation to a problem, and in many instances these did not coincide or connect well with the aspirations, views and perception of the problem by audiences, leading to rather limited responses.

   (b) Instrumental campaigns were often directed at individuals in a specific category only (e.g. female farmers), whereas the changes strived for depended on simultaneous change by several individuals and/or categories of actors (e.g. female farmers, male farmers, community leaders, veterinarians, traders, etc.). Typically, therefore, instrumental campaigns did not create conditions and opportunities for interdependent actors to engage in effective co-ordinated action.

   (c) The mechanical logic and emphasis on one-way communication that characterised many instrumental campaigns tended to reduce the learning capacity of such interventions. Feedback was not received and/or solicited, causing rigidity, inflexibility and lack of adaptive capacity.

Inspired by such experiences and insights, the idea of directing change and innovation towards pre-determined goals has been abandoned by many. It has been replaced by the idea that intervention should be a flexible process, in which goals and means are continuously adapted to ever-changing circumstances, insights and emergent dynamics. Initially, the nature of this process was still regarded as connected with '*planning*', as reflected in popular terms like 'process planning' (Brinkerhoff & Ingle, 1989) or 'participatory planning'. However, it has been proposed that terms like 'process *planning*' still suggest incorrectly that social processes can somehow be controlled (Aarts & Van Woerkum, 2002). Instead of talking about 'planning', we will refer in this book to *network building*, *social learning* and *negotiation* as the key processes to be supported in deliberate efforts to induce change and innovation (Aarts, 1998; Röling & Wagemakers, 1998; Leeuwis, 2000a). The term *network building* refers to the idea that change and innovation imply the establishment of new relationships between people, technical devices and natural phenomena (see sections 2.1.2 and 8.2). The idea of 'social learning' captures the fact that change is connected with individual and/or collective cognitive changes of various kinds (see section 5.3.1). Finally, the term 'negotiation' is used to indicate that meaningful changes of the status quo are likely to be accompanied by conflicts of interests between stakeholders, and that such conflicts need to be resolved (at least to some extent) by conflict management strategies such as negotiation, in order to make change possible. We further discuss social learning and negotiation in Chapters 9 and 10.

We speak about *'process management'* to indicate that social learning and negotiation processes cannot be 'planned' in advance. However, such human processes can, to a degree, be supported, guided and improved by facilitators (even if the process and its outcomes remain unpredictable), provided that the emerging dynamics are carefully monitored (see Chapter 14). The term 'planning', however, remains important in interactive approaches. First, although processes cannot be 'planned' in detail, they do require systematic thinking, and individual activities need to be well prepared and flexibly planned (see Chapter 15). Second, planning is important from an organisational management point of view. Organisations need to make sure they employ the right staff and create appropriate conditions (in terms of, for example, transport, communication facilities, secretarial support, time-management, equipment, etc.) for them to do their facilitation work. Finally, the components of the planning cycle can still be useful in interactive processes; not in the form of 'steps' or 'phases' but rather as 'tasks' and 'functions' that may orient discussions in a social learning and negotiation process. However, the enhancement of learning and negotiation processes involves a number of additional 'tasks' that are essential for forging productive outcomes, and which are generally overlooked in planning discourses (see Chapters 9, 10 and 14).

4.2.2   Arguments for an interactive model of communicative intervention

The process management approach to change and innovation is closely associated with an interactive, or 'participatory', model of communicative intervention. Here the role of communication is not so much to 'sell' or 'implement' pre-defined goals, policies and innovations, but rather to help *generate* and *design* appropriate goals, policies, and innovations in close interaction with societal stakeholders (Van Woerkum et al., 1999). More specifically, communication becomes an integral part of the *facilitation* strategies that aim to enhance learning and negotiation towards change.

In the development literature one can find several arguments for organising change processes in an interactive/participatory way (e.g. Chambers, 1994a; Pretty, 1994; Webler & Renn, 1995).

A first set of arguments is rather *pragmatic* in that it emphasises that an interactive approach is needed in order to be *effective* (Brinkerhoff & Ingle, 1989; Van Dusseldorp, 1990; Geurtz & Mayer, 1996). Examples of such arguments are that close interaction with stakeholders is needed:

- In order to gain *access to all sorts of relevant knowledge, insights, experiences and/or creativity* that stakeholders may have regarding, for example, history, the nature of problems, possible solutions, changing circumstances and capricious local dynamics, etc. (see Table 14.1). Such access is needed, then, in order to get proper information and feedback on which to base intervention initially and then adapt it continuously. In other words, interaction is needed to build in sufficient *learning capacity* in intervention processes.

- In order to gain *access to relevant networks, resources and people* that may be relevant to building effective links and support networks for innovations to materialise. Thus, interaction is needed for dealing with the collective nature of innovations and 'alignment' (see section 2.1.2 and Chapter 8.).
- Because it is theoretically inconceivable that people will change without some degree of mental, emotional and/or physical involvement. Interaction, then, is needed *to generate the required involvement and 'ownership'*.

Second, other authors advocate an interactive approach on the basis of *ideological* and/or *normative* grounds. Basically, the argument here is that *citizens have a wish, a moral right and/or a duty to be actively involved* in shaping their own future (Rousseau, 1968; Rahman, 1993).

Third, *political* considerations are often used to justify an interactive approach. The idea here is that interactive processes can help to *emancipate and empower particular groups* in society (e.g. Freire, 1972; Friedman, 1992; Rahman, 1993; Nelson & Wright, 1995). Through their involvement disadvantaged groups can build up the necessary skills, insights and resources (e.g. forms of organisation) that help them to strengthen their position in relation to others.

Finally, an interactive approach is often advocated to enhance the *accountability* of intervention activities. In this line of thinking, involvement of those who are supposed to benefit from external interventions helps to make projects and their staff more accountable to their 'clients'. By ensuring that prospective beneficiaries have a certain amount of control over project budgets and activities, interventions are expected not only to become more effective (the pragmatic argument) but also more legitimate from an ethical perspective. In the latter sense, accountability arguments can be seen to arise from a need felt by interventionists to justify and legitimise themselves and/or to create a favourable reputation (see also Hilhorst, 2000). To the extent that such aspirations are related to the desire of intervention organisations to 'stay in business', the accountability argument is also very much an 'organisational survival' argument.

4.2.3   In conclusion

We can conclude that the interactive approach towards communicative intervention is based on radically different ideas and assumptions from those underlying the instrumental model. In particular, it is characterised by far less confidence in the predictability and controllability of change. In addition, we have seen that different authors emphasise different reasons for advocating an interactive approach. In practice, the underlying rationale for using an interactive approach may vary from context to context, and may significantly affect the way processes are organised. In this respect, Nelson and Wright (1995) differentiate between the use of participatory methods and techniques *as a means* (that is, when an interactive approach is used mainly for pragmatic reasons in order to further goals that are still largely externally imposed), or *as an end* (when the process is used to empower participants so that they can determine their own future; that is, when political and ideological arguments prevail)[2].

4.3 Shortcomings and conditions: the relation between interactive and instrumental approaches

As mentioned earlier, instrumental forms of communicative intervention have become increasingly unpopular (if not politically incorrect) in intervention discourses due to their essentially 'top–down' nature and frequently badly adapted solutions, leading to ineffectiveness and/or detrimental consequences. Despite this unpopularity in communication for innovation rhetoric, we still see many instrumental intervention activities and campaigns in practice. On the one hand, this may be connected with political and/or organisational conditions that are unsuitable for an alternative, i.e. more participatory, style (see Chapter 14). On the other hand, we may have to acknowledge that it may not be possible and/or desirable to do away with instrumental forms of intervention altogether. In any case, it has become clear that interactive approaches are not a panacea and can run into many problems. The problems frequently mentioned include the fact that interactive processes:

- can be very time consuming and costly for both communication workers and clients;
- may generate little enthusiasm on the side of those who are supposed to participate;
- are frequently affected negatively by conflict, unequal power relations and/or unequal capacities to participate;
- at times do not help to generate tangible results or innovations;
- may end in compromises that nobody is really happy with;
- regularly produce outcomes that are ignored by those who are supposed to incorporate them;
- frequently raise high expectations which cannot be met;
- may reinforce the position of relatively powerful and well-to-do clients;
- often bypass regular democratic procedures;
- may prevent people from taking responsibility;
- etc.

Although these difficulties can be very real, we believe we should not throw away the baby with the bath water. Instead, we should focus attention not only on furthering the quality of process facilitation, but also on improving our insight into the factors that affect the productiveness of interactive processes, and on the conditions that may have to be in place for this to happen. We reserve this discussion for later in this book, after we have discussed the concepts of *social learning* and *negotiation* in more

---

[2] One could argue that at a higher level the use of interactive approaches always has pragmatic connotations; after all the wish to empower people is a goal usually set by some in order to affect the position of others.

detail (Chapters 9 and 10) as these are central to an interactive approach. Among other factors we will discuss important pre-conditions for learning (section 9.4) and negotiation (section 10.3). On the basis of this we conclude in Chapters 11 and 14 that a certain balance and iteration between instrumental and interactive intervention activities may be required in several instances.

Questions for discussion

(1) A government wants to reduce the occurrence of uncontrolled bush fires that are caused by farmers' land clearing practices. Discuss whether (and when) the government should adopt an instrumental or an interactive approach.

(2) Which argument for an interactive model of communicative intervention do you find most convincing? Where does this place you in the discussion (see section 4.2.3) about participation *as a means* or *as an end*?

(3) Many organisations, governments and politicians are reluctant to throw away the idea that the future can be planned, and many policy and project documents remain to be written with reference to the stages of planning (see section 4.1.1). In your view, why is the idea of planning so persistent?

PART 2
The relations between human practice, knowledge and communication

In Part 2 we first try to improve our understanding of human practices, and gain insight into why people do what they do (or do not do something) at a given point in time (Chapter 5). From this, it transpires that knowledge and perception play a vital role in shaping human practice. This implies at the same time that innovation and development (i.e. modification of human practice) require and/or go along with changes in knowledge and perception. Thus, in Chapter 6 we discuss several complexities regarding the notions of knowledge and perception.

   Communication, then, is an important process that people use to exchange experiences and ideas, and hence a vital trigger for altering knowledge and perception (Chapter 7). In this Part we take a closer look at these key concepts in the field of communication and innovation studies.

5 Understanding human practices: the example of farming

From the previous chapter we have learned that agricultural extension involves forms of communicative intervention that are aimed at facilitating and directing change processes to deal with the complex problems that face agriculture today. Moreover, we have argued that new solutions and innovations involve new patterns of co-ordination between people, technical devices and natural phenomena. For a sharper view of the potential and limitations of communication for innovation, a basic understanding is useful of the variables that play a role in shaping human practices regarding the natural and social environment. Without a proper understanding of why people do what they do (and do not do) at a given point in time, it will be impossible to contribute to change effectively.

Even though we have argued that rural change and innovation often do not involve only farmers, we will use farmers as an entry point for constructing a basic model for understanding human practices and/or responses to proposed changes. Thus, we will refer to farmers in order to *illustrate* and *explain* the model, but the model itself is meant to have a more general value, and the variables in it can also help us to better understand the practices of other categories of people, such as consumers, environmentalists, scientists, communication workers, etc. The model presented in section 5.2 draws on several disciplines, including agrarian and rural sociology, anthropology and social psychology. On the basis of this model, several important implications for communication for innovation are discussed in section 5.3. Before presenting the model, however, we will first discuss briefly what we understand by the term 'practices' in the context of farming.

5.1 Different levels and domains of farming practice

In this book we use the term 'practices' to refer to things people 'do' (and 'do not do') on a more or less regular basis (see also Giddens, 1984). Thus, one particular event or human action is not a 'practice', but when a person engages in similar actions over time and/or if many people act in a particular way we can speak of a 'practice'. Thus, practices are essentially *patterns of human action* or *regular activities*. It is important here to recognise that practices may or may not be easily recognisable in physical terms; one can, for example, easily observe that a farmer regularly feeds maize to his animals, but not that he or she avoids being dependent on input markets. Similarly, it is relevant to note that practices may or may not result from a *conscious ex-ante decision* to do something. In fact, many practices can be seen as *routines* that are not actively deliberated on as they take place, even if they may derive historically from deliberate decisions in the past. For example, a farmer does not deliberate every day whether or not to feed the cattle by hand or with the tractor, once he or she has decided to use a tractor some 10 years ago. Similarly, practices may be reasoned about (or rationalised) only *after* they have emerged; a young farmer may, for example, continue the farming practices of his or her parents without much active deliberation, and only later come to think and reason about them actively. Of course, human beings are involved in many practices over time, which are connected to each other in a complex way. Moreover, we can look at practices from different angles. We will illustrate this by discussing farming practices as a multi-dimensional phenomenon.

5.1.1   Farming practices at different hierarchical levels

When managing their farms farmers 'do' things at various levels. These levels can be looked at as different degrees of zooming in or zooming out; depending on one's position and focus one either sees 'the farm in relation to its wider environment', 'an individual animal' or something in between (see below). We will briefly explain some important levels, starting with the lowest level. In doing so, we will illustrate that different kinds of innovations (including technologies) exist at different levels.

The level of 'individual' production objects

When taking a 'close-up' of a farm one may, depending on the farm, see a cow, a section of land, a hoe, a piece of machinery, a bag of fertiliser, etc. These are the items that farmers directly manipulate and/or work with, and we will call them 'individual' production objects. There are many specific things that farmers 'do' with or to these production objects on a regular basis; they feed certain animals a particular mixture, they inseminate a particular cow with a special breed of bull, they use a specific fertiliser dosage in a particular corner of a field, etc. In other words, they engage in particular farming practices. Innovations at this level may, for example, include:

● an improved piece of equipment;
● an adapted way of handling existing equipment;
● a novel way of tilling a particular piece of land;
● a new crop variety;
● a change in fertiliser dosage or timing;
● an adapted way of making compost.

The level of 'aggregate' production objects

When looking at a greater distance, one does not see individual items, but aggregates. One sees 'a herd of animals', 'a grazing area consisting of several fields' or a 'machinery set-up'. Also at this level, farmers engage in farming practices; for example, they circulate a herd over a grazing area in a specific way, or they may have a particular way of maintaining, using or sharing a machinery set-up. Innovations at this level may, for example, include:

● a different pattern of grazing a herd (e.g. strip-grazing);
● a new fencing lay-out in the grazing land;
● a change over to a new cattle breed;
● a new machinery-sharing arrangement with neighbours.

The level of the farming system

When taking a bird's-eye view one sees the farm as a whole. One may see a mixed farm, characterised by complex relations and interconnections between the herd, the grazing area, the cropping area, the lay-out of fields, the machinery set-up, the cropping system, the farming family, etc. Again, farmers engage in organisational practices at this level which result in the maintainance of a particular order and balance between the various components. Innovations at this level typically include adaptation of the whole system, leading to a totally new mode of organisation, for example:

- a change from mixed farming to specialised farming (or the other way around);
- moving from conventional farming to biological farming;
- enlarging the farm;
- switching to a new cropping system;
- engaging in on-farm food processing.

The farm in its environment

When looking from a plane or satellite one might see a farm situated close to a river, far from a city, surrounded by a bad road and service infrastructure, among political unrest, and in a favourable climate zone. Such an environment produces opportunities and constraints. A farmer needs to forge and maintain balanced relations between the farming system and the wider environment. Innovations at this level may, for example, include:

- relocation of the farm;
- improved arrangements with input suppliers;
- shortening the chain between farmer and consumer;
- establishing co-operatives to enhance the bargaining power of farmers;
- developing suitable conditions for multi-functional agriculture.

As we have seen, farmers engage in practices at various *levels* of the farm, and in order to operate successfully these practices must be *co-ordinated* carefully. Similarly, *changes* in farming practice may emerge or be considered at each level, as innovations originating in one level are likely to have implications for another. For example, if a farmer aspires to improve the health of a herd, this may have repercussions for the way different categories of animals are fed, which again may lead to a conversion of grazing land into cropping area (or the other way round). Thus, innovations impinge on different levels and require careful co-ordination.

5.1.2   Different domains of farming practice

Instead of looking at a farm in a hierarchical way, one can also dissect it 'horizontally' and distinguish between different 'domains of farming' (Van der Ploeg, 1991). Such domains are essentially different *aspects* that need to be considered while farming. When managing a herd of cattle, for example, one needs to think about various *technical* aspects, such as animal health, feeding, fertility, production level, etc. At the same time, one may consider various *economic* variables such as income, cost effectiveness, cash flow, credit requirements, etc. In addition, farmers can take into account *social* issues, for example relationships with household or family members, with the community and/or with outside institutions. When thinking about farming practices at different levels, farmers will usually consider their implications for various domains. Thus, all farming practices can *at the same time* be seen as technical, economic and social practices, meaning that farmers do things in the technical, economic and social domain. Practices and innovations in various domains need to be co-ordinated carefully in order to arrive at a coherent farm. Important domains of farming practice, and hence innovation, that farmers are likely to take into account include[1]:

- *In the technical domain*: soil fertility, crop protection, animal health, production and yield, storage facilities, spatial organisation of the farm, regeneration of production potential, etc.
- *In the economic domain*: income, profitability, marketability, taxes, investments, cash flow, credit, fixed costs, variable costs, etc.
- *In the domain of social-organisational relationships*: relationships with input-providing organisations, organisations on the output side, state organisations, NGOs, members of the household, community members, farm labourers, ancestors, spirits, gods, etc.

5.1.3 Farming practices at different points in time

Farmers need to co-ordinate practices not only at different levels and in different domains, but also across different times. What a farmer does today needs to be co-ordinated with what is expected to happen tomorrow, or in the next month, season, year, decade and/or stage in the family cycle (Chayanov, 1966). Thus, farmers can do things with a different time horizon in mind. Management scientists speak of 'operational', 'tactical', and 'strategic' decisions, which are geared towards yielding short-term, medium-term or long-term consequences respectively (Davis & Olson, 1985). Farmers can, for example, invest in buying new land now, so that the farm can be taken over by two children in 15 years' time, or alternatively they can sell land because there will be no-one to inherit or continue farming (strategic decisions). Or they can plant an early variety of a crop, in the expectation that they can sell it at a higher price (tactical decision). Finally, farmers can postpone ploughing the land today, because they expect better weather conditions tomorrow (operational decision). Thus, we could say that farmers need to carefully match and co-ordinate past, present and future practices (at different levels and domains).

In all we see that farming entails a complex set of activities, in which distinct practices have implications in a three-dimensional space; that is, they have consequences

---

[1]   Note that Van der Ploeg (1991), who introduced the notion of domains of farming, speaks of slightly different domains. He distinguishes the domains as: production, reproduction, family and community, and economic and institutional relationships.

at different levels of the farm, in different domains of farming, and at different points in time. Managing a farm, then, is a process that involves the careful co-ordination and weaving together of practices involving each of these dimensions.

5.2   Understanding the social nature of technical practices

For a long time extensionists assumed that agriculture was a largely separate activity, carried out and/or decided upon by a single individual (i.e. 'the farmer'), and grounded primarily in rational technical and economic considerations. In the last few decades we have learned that such assumptions are seriously flawed. For many farming families, agriculture is only one of various income generating activities, implying that agricultural practices can only be understood in the context of practices in, at first sight, 'non-agricultural' domains (Hebinck & Ruben, 1998; Farrington et al., 1999; Ellis, 2000). Furthermore, it has become clear that agricultural decisions are not made solely by the individual 'head of the household', but extend to other household and/or community members (Maarse et al., 1998), and are also influenced by other actors in, or even outside, the agricultural production chain. Consequently, in addition to 'simple' technical and economic considerations, a range of other, often less tangible, issues play a role in shaping farmers' practices; these include issues of power, identity, culture, conflict, religion, risk and trust. In short, farming practices are shaped in a series of social interactions between different people at various points in time and in different locations, within the context of a wider social system.

   In developing a model that captures these complexities, we will start with a very simple model (derived from Röling & Kuiper, 1994) which identifies in laymen's terms which 'variables' can help to explain farmers' practices. This model suggests that what farmers (and other human beings) do or do not do depends on what they:

- *BELIEVE TO BE TRUE* about the biophysical and social world (i.e. what they '*KNOW*');
- *ASPIRE* to achieve (i.e. what they '*WANT*');
- (think they) are *ABLE* to do;
- (think they) are *ALLOWED* and/or *EXPECTED* to do.

A model like this merely offers a checklist for identifying factors that explain both what people *do* and what they *refuse to do*. In the latter case, for example, one can ask oneself for what *reasons* farmers do not use a particular chemical pesticide:

- Because they do NOT BELIEVE that the chemical pesticide may help control pests and improve yield?
- Because they do NOT ASPIRE to improve yield and/or use the chemical pesticide, e.g. in view of expected negative consequences and/or other priorities (even if they may 'believe')?
- Because they are NOT ABLE to use the chemical pesticide, e.g. since they lack skills or are unable to buy the pesticide and/or the necessary equipment due to lack of availability or money (even if they may 'believe' and 'aspire')?
- Because they are NOT ALLOWED/EXPECTED to use the chemical pesticides, e.g. because religious leaders or the government use pressures and sanctions to prevent them from doing so (even if they may 'believe', 'aspire' and 'be able to')?

In a similar way, the variables can help us to understand the things farmers already do. Thus, even if simplistic, these four variables can be quite helpful in understanding what farmers do and do not do at a given time. And by doing so, it also gives us some entry points for contributing to change and innovation; that is, based on an analysis of reasons, we can try to alter what farmers *believe to be true*, what they *aspire* to, what they are *able* to do and/or what they are *allowed* to do. As mentioned earlier (see section 5.1), such *reasons* should not be equalled to deliberate ex-ante decisions or intentions, and neither should they be confused with simple 'causes' of human action (even if, in retrospect, actors themselves at times present their behaviour as being explainable and/or 'caused' by certain views and perceptions[2]).

However, the model is not refined enough, and does not point to interrelations between the variables. Moreover, it is static rather than dynamic, and we would like to connect it with relevant insights and terminology from various academic disciplines. The model presented in Figure 5.1 is a more refined attempt to capture and

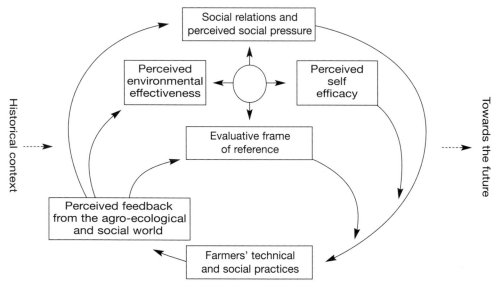

• *Social relations and perceived social pressure* connects with 'to be **allowed** and/or **expected** to'.
• *Perceived effectiveness of the social environment* connects with 'to be **able** to'.
• *Perceived self efficacy* connects with 'to be **able** to'.
• *Evaluative frame of reference* incorporates both 'to **believe**' and 'to **aspire**'.

Figure 5.1   Model of basic variables that are relevant to understanding individual farmers' practices and responses to proposed alternatives.

---

[2]   Whether or not it is useful and possible to interpret human behaviour as being 'caused' by internal or external factors is a big debate in the social sciences, and one which we will not enter at this point. Elsewhere in this book (Chapter 19) we discuss some differences between the social and natural sciences. Here we show that, even *if* one would like to speak of 'causes', they are extremely difficult to assess.

link a number of important variables for understanding 'individual' farmer practices and/or responses to proposed alternatives. In the text underneath the figure we have roughly indicated the relations between the two models.

In the following sections, we discuss, further refine and illustrate each variable separately. Subsequently, we discuss the links and dynamics through which these variables are connected, indicating the social shaping of farming practices.

5.2.1   Evaluative frame of reference: the basis for reasoning about practices

What farmers do or do not do depends in part on (a) their perceptions of the manifold consequences of certain practices; (b) the perceived likelihood that these consequences will emerge; and (c) their valuation of such consequences in relation to a set of aspirations. Taken together, we call this farmers' *evaluative frame of reference*, as it relates closely to their knowledge and mode of reasoning about the natural, economic and/or social world. Social psychologists call the overall inclination that results from reasoning about such aspects a person's *attitude* towards a specific farming practice (Fishbein & Ajzen, 1975; Ajzen & Fishbein, 1980).

Box 5.1   Overview of sub-variables constituting the 'evaluative frame of reference'.

Evaluative frame of reference

Perceived (technical and socio-economic) consequences
×
Perceptions of (un)certainty, likelihood and risk
×
Valuation of consequences and risks regarding aspirations

Perceived technical and socio-economic consequences (i.e. beliefs about consequences)

As farming is a complex and carefully co-ordinated activity (Van der Ploeg, 1991) even relatively 'minor' changes in agricultural practice may have a number of consequences, about which farmers usually have certain ideas beforehand. Farmers do not only consider possible technical consequences (e.g. yield expectations, required inputs, impact on quality, etc.), but also socio-economic effects (required labour organisation, income effect, impact on social relations, etc.). Take, for example, a new hybrid rice or maize variety. Based on various sources of information and experience, farmers may expect that using such a variety leads to, for example: improved or reduced yields of grains and crop-residues; increased or reduced vulnerability with regard to pests, diseases and weeds; changes in labour requirements at particular points in the growing season; increased or reduced dependence on external inputs like fertiliser and seed; a need to arrange a short term loan; changes in taste, storage requirements and marketability; etc. Underlying such perceived consequences there is usually a 'web' of interrelated *causal attributions* that actors perceive to be valid; that is, farmers tend to apply, either explicitly or implicitly, certain modes of causal reasoning ('x leads to y through mechanism z') in explaining why certain consequences happen, and in doing so they attribute consequences to specific causes. For example, they may think that hybrid maize is more susceptable to pests than traditional varieties either because 'insects like the taste better', 'the plants are weaker', 'they make the soil tired', 'the ancestors are angry', or a combination of these.

Farmers' knowledge and beliefs can originate from various sources, for example, from local and/or own experiences and experimentation (see Chapter 13, section 13.5.8), from experiences elsewhere (as passed on, for example, by communication workers, traders or migrants), or from formal agricultural research. In many cases, the precise origin can no longer be established. We often speak of a mixture of sources, and hence of 'creolised' (Hannerz, 1996) knowledge and perceptions that include both 'local', 'external' and/or 'scientific' elements (see Chapter 6). At the level of a community or society, such knowledge is often part of a shared knowledge base that Giddens (1976) calls 'mutual knowledge'. As we discuss later in Chapter 6, mutual knowledge is not always explicit, but in many instances remains implicit or taken-for-granted (Schutz & Luckmann, 1974; Giddens, 1976). Nevertheless, such 'tacit' knowledge (Nonaka & Takeuchi, 1995) can play an important role in shaping farmers' practices.

Despite farmers' knowledgeability and expertise, they are usually not all knowing and may at times lack the insight to draw appropriate inferences. This is especially so when phenomena are difficult to observe (Richards, pers. comm.) either because of their inherent characteristics (e.g. viruses and nutrient flows), gradual nature (e.g. decline of soil fertility) or spatial distance (e.g. migration of insects from other regions). Hence, there is scope for learning and expansion of their knowledge base. However, for understanding what farmers do and do not do, it is *irrelevant* whether their beliefs about consequences are – in the eyes of, for example, scientists – valid, correct or complete. What is important here is that farmers' practices and/or their rejection of alternatives are, to a degree, associated with their perception of the consequences of such practices at various levels and domains of farming (see section 5.1). In this section we have emphasised consequences in the technical and economic domain. The consequences that farmers may consider in the social-organisational domain, i.e. social-organisational consequences, will be discussed in more detail in section 5.2.4, when we discuss issues of social pressure.

Perceptions of (un)certainty, likelihood and risk

Being exposed to the capriciousness of weather conditions, market prices and the like, farmers tend to be aware of the importance of risk and uncertainty (see also Huirne et al., 1997). Thus, ideas concerning possible consequences of using, for example, a new rice variety are often associated with perceptions about the likelihood of certain consequences (e.g. higher yields), and the risk that they do not occur

(Fishbein & Ajzen, 1975)[3]. For example, farmers may expect that higher yields will, depending on weather conditions, be obtained in 3 out of 5 years, and may foresee total crop failure in the remaining two years.

When reasoning about certain innovations, farmers may consider risks in each of the domains mentioned in the previous section, so that we can speak of 'technical', 'economic' and 'social-organisational' risks (this latter type of risk will be discussed in more detail in sections 5.2.2 and 5.2.4). Many of the points made about the origin and validity of farmers' knowledge and perception about consequences, are also true for their knowledge and perception of chances and risks. Risks are notoriously difficult to assess and predict, so risk perceptions can be 'distorted' in various ways. However, once again, for understanding farming practices one has to take into account farmers' perceptions, rather than other people's evaluations of these. The literature on risk perception (e.g. Slovic, 1987; Nelkin, 1989; Uitdewilligen et al., 1993) suggests that the awareness of 'technical' risks (and also their over- or under-estimation) depends on, among other factors, their frequency of occurrence (e.g. frequent gales versus one serious gale every 50 years), visibility (e.g. diseases caused by visible insects or invisible viruses), magnitude (e.g. low or high number of plants affected), and the directness and duration of consequences (e.g. the immediate damage from a tropical cyclone, versus the gradual effects of declining soil fertility). Moreover, it is important to mention here that risk perceptions may vary across cultures depending, for example, on the extent to which nature is regarded as 'robust', 'fragile', 'capricious' or 'robust within limits' (Thompson et al., 1990; Douglas, 1992). In cultures and circumstances where nature is experienced as capricious, for example, farmers may regard nature as threatening, and may engage in all sorts of practices to control and reduce risk (e.g. build dykes, organise rain-making ceremonies, apply intercropping). Similarly, where nature is regarded as fragile, there may be an inclination to protect and conserve the natural environment (Aarts, 1998).

Valuation of consequences and risks: the importance of different sets of aspirations

In processes of reasoning about the pros and cons of particular farming practices, perceptions about consequences, likelihoods and risks are linked to subjective preferences and aspirations. In this way they are evaluated as being 'positive' or 'negative' (i.e. as 'advantages' or 'disadvantages'). If, for example, using a new rice variety is expected to involve shifting labour peaks, this may be regarded as desirable or

---

[3]   As demonstrated in this example, the terms uncertainty, likelihood and risk are closely related. 'Uncertainty' is a more general concept, referring to a mental state in which people do not know what may happen and/or how to define or deal with a situation. The terms 'likelihood' and 'risk' are more particular in that they refer to the chances that something *specific* may happen, or not. The term 'risk' is associated usually with things that one would *not* like to happen (e.g. having an accident), while 'likelihood' is used more often in connection with things one would like to happen (e.g. winning the lottery). Moreover, the terms 'risk' and 'likelihood' are used frequently with the connotation that they can be *calculated* or *estimated*. However, we use the terms more broadly, since we are interested mainly in human *perceptions* regarding likelihood and risk, which are usually more qualitative in nature and far from precise.

non-desirable depending on the existing labour calendar and preferences with regard to household labour organisation. If more labour is required in periods when there are already labour shortages, such a change is likely to be evaluated negatively.

It is important to note here that several types of aspirations play a role in the valuation process. We mention four basic types, of which the second and third are a partial prelude to the issue of *social influence* that we discuss in section 5.2.4.

First, farmers may have '*technical/economic' goals and interests* for evaluating certain practices. A farmer may, for example, strive to: reduce labour requirements during certain periods (and choose a late harvesting variety); minimise the use of pesticides and fertiliser (and opt for a traditional maize variety); and/or spread risks (and use intercropping in scattered fields across different agro-ecological zones). Note that, at any one time, there will usually be a variety of such interests and goals, which relate to different domains of farming (see earlier list of important domains).

A second type of aspiration involves '*relational' (including 'political') goals and interests*. Farmers may opt for certain farming practices because they want to maintain good relations with family members, neighbours, traders, farm labourers and/or others. For example, they may prefer to hire a neighbour's ox-plough rather than use the services of a mechanised contractor, in order not to disrupt existing labour exchange and support arrangements between families. Similarly, farmers may opt for composite maize varieties rather than hybrids in order to maintain a level of independence from input suppliers. Alternatively, they may use a hybrid variety mainly to maintain good relations with change agents or agricultural administrators.

A third type of aspiration is more *cultural*, and involves *social norms and values*. Values, here, are culturally specific notions about what is 'good' or 'bad', 'right' or 'wrong', 'fair' or 'unfair', and 'important' or 'insignificant'. Such notions are often 'translated' into norms; that is, into concrete collective expectations and prescriptions with regard to human behaviour in specific circumstances. In Bhutan, for example, it is unacceptable in some communities to use even minimal amounts of pesticides since religion forbids the deliberate killing of animals, including insects (Van Schoubroeck, 1999). Here, the underlying values include 'respect for life' while the norm is 'not to kill insects'. In a similar vein, it has for a long time been culturally acceptable in industrialised countries to use agricultural practices with limited consideration for animal welfare or the long-term consequences for the environment. Clearly, such cultural preferences are linked with earlier mentioned 'relational' aspirations in that following or not following norms and values is perceived to affect or not affect relationships with community members, religious leaders, future generations, ancestors, spirits, etc. Moreover, both relational and cultural aspirations are closely linked with people's understandings of their *responsibility* in a given situation; that is, they may influence the extent to which people feel that they or others should take action, or not, in a given situation. For example, although farmers in the Netherlands feel some responsibility to contribute to animal welfare, nature and landscape conservation, many of them feel that government bodies and consumers should take the lead in creating better conditions for farmers to make such a contribution (e.g. by paying them to provide such services). In essence, this example shows that, in view of relational and moral considerations, farmers shift the prime responsibility for conservation to others. Consumers and government bodies tend to do the same, so that eventually none of the parties feel sufficiently responsible to take the initiative, leading to a vacuum in responsibility and effective action (Aarts & Van Woerkum, 2002).

Finally, the fourth type of aspiration can be referred to as *'emotional' interests.* The term 'emotional' refers to human feelings like fear, safety, happiness, love, guilt, appreciation, regret, disgust, peace of mind, anger, etc. Depending on the context, we tend to experience certain emotions as positive and others as negative. Consequences and risks are evaluated at the emotional level as well, and in some cases practices are directly associated with emotions. For example, farmers may not like personally to kill animals they have known for a long time, and may prefer to leave it to others. In many cases, the relationships are more indirect. When certain practices are seen as jeopardising relationships with neighbours (i.e. to go against relational interests) this may at the same time invoke an emotion of regret. Similarly, risky investments may arouse feelings of fear. Thus, emotional and more rational evaluations influence each other and go hand in hand (Van Woerkum, 1991).

In the valuation process farmers may take into account the perceived likelihood or certainty that consequences occur and contribute to the realisation of the various types of aspirations. With respect to the valuation and acceptance of perceived risks, the expected impact on the various aspirations (i.e. the magnitude of potential 'gains' and 'losses') will play a role as well. Here, a number of additional issues are also important, which relate mainly to farmers' perceptions of whether or not such risks can be controlled. We discuss this in more detail in section 5.2.3.

Finding a balance

From the above it can be concluded that, even if we limit ourselves to matters of cognition and evaluative reasoning, understanding farmers' practices is highly complex. Farmers' decisions may involve perceptions about the consequences of such practices in a large number of distinct domains, and are linked with an even higher number of perceptions regarding (un)certainty, likelihood and risk. Moreover, such consequences and risks are being evaluated against a diverse set of aspirations, each with a different priority. This inevitably poses dilemmas and contradictions for farmers, as the consequences in one domain (e.g. production) may be perceived as positive, while the consequences in others (e.g. labour organisation) may be regarded as negative. Another complicating factor here is that we have so far talked mainly about farmers from a single *identity*, namely as 'a farmer'. However, people can approach a situation from various identities (e.g. as 'a farmer', 'a husband', 'a parent', 'a citizen', 'a student', etc.), and depending on their identity in a given situation, the beliefs and aspirations that are salient and relevant to them may change. In short, deciding about farming practices requires a careful balance of numerous considerations and trade-offs. Given this complexity, one wonders how farmers are able to take decisions at all. This is an issue we discuss further in Chapter 9.

5.2.2   Perceived effectiveness of the social environment

For the successful application of farming practices, farmers often depend on others. Thus, an important factor that influences farmers' practices is their perception of whether or not their socio-economic environment is able to support these adequately. Below we will highlight two, closely related dimensions of this perceived environmental effectiveness: perceived effectiveness of the agro-support network, and perceived effectiveness of (inter)community organisation.

Box 5.2 Overview of sub-variables constituting the 'Perceived effectiveness of the social environment'.

Perceived effectiveness of the social environment

Perceived effectiveness of the agro-support network
+
Perceived effectiveness of (inter)community organisation

Perceived effectiveness of the agro-support network

Implicit to each farm technology or practice is a network of support relationships that makes it possible. This is most obvious for high-tech farming practices, but is equally true for technically less-sophisticated practices. For example, it is clear that mechanised farming requires a physical and organisational infrastructure that: (a) makes it possible to either buy tractors and corresponding farm implements, or hire them at the appropriate time; (b) ensures continuous availability of petrol, oil, repair services and spare parts; (c) allows machinery to move around easily; (d) guarantees sufficient access to land over the years (to justify the investment), etc. In other words, one needs well-functioning and reliable dealers, contractors, repair workshops, credit systems, land-tenure arrangements, petrol providers and road infrastructure. In many ways, the use of draught animals as a source of power requires similar arrangements, to ensure the availability of animals, implements, animal feed, grazing lands, etc. Similarly, high-yielding seed varieties often require an adequate supply of both fertilisers and means for biological and/or chemical pest control, and sufficient storage, transport and marketing facilities. The presence of these kinds of necessary infrastructure and services is far from self-evident, and even if they exist they may not operate in a reliable way. Here too, farmers' *perception* of the availability, quality and reliability of physical and organisational infrastructures is what counts. Such perceptions are bound to shape farmers' practices, as farmers are likely to opt only for those practices that they expect to be adequately supported by infrastructure. Important infrastructure that farmers may consider includes:

- organisation of input supply;
- organisation of marketing (traders, transport, storage, etc.);
- governmental price policy;
- road system;
- organisation of water delivery and/or drainage;
- availability of special facilities and services;
- credit systems;
- arrangements for land tenure.

*Perceived effectiveness of (inter)community organisation:
dealing with social dilemmas*

In relation to various farming practices, farmers depend not only on the effectiveness of the wider support network, but also on the practices and behaviour of their colleague farmers. Certain modes of crop-protection, for example, only work effectively if all members of a farming community use them at the same time (e.g. Van Schoubroeck, 1999). The same may be true for the growing of certain newly introduced crops, which may be viable only if other farmers grow them as well, so that overhead costs (e.g. equipment, transport, storage) can be shared. Similarly, the way in which a farmer uses collective resources such as water and grazing lands will partly depend on the way in which other farmers are expected to use these resources. For example, if farmers feel that their colleagues will use whatever water is available, without considering the needs of others, they themselves may also engage in such 'selfish' practices.

Here we touch on an issue that often plays a role in the use of collective (natural or other) resources and services, namely the existence of *social dilemma situations*. These are situations in which a tension exists between short-term individual interests and long-term collective interests (Messick & Brewer, 1983). Communal grazing lands are a classic example of a so-called 'commons dilemma' (Hardin, 1968). Here, the individual farmer tends to have an interest in letting as many cattle as possible graze on the communal land. However, if all farmers let such an individual interest prevail, this leads to serious overstocking, causing animals to die or get sick. These negative consequences can assume such proportions that each farmer would have been better off if they had restrained themselves and grazed fewer cattle on the common grazing land. The basic difficulty here is that interdependent farmers will only decide to restrain themselves if they *trust* that other community members will do so as well. Whether or not such trust exists may depend, among other factors, on the size of the community, the existence, nature and strength of a group identity, and the visibility of co-operative and non-co-operative behaviour (Koelen & Röling, 1994)[4]. Moreover, trust is usually enhanced when agreed rules exist, with the implementation of them supported and controlled by well-functioning organisation (Ostrom, 1990; Baland & Platteau, 1996).

The above is also relevant to a second type of social dilemma situation called a 'public good' dilemma. Here the issue is not so much about 'taking' (e.g. from the common grazing land), but about 'giving', for example for the maintenance of 'public' services such as irrigation canals, roads, etc. Here too farmers are most likely to contribute if they are reasonably sure that others will do so too. It is relevant to note here that, when analysing multi-actor problematic situations, one often recognises social dilemma-like dimensions.

---

[4] Based on social-psychological literature, Koelen and Röling (1994) argue that people are less likely to act in the common interest when communities are large, heterogeneous and/or lack a sense of belonging together. Moreover, if making a sacrifice in the interest of the collective is highly visible, community members may be more inclined to do so because they may like to make a good impression on others (i.e. serve self-presentational interests; see section 6.3). For similar reasons, community members may be deterred from selfish choices.

The most important conclusion at this point is that farmers' practices are shaped not only by farmers' perceptions of the quality of the wider support network, but also by their views on the effectiveness of local forms of organisation in relation to specific issues. Important aspects of (inter)community organisation that farmers may consider include:

- the use of collective resources (e.g. water, land);
- the maintenance of collective services (e.g. irrigation canals, roads);
- input supply;
- storage, marketing and/or transport of produce;
- credit provision;
- land-tenure;
- crop-protection;
- labour exchange.

Summary: the significance of social-organisational risks and trust

From the above we have learned that perceived environmental effectiveness in connection with certain farming practices is closely connected with farmers' *trust* in the functioning of organisations and institutions – inside, outside and between communities – on which the success of farming practices depends; or, in other words, with their perception of 'social-organisational risks' that exist in relation to specific farming practices. Of course, farmers cannot and do not avoid all the organisational or other risks that they face. We discuss this further in the next section.

5.2.3   Perceived self-efficacy

Farmers' practices are not only shaped by their level of trust in the functioning of organisations (see section 5.2.2), but also by their confidence in their own capabilities. In some cases farmers may refrain from practices that they regard as beneficial solely because they think they cannot properly and/or realistically apply them. Alternatively, they may continue with practices that they regard as sub-optimal because they feel that they are 'good at them'. Social psychologists have called this variable 'perceived self-efficacy' (Bandura, 1977, 1986) or 'perceived behavioural control' (Ajzen & Madden, 1986). Below, we discuss several dimensions of perceived self-efficacy.

Box 5.3   Overview of sub-variables constituting 'Perceived self-efficacy'.

Perceived self-efficacy

Perceived ability to mobilise resources
+
Perceived availability of skills and competence
+
Perceived validity of the evaluative frame of reference
+
Perceived ability to control or accommodate risks

Perceived ability to mobilise resources

Applying a particular practice usually requires a mixture of resources in terms of labour, cash, land, etc., not only in a narrow sense (e.g. for buying a hybrid seed), but also for dealing with the consequences that may follow from such a decision (see our earlier discussion in section 5.2.1). If, for example, farmers feel that they may not be able to raise the money to buy fertilisers and pesticides, or feel unable to mobilise sufficient labour to process a bigger harvest, they may abandon the idea of using hybrid varieties.

Perceived availability of skills and competence

A special category of resources needed to perform specific practices is 'skills and competence'. For implementing forms of biological pest control, for example, farmers may need to be able to (a) prepare biological pesticides; (b) distinguish between 'good bugs' and 'bad bugs'; (c) use tools for evaluating the balance between such bugs; and (d) draw appropriate conclusions about when to take which measures. Farmers may well believe that they have insufficient competence and skills for applying, in this case, biological pest control, and therefore resort to more 'traditional' means of crop protection (i.e. chemical pesticides).

Perceived validity of the evaluative frame of reference

In section 5.2.1 we argued that farmers may consider all sorts of consequences, risks and aspirations in evaluating farming practices. In doing so, they draw upon knowledge derived from many sources. However, farmers may well be insecure about the relevance and validity of their knowledge to a specific situation. They may, for example, expect higher yields by using hybrid varieties, but at the same time have serious doubts about whether this will prove correct. We could call this a 'knowledge risk', i.e. the perceived risk that the knowledge of consequences and their likely occurrence is flawed. When farmers distrust their own knowledge of certain practices they may reconsider such practices and/or further check or test their assumptions, for example by on-farm experimentation (see section 13.5.8).

Perceived ability to control or accommodate risks

In the preceding sections we have come across various types of risk that farmers take into account when evaluating farming practices; i.e. technical risks, economic risks, social–organisational risks (section 5.2.2, and also 5.2.4) and knowledge risks (see above). As mentioned earlier, farmers cannot and do not avoid all risks. Whether or not certain risks are accepted or 'taken for granted' can depend on a number of factors, of which we have already mentioned (in section 5.2.1) the chance that a negative consequence occurs (e.g. 5% versus 90%), and the perceived balance between potential 'gains' and 'losses'. In addition to these more obvious factors, there are a few related to the perceived ability to control or accommodate risks, which relates to the issue of self-efficacy.

In order to place this discussion in its appropriate context, it is important to recall that farming is by definition a risky business, due to the need to deal with living materials, weather conditions, etc. In other words, farmers cannot run away from certain risks, and have to 'accept' certain risks as a consequence of being a farmer. Thus, for farmers the discussion is not so much whether or not to take risks, but rather how to deal with risk and how to choose between alternative risks: what risks can be accepted, and which risks must be minimised. For understanding farmers' responses in this respect, it is not only their risk perceptions that are important (see section 5.2.1), but also their perceptions of options to control risks. We will mention five important dimensions.

The first is farmers' *perceived ability to prevent risks from happening*. In order to prevent water shortages from happening, for example, farmers can build water storage facilities, provided that they have sufficient resources, skills and competence to do so (see above). A second element is farmers' *perceived ability to pass on the consequences of risks to (or share them with) others*. Such opportunities exist, for example, when risks can be insured through insurance companies, when there are arrangements within a community to share risks, or when governments or donors are willing to compensate for risks. The perception of whether or not reliable options exist is closely related with the issue of perceived environmental effectiveness (see section 5.2.2). Third, farmers' ways of dealing with risks can be understood through their *perceived ability to accommodate risks*. Some farmers find themselves in the position of simply accepting a particular risk (e.g. total crop failure in case of drought), because they feel they can carry the consequences (e.g. buy additional food), whereas others do not have the capital to do so. Fourth, farmers' *perceived ability to reduce risks* can play a role. Well-known strategies to reduce risks are, for example, diversification of activities and/or crops, intercropping, and using land in different agro-ecological zones. In fact, these are all strategies to *spread* risks, which can – at least for some farmers – be seen as a 'last resort' if they feel that there are no other viable strategies to control or accommodate risks. But here too, some farmers find themselves in a better position to spread risks than others. Finally, farmers' responses to risks may be shaped by their *perceived ability to predict risks in advance*. If farmers feel relatively certain about the chances that risks occur (e.g. when they know that a crop failure normally happens once in three years), or are able to 'see risks coming', such as by means of early warning indicators (e.g. when they know that certain diseases are more likely to occur if the early growing season is very wet), they may be more inclined to take such a risk since they feel they can at least *prepare* for it.

5.2.4   Social relationships and perceived social pressure

Farmers' practices are shaped also by pressures that farmers experience from other people with whom they relate. This influence is partly 'direct', but also 'indirect' in the sense that social pressures can influence earlier mentioned variables such as farmers' evaluative frame of reference, perceived self-efficacy and perceived environmental effectiveness (for a discussion of these 'indirect' influences see section 5.2.5). Social pressure can be divided into three dimensions: (a) the perceived desires and

Box 5.4 Overview of sub-variables constituting 'Social relations and perceived social pressure'.

Social relations and perceived social pressure

Perceived desires and expectations from other actors

×

Resources that others are perceived to mobilise in order to persuade

×

Valuation of expectations, resources and relationships

---

expectations from other actors (in relation to farming practices); (b) the resources, including rewards and sanctions, that others are perceived to mobilise in order to persuade; and (c) the valuation of expectations, resources and relationships.

Perceived desires and expectations from other actors

Farmers are not isolated actors operating in a neutral environment. They have direct and indirect relationships with other people who often have certain explicit or implicit ideas about what they would like a farmer to do in a specific context. Such actors can include spouses, children, relatives, village leaders, donors, international organisations, agro-industry, communication workers, politicians, scientists, governments, etc[5]. For example, female spouses may reject certain farming practices proposed by their husbands because they do not like the consequences for them (e.g. increased workload); male spouses may oppose their wives engaging in cash crop production because they fear the wives' increased independence; governments may try to convince farmers to grow cotton rather than maize; foreign donors may push farmers to produce in an environmentally friendly way because of political priorities in their own country; religious leaders may discourage the use of pesticides in view of religious prescriptions, etc.

As the examples show, there are many reasons for others to try and influence farmers. These can range from an aspiration to maintain power positions, a desire to meet national economic needs, or a wish to maintain cultural norms and values. Such underlying aspirations originate largely from the evaluative frame of reference (see section 5.2.1) of the *other*. Again, what matters for understanding farmers' practices is farmers' *perception* of the desires and expectations of others (which, of course, does not necessarily coincide with the way the others experience their own positions). This perception can be based on various grounds, including overt statements by others, non-verbal practices and signals (see Chapter 7), observations of other people's practices and responses, knowledge of other farmers' experiences,

---

[5] The latter five categories belong to what Benvenuti (1982, 1991) has called the Technological-Administrative Task Environment (TATE) in which farmers operate, which tends to have a considerable 'prescriptive' influence on what farmers do or do not do. A related term is that of the 'technological regime' (Kemp et al., 2001) in which farmers operate. Like TATE, a 'regime' expresses the idea that there is often a dominant set of rules and principles, including technological ones, that farmers are expected to follow and from which they cannot easily deviate.

information provided by the media, common knowledge of social norms and values, empathy, imagination, etc. (e.g. Fulk et al., 1990; Van Meegeren, 1997).

Resources that others are perceived to mobilise in order to persuade

When others want farmers to perform or not perform certain practices, they usually mobilise a number of resources to strengthen their claim. This is because they feel, rightly, that just *telling or informing* farmers about their desires will often not have sufficient persuasive power. Thus, they try to make certain options more attractive than others by employing 'rewards' and 'sanctions'. Such 'carrots and sticks' usually take the form of the granting or withholding of assets that farmers find valuable; in other words, items that are regarded as resources. For example, a government may want farmers to produce cotton, and reward those who do so with cheap fertiliser, withholding such fertiliser from those who refuse to grow cotton. Alternatively, the government may make a law or rule that prescribes the growing of cotton, and can try to enforce it with a system of policing and fines. Similarly, a Bhutanese lama (priest) may not want the members of his community to use pesticide, and may withhold certain religious ceremonies from those who do. These examples show that what counts as a 'resource' can vary considerably in different situations and between different people; after all, some Bhutanese farmers may care more about religious ceremonies than others. This implies that what some farmers regard as a 'reward' (i.e. positive) others may regard as a 'sanction' (i.e. negative). Also, rewards and sanctions are not always applied immediately, but can include *promises* or *threats* that extend into the future (i.e. future expectations). Rewards and sanctions frequently employed to persuade farmers include:

- (non) access to agricultural inputs;
- (non) access to subsidies;
- (non) access to output channels;
- increased or reduced social status and personal recognition;
- (non) access to networks of people (e.g. foreign donors, government officials);
- money in the form of bribes, subsidies or fines;
- (non) access to land;
- (non) access to information and knowledge;
- improved or reduced social security;
- improved or reduced safety and protection;
- (non) access to jobs.

Some of the pressures imposed on farmers are explicit, while others remain much more implicit and hidden. Farmers may have a more or less articulate perception of the various rewards and sanctions they are facing, which again need not necessarily conform with other actors' views and intentions. These perceptions can in many ways be regarded as being about *social–organisational consequences* associated with the use of certain farming practices, which we touched on in section 5.2.1. As is the case with technical and economic consequences, social–organisational consequences are associated with perceptions of likelihood and risk (see section 5.2.1). For example, farmers may feel that there is a high or low chance that they will face sanctions if they refuse to grow cotton.

Valuation of expectations, resources and relationships

Like technical and economic consequences, expected social–organisational consequences are evaluated in relation to farmers' aspirations (see section 5.2.1); that is,
individual farmers consider the balance between likely positive rewards and negative
sanctions in view of a variety of economic, technical, relational and/or cultural goal
orientations (see section 5.2.1).

5.2.5   The dynamics within the model

So far we have dealt with various important variables in a largely 'static' and separate way. We have, so to speak, taken a photograph of relevant factors that shape
farmers' practices at a given time. We now have a more refined picture of variables
that may help to explain farmers' practices. We can now see that what farmers do or
do not do is shaped by:

(1)   What they *believe to be true* about the agro-ecological and social world (as part
      of an evaluative frame of reference) which includes multiple:
      (a)   beliefs about consequences, including causal attributions;
      (b)   perceptions of (un)certainty, likelihood and risk.

(2)   What they *aspire* to achieve (as part of an evaluative frame of reference) as
      expressed through interrelated aspirations of various kinds, including:
      (a)   technical or economic goals and interests;
      (b)   relational (including 'political') goals and interests;
      (c)   cultural aspirations, including also responsibility considerations;
      (d)   emotional aspirations.
      Such aspirations play a role in the evaluation of multiple consequences and
      risks associated with a particular practice, resulting in a complex set of '*advant-
      ages*' and '*disadvantages*' of various kinds, and at various levels, domains and
      points in time. The overall balancing of these is often called a person's positive,
      negative or indifferent *attitude* towards the specific practice.

(3)   What they think they are *able* to do (as connected with perceived self-efficacy
      and perceived effectiveness of the social environment) given their perceived:
      (a)   ability to mobilise resources;
      (b)   availability of skills and competence;
      (c)   trust in the validity of their knowledge;
      (d)   ability to control or accommodate risks;
      and given their expectations regarding the:
      (e)   effectiveness of the agro-support network;
      (f)   effectiveness of (inter)community organisation.

(4)   What they think they are *allowed* and/or *expected* to do (as connected with
      social relations and perceived social pressure) in view of the perceived:
      (a)   desires and expectations from others;
      (b)   'rewards' and 'sanctions' (resources) mobilised by others;
      (c)   importance of, and balance between, rewards and sanctions (vis-à-vis aspira-
            tions of various kinds).

All of this, of course, involves taking into account and co-ordinating practices at different *levels*, *domains* and *points in time*, while also taking into account viewpoints from non-farming identities. As we illustrated at the beginning of section 5.2, we can use this refined photograph to identify reasons why farmers may or may not engage in certain practices. These reasons and variables have affinity with what others have called different forms of 'capital'[6]. Similarly, we have now arrived at a more refined overview of entry points for change.

Despite all this refinement, our model is still incomplete. This is because, in everyday life, we are dealing with 'moving pictures' and highly dynamic situations rather than with 'photographs'. In addition, we are interested in *change*, which is something the model has not captured so far. In this section we therefore discuss three different forms of dynamics that are important. First, we discuss the matter of time; that is, the relationships between past, present and future from the perspective of the model. Then, we discuss the dynamic interrelations between the variables that we have discussed; that is the way in which they influence each other over time. Finally, we identify a number of social processes that take place within the model, which link to other bodies of social theory that may improve our understanding of the dynamics in the model.

Dynamics over time: the importance of feedback, routine and path dependence

The farming practices that farmers employ today need to be understood in their historical context. Obviously, farmers do not start from scratch. They are building on a historically grown farming system and body of knowledge, that operates in a society and agro-ecological environment that over time has taken on certain characteristics – called 'structural properties' by Giddens (1984) (see Chapter 6). Thus, for specific actors the set of variables discussed so far is, at a given time, configured in a particular way, influenced by previous events, practices, experiences and historically grown structural features. For example, in the present a farmer may opt for a new 'hybrid' rice variety in view of his or her current evaluative frame of reference (e.g. shaped by a positive experience of a neighbour), a favourable perception of environmental effectiveness (e.g. based on a well-functioning agro-support network in the last few years), a positive feeling about his or her capability to manage the variety (high self-efficacy created, for example, through observation of the neighbour's practices), and considerable pressures from a change agent, who promises increased access to subsidised inputs in the future if the farmer changes to the new variety. Adopting the new variety means new farming practices in a variety of domains (see section 5.1.2). In the course of the season, these practices lead to particular, expected or unexpected, outcomes in each of these domains (e.g. effect on the labour calendar,

---

[6]   Different authors speak of different kinds of capital – such as 'natural', 'financial', 'human' and 'social' capital – as factors that may shape development, including changes in agricultural practice (see Putnam, 1995; Uphoff, 2000). The latter two categories clearly overlap with what is mentioned under points (1) and (3) in particular, in the list on page 79. Although we mention similar issues (e.g. knowledge, skills, social organisation, trust), a clear difference is that we speak of them in terms of *perceptions* rather than as factual and measurable 'stocks' of capital.

changes in yield and income, altered dependency on external inputs, etc.). Such outcomes can be regarded as *feedback*, which is a crucial mechanism in the model as it forms the connection between different points in time (see also section 9.3).

As shown in Figure 5.1, we can distinguish between feedback from the agro-ecological world, and feedback from the social–organisational world. The first refers to biophysical responses, which in the case of the hybrid rice variety may include a higher yield of grains, a reduced yield of crop residues, changes in taste and palatability, the emergence of new pests, changes in the nutrient balance within the soil, etc. Feedback from the social–organisational world refers to the human responses experienced in relation to the adoption of, in this case, the new rice variety. These can include, for example, the ease with which the product is sold, the extent to which the change agent keeps his promise, the satisfaction expressed by household members with the new labour arrangements and income situation, etc. Through such feedback, the original perceptions of farmers (i.e. their evaluative frame, perceived self-efficacy, perceived environmental effectiveness and perceived social pressure) can be altered, resulting in a new situation which, in this example, may or may not be conducive to the continued use of hybrid rice. In any case, farmers are likely to adapt certain farming practices on the basis of their experiences, i.e. on the basis of the feedback encountered. It is important to note, however, that the space for change at any point is, at least to some extent, limited by previous practices, decisions and happenings, implying that future practices are in part *'path dependent'* (Rip, 1995; Garud & Karnoe, 2001). Once one has started to go in a certain direction (e.g. through earlier investment decisions or social commitments) there is no easy way back or sideways.

Apart from the feedback that results directly from the use of new farming practices, the agro-ecological and social–organisational world may provide all sorts of other 'inputs' which may affect farmers' perceptions of a particular issue. Such 'external' triggers can take many forms, including for example, the outbreak of a war, a change in world trade arrangements, a severe drought, etc. The outbreak of a war, for example, may alter farmers' trust in the agro-support network (perceived environmental effectiveness), and lead them to reconsider the use of hybrid rice.

The model can be regarded as a cycle that is in continuous motion, and through which social actors connect their past, present and future[7]. Practices (as regular patterns of action, see section 5.1) provide some stability to this cycle. Usually, such practices are no longer deliberated on as they take place; they have become habits or *routines*. However, as Giddens (1984:5) points out, social actors continuously and routinely monitor their practices as an integral part of the ongoing flow of life and experience. This means that, even if actors are not deliberately seeking feedback, they are involved in what Giddens calls 'reflexive monitoring of action'. As a result, the relative stability of practices may be 'disturbed' by changes in feedback from the

---

[7] An interesting observation by Van der Ploeg (1999) is that the extent to which farmers' aspirations are linked to the past, present or future may differ. In some cultures or historical episodes farmers may be guided by a wish to reproduce the past, whereas in others they may strive for immediate outcomes in the present, or aspire to arrive at an abstract idea or model of farming in the future (Van der Ploeg, 1999).

social–organisational or biophysical world. In view of such feedback, current practices may become regarded as problematic which may, in turn, lead various people
to actively consider and/or perform alternatives modes of action. One could say that
the existing configuration of variables and practices is 'shaken-up', resulting in a
certain amount of 'turmoil' until a new 'balance' emerges. Such a new 'balance'
includes new forms of co-ordinated action, which may over time become a regular
pattern again. In this way, new practices and routines come about.

Dynamic interrelations between variables: the social construction of perceptions

Apart from the dynamics created through the performing of farming practices, and
the emergence and incorporation of feedback as a result of this, there are dynamic
processes which occur while practices are 'in the making'. This is because the
variables discussed in sections 5.2.1 to 5.2.4 do not only shape farming practices
'directly', but also 'indirectly' in the sense that they influence *each other* (see Figure
5.1). For example, perceived social pressure does not only have a direct bearing on
farming practices, but also tends to have an influence on the evaluative frame of
reference (and vice versa), so that it has indirect consequences as well. Likewise, the
evaluative frame of reference can shape farmers' perceived self-efficacy or perceived
environmental effectiveness.

Let us clarify these complex interrelations with a few examples. Take farmers'
perceptions of the technical consequences of adopting a hybrid maize variety
(as part of their evaluative frame of reference, see section 5.2.1). These perceptions
(e.g. the idea that it can double the yield) are not necessarily an independent and
neutral reflection on certain observations or statements of others, but these ideas
are *themselves* subject to social pressure. For example, a local leader, communication worker or an applied research organisation may well exert pressure on farmers
in order to influence their belief. They may mobilise certain rewards (e.g. to be
recognised as a 'good farmer') or sanctions (e.g. to be ridiculed as a 'stupid farmer')
for thinking along particular lines. Thus, farmers' convictions and thoughts are
not simply neutral reflections of 'reality', but tend to be moulded by the pressures
imposed by various others who can effectively mobilise resources to support adherence to their views (e.g. teachers, change agents, agro-industry, government,
etc.). This has been called the *social construction of knowledge*. This idea essentially
means that, in a given situation, a person's thoughts and beliefs are not neutral
but are influenced by social pressures and perceived political, economic, relational
or normative interests (see also section 6.3). In short, people tend to believe what
they think it is in their interest to believe, given experienced social pressures. If a
farmer experiences pressure from everybody around to believe that hybrid maize is a
miracle solution, it is quite hard to resist that view. It is relevant to recall here
perhaps the history of Copernicus and Galileo who in their times (the 16th and 17th
centuries respectively) had great difficulty convincing the clergy that the earth was
circulating around the sun; among other reasons, it was in the interest of the church
to defend the idea that the sun circulated around the earth as this was suggested in
the bible.

It is important to note here that farmers are not 'passive receivers' of other actors' social pressures, but are themselves often actively involved in exerting social pressures, or sometimes counter-pressures, as well; that is, they too engage in constructing support networks for their views of reality.

Similarly, social pressure can be exerted to influence, positively or negatively, farmers' perceived self-efficacy or perceived environmental effectiveness. For example, to keep the benefits of a certain innovation within a limited group, those who have adopted it may well emphasise to others how difficult it is to make the innovation work, so as to undermine such farmers' self-confidence. Moreover, farmers' perceptions of environmental effectiveness may well have an impact on their technical risk perceptions, while in turn such technical risk perceptions affect a person's perceived self-efficacy. For example, when farmers perceive that the supply of chemical pesticides is unreliable (i.e. perceived social–organisational risks are high), this may lead them to believe that there is an increased risk of severe crop damage due to pest attacks (i.e. perceived technical risks are high), and – in the absence of training in biological control strategies – may reduce their perceived self-efficacy in dealing with a particular crop (e.g. hybrid rice).

The variables mentioned are interconnected in complex ways, as shown by the small circle in Figure 5.1. In a way this is not surprising, as all the variables mentioned are perceptual, so in essence the conclusion is that an actor's 'cognitive system' (see also Röling, 2002) is made up of a complex and dynamic web of interrelated perceptions.

Social processes at work: linking the model with other theories

The basic model for understanding farming practices, presented in this chapter, provides a particular analytical perspective, focusing on cognition, and is inspired mainly by social psychology, agrarian and rural sociology and anthropology. However, it does not claim to be exclusive in the sense that farmers' practices can only be understood through this model. In fact, like any model, it has biases and shortcomings that may render it less useful in specific situations. An important limitation of the model, for example, is that, although it indicates how practices of different actors can be interlocked (e.g. through social pressure, feedback and (dis)trust), its main focus is on the *individual actor* and his or her practices at a *particular* time. This would imply that, for an understanding of the outcomes of a series of multi-actor interactions (e.g. meetings between different stakeholders, or interactions within composite actors, such as organisations), one would have to look at all the individual perspectives, and the way in which they are interlocked (also over time). In other words, one would have to apply the model to a large number of actors who operate in subsequent and simultaneously occurring action contexts.

Although this may at times be an interesting, but rather demanding, exercise, there are social theories that give insight into multi-actor processes more transparently and comprehensively. Elias' 'established/outsider' perspective (Elias & Scotson, 1965), for example, is a powerful and – compared to our model – relatively transparent aid to understanding how and why 'established' groups (e.g. farmers living at the end of a water catchment) think about themselves (i.e. take on a certain identity) and stigmatise 'outsiders' who somehow pose a threat (e.g. 'upstream' farmers or industries who compete for water). More generally, social science provides numerous additional perspectives which can improve our insight into the logic of farmers' practices and/or enhance our understanding of the dynamics within the model. More specifically, theories that deal with dynamic social processes can often add to our understanding of the interrelations between variables in the model (e.g. communication theory, learning theory, negotiation theory) and/or can provide a particular focus to the model in terms of possible outcomes that one is interested in (e.g. adoption of technology, conflict resolution, new structural properties, as connected to, respectively, adoption theory, negotiation theory and structuration theory). Alternatively, and especially if theories include cognitive or perceptual variables, the model can contribute specific insights to other perspectives. For example, the model may add to economic theory by shedding more light on how certain 'preferences' (an important variable in economics that is usually treated as 'given') come about. In all, the model can be enriched by other bodies of theory and vice versa. Below, we briefly highlight some relevant theories on social dynamics that may be fruitfully linked to the model. Many of these will be discussed in more detail in later chapters of the book.

- *Communication theory*: Communication refers to the process of exchanging messages and signals between social actors, as an important element in the construction of meaning and perception, which is a central concern in the model. Thus, communication theory (e.g. Dervin, 1981) is likely to enrich our understanding of the dynamics in the model, and vice versa (see Chapter 7).

- *Learning theory*: Theories of individual learning (e.g. Kolb, 1984) deal with the ways in which individuals acquire the ability to perform new practices and patterns of action, as related to new knowledge (perceptions) and skills. This issue of changes in practice and cognition is again a central concern in the model, so that learning theory too can form an important source of improved understanding (see Chapter 9).

- *Social learning theory*: Social learning theory deals specifically with learning processes that occur between different social groups (Leeuwis & Pyburn, 2002; Röling, 2002). Important issues here are, for example, processes through which groups with different cultural backgrounds and/or conflicting interests gain insight (or not) into each other's perspectives, develop mutual trust (or not) and are able to arrive at more convergent views. Since our model indicates the importance of such different backgrounds and relationships of trust for understanding farming practices, social learning theory can form a further source of inspiration (see Chapter 9).

- *Cognitive dissonance theory*: The theory of cognitive dissonance (Festinger, 1957) gives insight into the *relationships* that may exist between cognitions, and the ways in which actors deal with these. In particular, it helps to understand how actors deal with cognitions/practices that are somehow in conflict with each other; they may, for example, know that smoking cigarettes increases the chances of getting lung cancer, but also know that they love to smoke cigarettes. Cognitive dissonance theory proposes that actors adapt their cognitions in an effort to avoid such contradictions, i.e. they reduce cognitive dissonance. This may be relevant to the understanding of farmers' perceptions and practices (see Chapter 9).

- *Autopoietic systems theory*: Drawing upon biological theories (Maturana & Varela, 1984) of cognition and self-reproduction (or 'autopoiesis'; 'autos' is Greek for 'self', and 'poiein' means 'to make'), social scientists have proposed that human minds tend to be 'operationally closed' (Luhmann, 1984:346 onwards). By this it is meant that people, or organisations of people, tend to interpret the world around them in terms of already existing concepts and rules, implying that they tend to be 'blind' and find it difficult to take on board new ways of looking at the world. Thus, the theory deals with fundamental difficulties in human learning (see Chapters 9 and 16).

- *Theories of innovation and adoption*: A wide variety of disciplines (ranging from management science to the philosophy of science) are concerned with the way innovations are created and become widely adopted in society (or not). Such theories include actor network theory (Callon et al., 1986; Law & Hassard, 1999; see Chapter 20), knowledge systems and network theory (Hannerz, 1980; Röling, 1992; Engel, 1995; see Chapter 17), strategic niche management (Rip, 1995; Kemp et al., 2001) and adoption and diffusion of innovations theory (Rogers, 1983) (see Chapter 8). To the extent that we are interested in understanding *changes* in farming practice, these bodies of literature have much to add to the model presented.

- *Negotiation theory*: Negotiation theory gives insight into the way conflicts between social actors emerge, are resolved or continue to exist (Pruitt & Carnevale, 1993). Also, there are normative theories on how conflicts can be resolved (Fisher & Ury, 1981). As our model indicates, farming practices emerge in a complex social context in which conflicts of interest are likely to exist and affect the way farmers go about farming. We will further discuss the relevance of negotiation theory in Chapter 10.

- *Structuration theory*: Structuration theory (Giddens, 1984) helps us to understand how structural conditions (in Giddens' terms, structural properties) constrain and enable human action, and are reproduced and adapted through time as a result of the interactions between active agents. Since farmers' practices are shaped to a considerable degree by seemingly stable structural conditions, theories on how such properties emerge and change are of considerable interest (see Chapter 6).

Clearly, there are many more theories on social dynamics that may in a given context be linked to our model and may help to shed light on farming practice and changes to it, for example theories on exploitation (Marx & Engels, 1973; Wallerstein, 1974; De Janvry, 1981), identity formation (Wetherell, 1996), globalisation (Hannerz, 1996), etc. However, this book builds largely on the perspectives outlined above.

Summary

The model in its 'static' form (i.e. a checklist of variables) helps us to understand better why (i.e. for what reasons) people engage in certain practices (or not) at a given time. Also, it enables us to identify entry points for change. However, it does *not* give us much insight into *change processes*. Our discussion of the dynamics in the model allows us to conclude that our understanding of change processes can be enhanced by looking at:

● the nature and quality of feedback;
● social construction processes;
● various dynamic theories, related to, among others, learning and negotiation.

We will discuss these issues in the subsequent chapters of this book.

5.3   Implications for communication for innovation

From the above reflections into the logic of farming practices, we can derive several important insights into the potential, limitations and principles of communication for innovation. In this section we will briefly highlight some of these. They will be discussed further in subsequent chapters.

5.3.1   The central role of knowledge and the need to be modest

The model expresses clearly the idea that different forms of knowledge and cognition play a central role in shaping farmers' practices, and vice versa. Except for practices, all variables in the model are essentially perceptual in nature, perceptions that are shaped through experiences from both the past and present. In other words, we assume that knowledge and practice pre-suppose each other, and that farmers and other actors are essentially 'cognitive agents' (Röling, 2002). From this it follows that, to a considerable extent, 'the world can change if only (and only if) sufficient people change their way of thinking about it'. As communication is an important process through which experiences are exchanged and through which knowledge and perceptions are moulded, professional communication can, in principle, be a powerful aid to achieving change (see Chapter 7).

An implication of the above, however, is that communication workers are not the only ones who may have relevant knowledge and information with respect to farming and the solving of problems. We have seen that managing a farm is a complex process of co-ordinating different levels, domains and aspirations; through this process much relevant contextual and experiential knowledge is generated, into which external agents (e.g. communication workers) can at best have partial insight. Thus, in many ways farmers are 'experts', even if they are not all-knowing. The challenge for communication workers, then, is to offer a different 'expertise' that recognises and adds to the one farmers already have, and/or to enhance the experiential learning processes that take place in farming (see also Chapters 9 and 13).

5.3.2   The relationship with different communication strategies and functions

In essence, all the variables mentioned in the model can serve as entry points for inducing change in the context of a wider dynamic process. Thus, the model shows that communication for innovation can be geared towards changing perceptions and practices of various kinds. For example, it can focus on enhancing perceived self-efficacy (e.g. by providing training in certain skills), on improving perceived environmental effectiveness (e.g. through strengthening forms of community organisation), on organising relevant forms of feedback (e.g. by introducing new indicators for soil depletion), or on altering farmers' evaluative frame of reference (e.g. by supporting on-farm experimentation). In other words, there is a fairly clear connection between the different communication strategies and functions discussed in Chapter 2, and the variables in the model (see Tables 5.1, 5.2). Each communication strategy or function works on a particular variable (or variables), identified by an analysis of why problematic practices occur, and so it can be seen which variables can serve best as entry points for change.

From an instrumental perspective (see Chapter 4) there are similar associations between policy instruments other than communicative intervention, and the variables in the model (see Table 5.3). In discussions on the usefulness of policy instruments (see Chapter 4), the use of communicative intervention as an instrument is frequently restricted only to influencing people's *reasoning* and *evaluative frame of reference* by providing novel arguments, insights and feedback. However, as indicated

Table 5.1   The connection between different communication for innovation services or strategies and the variables in the model for understanding human practices.

	Logic of communication for innovation in terms of the main variables that are 'worked on'
Focus on 'individual' change/farm management communication	
Advisory communication	• widening of individuals' *evaluative frame of reference*
Supporting horizontal knowledge exchange	• widening of individuals' *evaluative frame of reference*
focus on collective change/co-ordinated action	
Generating (policy and/or technological) innovations	• widening/connecting stakeholders' *evaluative frame of reference* • changing *perceived environmental effectiveness* through new arrangements • altering *perceived social pressures* from the policy/technology environment
Conflict management	• widening/connecting stakeholders' *evaluative frame of reference* • changing *perceived environmental effectiveness* through newly negotiated arrangements • altering *perceived social pressures* from other stakeholders
Supporting organisation development and capacity building	• improving *perceived environmental effectiveness* within a particular interest group
Focus can be individual or collective persuasion	
Persuasive transfer of (policy and/or technological) innovations	• exerting *social pressure* • altering people's *evaluative frame of reference*

Table 5.2 The connection between general communication for innovation functions and the variables in the model for understanding human practices.

	Logic of communication for innovation in terms of the main variables that are 'worked on'
Awareness raising and making people conscious of pre-defined issues	• altering people's *evaluative frame of reference* in order to stimulate acceptance of specific problem definitions
Exploring views and issues	• exploring and analysing alternative *evaluative frames of reference* that are (or may be) relevant to a situation
Information provision	• widening people's *evaluative frame of reference* • improving *perceived self-efficacy* for dealing with a specific situation
Training	• enhancing perceived *self-efficacy* on specific issues

Table 5.3 The connection between various policy instruments and the variables in the model for understanding farmers' practices.

	Logic of policy instruments in terms of the main variables that are 'worked on'
Regulations	exerting *social pressure* in the form of sanctions for those who do not perform according to the regulations
Provisions	enhancing *perceived self-efficacy* or *perceived environmental effectiveness* by the establishment of enabling facilities
Group pressure	mobilising other citizens/peers to exert *social pressure* in order to encourage or discourage certain practices
Subsidies/fines	exerting *social pressure* in the form of positive or negative financial stimuli with respect to certain practices

by Chapter 4 and Tables 5.1 and 5.2, we feel that communication for innovation can serve wider purposes; that is, it can be used to work on a broader range of variables (i.e. all variables in Figure 5.1), and even contribute to the *emergence* of other policy instruments such as regulations, provisions and subsidies/fines. In any case, the above suggests that deciding on appropriate communication strategies and other policy instruments requires a thorough situation analysis (see section 5.3.8) that explores which bottlenecks exist (e.g. inadequate knowledge, lack of trust, high social–organisational risks, low self-efficacy, inadequate provisions, etc.) in relation to a specific problem.

5.3.3   Communicative intervention must be 'tuned' to other communication processes

As we have seen, farmers are not isolated individuals. They are part of multiple social networks, and in that context they are likely to communicate, directly or indirectly, with a variety of actors (neighbours, religious leaders, dealers, spouses, researchers, etc.) in relation to specific issues. From these sources they may well receive contradictory information and experience different pressures. To be effective, it is crucial that communication workers understand they are just one of the communicating parties, and that they are aware of and/or anticipate the way in which farmers communicate about a specific issue with others (see Chapter 7). When communication workers fail to tune in to the way farmers speak about certain issues, they are likely to talk about topics that are not the main concern of farmers (Van Woerkum, 2002).

5.3.4   The need to anticipate diversity among farmers

Our model indicates that a large variety of cultural, technical, economic and relational aspirations and preferences may play a role in shaping individual farmer practices. Typically, such aspirations may vary widely across individuals and farming households. Sociological research suggests that differences exist even among farmers in the same region, and with similar farm lay-outs, household composition, age and education levels (Leeuwis, 1989, 1993; Van der Ploeg, 1990). From these different sets of aspirations, distinct patterns of farming emerge, called *styles of farming* (Van der Ploeg, 1994; see also Figure 15.2, Table 15.3). Each farming style has its own logic and different social relations, cropping systems, technology use, labour organisation, dependency on markets and off-farm income, etc. (see Leeuwis, 1989, 1993; Van der Ploeg, 1994). Thus, each style represents a different way of connecting and ordering the agro-ecological, the technical and the social world (Roep, 2000). It is significant that different patterns of farming can do well (be viable) in a given context. In other words, in farming there are 'different ways of doing things right'. This insight is at odds with mainstream micro-economic theory which essentially proposes that in a specific context there exists only one rationally optimal way of allocating production means and organising farming. Communication for innovation scientists and practitioners have long supported the idea that there was basically one 'right' direction in which farming should develop (see Chapter 8). Such an approach does not do justice to the existing diversity and runs the risk of failing to capture promising development opportunities.

Apart from a classification into farming styles there may, in a given situation, be other relevant (e.g. problem-specific) ways of describing and classifying diversity among farmers, for example in terms of their media use, attitude to nature, etc. Thus, in each problem situation, it is important to consider carefully what are the most relevant dimensions of diversity among farmers (see Chapters 7 and 15).

5.3.5   Linking multiple socio-technical innovation processes

During our discussion of the model it has become clear that changes 'never come alone'. As farming is a carefully co-ordinated activity, a change in one domain (e.g. pest management) has repercussions for other domains (e.g. labour organisation, weeding, storage, etc.). Thus, one is always dealing with multiple changes in a complex farming system. As mentioned in Chapter 1, it is important to acknowledge that most innovations in agriculture have two basic dimensions: technical and social–organisational. The first dimension refers to all sorts of biotic and abiotic artefacts and practices (e.g. new seeds, animal breeds, machinery, rotation systems, etc.),while the second involves novel social arrangements (e.g. new forms of labour organisation, marketing arrangements, community action, legal arrangements, etc.). The two dimensions are closely connected in the sense that both provide space, conditions and limitations for the other. For example, the existing division of labour between men and women in the weeding of crops may pose restrictions on a new ox-weeding device as men may pose different demands from women on such a piece of equipment. Alternatively, any weeding technology based on the use of draught animals as a source of power requires all sorts of organisational innovations (arrangements for exchanging oxen and ox-weeders, herding of animals, training of oxen, etc.) in places where the use of animals is not common. In innovation design, this mutual inter-dependency requires that both aspects of innovation are dealt with simultaneously in order to arrive at a viable combination of technical and social–organisational solutions (see Chapter 8).

It is important that change agents are aware of the multi-dimensional character of innovation, and incorporate this in their communication strategy. In doing so, however, it will often be unhelpful to develop one rigid package of innovations, as this tends to undermine the capacity to deal with diversity. What is needed instead is a careful evaluation of how certain innovations can be adapted and connected with other changes, in a way that suits the needs and requirements of different farmers.

5.3.6   The multi-layered character of technology and policy acceptance

The realisation that change processes usually involve numerous interconnected changes at the same time also implies that for a proper understanding of the degree of 'adoption' or 'acceptance' of pre-defined innovations (see Chapter 8), we need to look at 'acceptance' of different components that together constitute an innovation. This means that processes of 'adoption' and 'acceptance' are more multi-faceted than has been suggested previously (e.g. by Rogers, 1983; see also Chapter 8). This is compounded by the fact that our model for understanding farmers' practices indicates that innovations are developed, proposed and advocated in a complex social setting, and cannot be regarded as 'neutral'. In relation to externally introduced innovations, this leads us to separate various *layers* of policy or technology acceptance (adapted from Aarts, 1998):

(1) *Acceptance of the (perceived) underlying problem definition*
    The first layer of acceptance/adoption relates to whether or not farmers agree and personally identify with the, often implicitly, proposed problem definition for which a technology or policy is supposed to be a solution. Farmers are less likely to accept solutions for problems that they do not recognise.

(2) *Acceptance of the legitimacy of intervention*
    This second layer of acceptance/adoption relates to whether or not farmers find intervention/social pressures (see section 5.2.4) from specific outside

agents justified and acceptable. Farmers are less likely to accept solutions proposed by agents who, in their view, have no 'right' to interfere in particular affairs.

(3) *Acceptance of the credibility and trustworthiness of an intervening agent*
An important layer of acceptance/adoption relates to whether or not farmers find the intervening agent a credible and trustworthy source of information. Farmers are less likely to seriously consider solutions introduced by agents who are not trusted and/or lack credibility for other reasons (e.g. lack of affinity, seniority, capability, etc.).

(4) *Acceptance of the diverse (perceived) consequences of the composite innovation*
The fourth layer of acceptance/adoption relates to whether or not farmers accept the manifold technical, socio-economic and social–organisational consequences of proposed solutions. Such acceptance involves evaluation of perceived advantages and disadvantages in relation to technical, economic, relational, cultural and emotional aspirations (see section 5.2.1). Thus, considerations that may enter the evaluation include the perceived:
- technical and economic *effectiveness*, and *efficiency* of proposed solutions;
- *fairness* of the consequences;
- political *desirability* of proposed solutions and their consequences;
- *cultural acceptability* of proposed solutions and their consequences;
- *practical feasibility* of specific solutions in view of, among others, efficacy issues (see sections 5.2.2 and 5.2.3).

(5) *Acceptance of (perceived) risks*
The fifth layer of acceptance/adoption relates to whether or not farmers accept the technical, socio-economic, social–organisational and knowledge risks that may be associated with technology or policy solutions (see sections 5.2.1, 5.2.2, 5.2.3 and 5.2.4).

From the above it is clear that acceptance and adoption are not only shaped by the multi-faceted properties of an innovation, but also by the characteristics of the social dynamics surrounding it. To be effective, it is important for change agents to acknowledge and monitor the various layers of acceptance, including their own acceptability as an intervening agent (layers 2 and 3).

5.3.7 The illusion of supporting rational decision-making

Our model for understanding farmers' practices makes clear that, even if farmers engage in reasoning about certain practices, it is misleading to regard farmers' practices as the outcome of a rational decision-making process in the sense that farmers deliberately and objectively analyse problems, set goals, analyse causes of problems, review a range of alternative solutions, and choose the best possible solution (see Chapter 2). Rather, farmers' practices are shaped over time by routine, ever-changing and often implicit aspirations, and social pressures and continuous feedback from the natural and the social world, which only on specific occasions become part of a deliberate and explicit attempt to make a decision. And even then, considerations other than objective facts are likely to influence the outcome (recall the different sets of aspirations mentioned in section 5.2.1).

This raises a number of questions with regard to the usefulness of models for rational decision-making as a basis for communication for innovation activities. In the past, communication workers have made extensive use of normative decision-making models (i.e. models that prescribe how 'good' decisions should be made) as they had a key interest in improving farmers' decision-making. One can argue that such models remain useful, especially because so many decisions are made implicitly and routinely. On the other hand, one could take the position that such models deviate from practice to such an extent that they are of limited value, so more realistic and practical modes are necessary for assisting farmers in managing their farms. In this book we are inclined towards the latter view (albeit with some nuances, see section 9.2), not least because our explorations have so far shown that evaluating possible changes in farming practices is highly complex. This is because of the interconnections between different levels and domains of farming, the large variety and different nature of possible consequences, the multiplicity of goals and aspirations that play a role, the many risks and unpredictabilities involved, and the multi-faceted character of innovation itself. In such a context, following a rational decision-making procedure would involve a huge number of variables and 'calculations', and require so much predictive knowledge and information that it simply cannot be considered as a realistic option in many instances (Simon, 1976). In Chapter 13 we will return to the issue of alternative modes of farm management support.

5.3.8   The need for analytical capacity in communication for innovation organisations

We have argued in section 5.3.2 that, in order to decide on appropriate communication strategies, organisations that apply communicative intervention need to have a fairly thorough understanding of the social and technical bottlenecks that may exist in relation to specific issues. Moreover, we have proposed that, to play a stimulating role in change processes, such organisations need to monitor the dynamics of learning and negotiation processes. In Chapters 14 and 16 we discuss the kinds of investigation that may be helpful. At this point suffice it to say that organisations need the analytical capacity to make both types of analysis, and that this capacity is often lacking in present day communication for innovation institutions.

Questions for discussion

(1)   A government change agent advises a farmer to apply X amount of fertiliser in order to obtain higher yields for a certain crop.

    (a)   The farmer decides to apply less fertiliser than recommended by the change agent because he or she has other priorities on which to spend the money.

(b)   The farmer decides to apply less fertiliser than recommended by the change agent because he or she is not convinced that adding more fertiliser will lead to higher yields and thinks fertility is not the limiting factor.

Discuss which variables and sub-variables in the model for understanding farmers' practices are involved in these two examples.

(2)   A farmer is considering changing to a more intensive and profitable farming system which, however, will require a greater labour input from children, relatives and/or hired labourers. Discuss how a person's considerations and evaluation of this farming system may differ when looking at it from the identity of 'a farmer' or 'a parent'.

(3)   Governments in Europe and elsewhere stimulate ecologically sustainable farming but they are not very successful. Discuss which stakeholders are involved, and how different variables in the model may characterise their position. Which stakeholders and variables do you think are the most significant limiting factor? What communication for innovation strategies (see Chapter 2) would be most appropriate for the stimulation of ecologically sustainable agriculture?

(4)   Discuss the role that path dependency (see section 5.2.5) has played in your lives.

6 Knowledge and perception

In the previous chapter we have seen that different forms of perception play a vital role in shaping human practice. This implies that innovation and development (i.e. modification of human practice) require and/or are accompanied by changes in perception. In everyday language, perceptions and beliefs relating to the functioning of the biophysical and social world, including also the causal processes involved, are usually referred to as 'knowledge'. In this chapter we elaborate on several complexities related to this specific kind of perception (or form of cognition). We single out this particular kind of human cognition because – particularly in the field of agriculture and rural resource management – knowledge is a relevant and much debated concept. More generally, we have seen in Chapter 1 that knowledge and information are thought to be increasingly important and significant for society at large. An in-depth discussion of these concepts is necessary not just because knowledge and perception are closely intertwined with human action and changes in it (see Chapter 5), but especially since *differential* and at times *contradictory* perceptions, arguments and knowledge claims tend to play a role in processes of change. For example, we frequently witness disputes between scientists and farmers over the correctness and validity of farmers' convictions, and vice versa. More generally, the various stakeholders in a complex problem tend to understand, define and present the situation differently.

In order to come to grips with this, we introduce some theoretical ideas regarding the concepts of knowledge and perception. In the closing paragraph of this chapter we illustrate the usefulness of these insights for understanding controversies regarding knowledge in an intervention context.

6.1 Knowledge, perception, information and wisdom

Knowledge can be seen as the basic means through which we understand and give meaning to the world around us. Concepts like 'meaning', 'interpretation' and 'perception' are largely synonymous, and all refer to the outcome of applying our knowledge to a particular situation. If, for example, we look at the sky and see a dark cloud, we have already used knowledge; that is, we give meaning to that grey thing above us by identifying it as a 'cloud' (instead of, for example, a strangely shaped hole in the air). Depending on what we *know* about clouds, we may even infer that it may soon start to rain, and conclude that we need to prepare for sowing maize; that is, our perception/interpretation of the situation is that it will possibly rain, and that this is a good time for sowing maize. Knowledge can perhaps be most easily understood as a collection of interconnected 'schemes of interpretation' (e.g. 'a grey thing in the air is a cloud', 'a cloud consists of fine drops of water or ice', 'under certain conditions a cloud may drop its water, and produce rain', 'seeds need water to germinate', etc.) that we have available in our heads, and that we can mobilise to give meaning to a particular situation. As illustrated above, some of these schemes are 'definitional' (e.g. 'a cloud is . . .') while others include 'associative' or even 'causal' components (i.e. 'if the cloud is grey, then . . .'). By connecting such schemes to each other we can develop a sophisticated understanding of a situation, and can act accordingly, and more or less effectively depending on the quality of our knowledge. As explained in Chapter 5, we do not have knowledge only about the natural and biophysical world (e.g. the weather), but also about the social world. On the basis of social knowledge, for example, we impute meaning to the way a person dresses ('a person dressed in black and with a white collar may be a Roman Catholic priest'), a language or facial expression ('if a person curses while frowning he may be really angry'), or a particular social position in a community ('in this community the chief is the recognised authority for solving conflicts about land').

The above shows that knowledge and perception are closely intertwined with the concept of *information*. Perceptions or meanings inform us about a particular state of affairs, and thus constitute information. With the help of information and related terms (perception, meaning, interpretation), human beings reduce uncertainty and bring order to the world around them. Contrary to notions like perception and meaning, the term information is, in everyday language, often associated with knowledge that has been captured and stored in a physical (or nowadays electronic) form such as a book, leaflet, file, newspaper, picture, sound, website, etc. Thus, while most authors would argue that perceptions and meanings, like the knowledge on which they are based, remain essentially in people's heads (Berlo, 1960; Schutz & Luckmann, 1974; Röling, 1988), the idea is that they can be made tangible as information. When we deal with the notion of communication in section 7.2, we will see that it nevertheless remains difficult to speak of information as something that has a fixed and easily transferrable meaning.

Another important category related to knowledge is that of '*wisdom*'. According to Bierly et al. (2000:597), 'wisdom relates to the ability to effectively choose and apply the appropriate knowledge in a given situation'. Thus, wisdom is about judging and selecting relevant schemes of interpretation, and translating them into action (or inaction). Because situations can always be interpreted differently, wisdom is an important human capacity. When faced with a conflict situation, for example, there are usually several alternative explanations of what the conflict is about (see section 10.2.3), as well as various possible interpretations regarding, for example, the seriousness of the situation, whether or not (and which) action has to be taken, and who is in the best position to do so.

Our description of knowledge as schemes of interpretation, and information as knowledge expressed in a tangible form, deviates slightly from conventional definitions. More conventionally, a separation is made between three categories: data, information and knowledge (e.g. Harrington, 1991; Jorna, 1992; Weggeman, 2000). Data are considered to be facts or forms of sensory input (visual observations, smell, etc.), which constitute the raw material for information, which is looked upon as interpreted data (Röling & Engel, 1990; Jorna, 1992). Knowledge is regarded as the body of mental inferences and conclusions that people build from different elements of information, and which allows them to take action in a given context. In this view, a long list of figures (e.g. on temperature and rainfall) produced by an automatic weather station is considered data. When these data are fed into a computer and analysed, it may be concluded that average temperature over the years is rising. This is regarded as information, as it provides an interpretation or pattern to the data, and reduces uncertainty about something. When this information is connected to other information on, for example, carbon dioxide levels, one may arrive at the conclusion (i.e. knowledge) that rising temperatures are related to higher levels of carbon dioxide.

The difficulty with this conventional view is that data are regarded as an independent and objective category, while we feel that a lot of human knowledge and interpretation is already used to produce data. The concept of 'temperature' is essentially a human idea, and it is clear that each measurement (data) already *means* something, at least to the people who have invented the measurement device. Thus, the suggested distinction between data and information is difficult to maintain. The proposed separation between knowledge and information is equally fluid and questionable, because the information that average temperatures are rising over the years is already an inference (i.e. knowledge) in itself, while the process of arriving at this information presupposes the use of a lot of additional knowledge, as incorporated in statistical and analytical procedures. Therefore, we find the separation between data, information and knowledge unnecessarily complex and misleading. In terms of the definitions suggested earlier (i.e. that knowledge is constituted by mental schemes of interpretation, and information by tangible expression of that knowledge), this example of measuring temperature merely suggests that different elements of knowledge can be connected to each other, and used to compile new knowledge. This knowledge can be converted into information through speech, written language expressions, graphic representations, etc.

It is important to note here that information (e.g. the symbolic representation of knowledge, see Chapter 7) is not the only form in which knowledge can be made tangible. In many ways human actions and practices, as well as technologies and other material artefacts (e.g. machines, seed varieties, roads, bridges) can be seen as tangible expressions of knowledge. At this point, however, we shall focus mainly on the mental and symbolic side of knowledge. When discussing the relationships between knowledge and power (in section 6.6) we touch upon the more physical dimensions again.

6.2   Life-worlds: the locus of discursive (explicit) and practical (tacit) knowledge

Human beings possess a lot of knowledge of which they are not immediately aware. In relation to this, Giddens (1984:374–5) speaks of discursive consciousness, practical consciousness and unconsciousness. Discursive consciousness is described as:

> 'What actors are able to say, or to give verbal expression to, about social conditions, including especially the conditions of their own action; awareness which has a discursive form.' (Giddens, 1984:374)

Practical consciousness, as distinguished from unconsciousness, is:

> 'What actors know (believe) about social conditions, including especially conditions of their own action, but cannot express discursively; no bar of repression,

however, protects practical consciousness as is the case with the unconsciousness'.
(Giddens, 1984:375)

When we equate consciousness with knowledge we can see that discursive knowledge refers to knowledge that we are aware of, have reflected upon, and can easily capture in language (i.e. it can be converted into information). Van Woerkum et al. (1999:3) writes of explicit knowledge. This might be, for example, the knowledge that farmers are presented with on a course on pest management. Practical knowledge, however, is something we know and apply, but find it difficult to talk about. Most people, as Giddens argues, are able to produce a grammatically correct sentence in their native language. They *know* the grammatical rules of the language, even if most people would find it difficult to explain them. In farming there is much practical knowledge; many farmers just know the best time to sow a particular maize variety, even if they cannot always explain the underlying principles and laws of nature. Similarly, they can effectively use a hoe or tractor on a heavy type of soil in a particular field, but would find it difficult to explain the perfect motions to an outsider. Thus, a lot of knowledge is embedded in practical routines, contextual experience, skills and physical memory.

In principle, such practical knowledge – others speak of 'tacit' knowledge (Nonaka & Takeuchi, 1995), 'implicit' knowledge (Van Woerkum et al., 1999) or 'mētis' (Scott, 1998) – can be made partly explicit and/or transferrable to others, but this usually requires considerable effort and energy, and obviously requires the co-operation of the person with the knowledge. One can, for example, elicit farmers' rationale and tacit knowledge in in-depth group discussions, using various entry points for debate, such as observations regarding current practices, changes in practice over time, local proverbs and taboos (Sadomba, 1996), religious prescriptions and rituals, etc. This is less feasible in regard to our 'unconsciousness', which is an area of knowledge characterised by perceptions and motives that we are not aware of and which is 'sealed off' by psychological processes of repression (e.g. caused by traumatic experiences in the past), which means that emotional barriers have to be overcome in order to gain access.

As we have argued in section 6.1, discursive and practical knowledge can be seen as the 'reservoirs' or 'repertoires' of interpretative schemes that actors can draw upon in assigning meaning to their environment. Explicit knowledge can be seen as only the 'tip of an iceberg'. In sociological terms, this 'iceberg' can be called an actors's life-world:

'The concept of life-worlds derives from Schutz and implies simultaneously both action and meaning. It is a "lived-in and largely taken-for-granted world" (Schutz and Luckmann, 1974). It is constituted of various forms of social knowledge, intentions and evaluative modes, and types of discourse and social action, through which actors attempt to order their worlds. Such life-worlds are the products of past experiences and personal and shared understandings, and are continuously reshaped by new encounters with people and things. Although the researcher attempts to come to understand the make-up of different life-worlds, they are essentially actor rather than observer defined.' (Leeuwis et al., 1990:26, note 3)

As indicated in this quote, a person's knowledge base or life-world is subject to continuous change. Even if it may have specific individual characteristics, a considerable part of a person's life-world tends to be shared with others, i.e. with the members of a 'community'. Giddens refers to this type of shared knowledge as mutual knowledge, which he defines as:

'taken-for-granted "knowledge" which actors assume others possess, if they are "competent" members of society, and which is drawn upon to sustain communication in interaction.' (Giddens, 1976:107)

It is this mutual knowledge that provides opportunities for effective communication between people (see Chapter 7).

6.3   Multiple realities and knowledge construction

As we have already indicated in section 5.2.5, it is important to realise that knowledge and perceptions are not neutral, but are subject to social influences and related to social interests. We often see that different actors have different perceptions about a particular phenomenon or situation. They define the same situation differently, use different knowledge to interpret it, apply different arguments, etc. The social influences underlying these different realities can originate from: (a) people's wider social background and history; (b) concrete 'political' contexts and group interests; or (c) individual interests in specific interaction settings.

Culture

Culture is an important background influence, which relates closely to the notion of mutual knowledge discussed in the previous section. Cook & Yanov (1993:379) describe culture as:

'a set of values, beliefs and feelings, together with the artifacts of their expression and transmission (such as myths, symbols, metaphors, rituals), that are created, inherited, shared and transmitted within one group of people, and that, in part, distinguish that group from others.'

When we talk of cultural diversity, then, we refer to the differences in values (what is important, unimportant, good, bad, right, wrong, etc.), norms (behavioural prescriptions based on values), behavioural patterns (e.g. routines and rituals), concepts (e.g. the things expressed in language), beliefs (e.g. as expressed in religion), artefacts and technology that exist between certain communities of actors (adapted from De Jager & Mok, 1974; Gouldner & Gouldner, 1963). There is huge cultural diversity between and within different regions in the world (e.g. Africa, Asia, Europe, etc.) even if some authors see homogenising tendencies (e.g. branches of McDonald's throughout the world; see Ritzer, 1993). Based on their cultural background, people can have a totally different outlook on the world, and it is not surprising that this extends to agriculture. Many African farmers, for example, choose agricultural practices on the basis of their wish to appease spirits and maintain relationships with their ancestors (Sadomba, 1999). Thus, their agricultural knowledge is closely related to religious beliefs and requirements (which may again vary by tribe, region, etc.). For many Western European farmers, farming is more about staying in business and applying scientific knowledge, but here too religious connotations can play a role. In the Netherlands, for example, farmers of different denominations (Roman Catholics, various branches of Protestants, atheists) hold different views about the acceptability of working on religious holidays (including Sundays), which affects the organisation and planning of agricultural work. Also, we can see important cultural differences between conventional and biological farmers in the Netherlands. For example, biological farmers – from a holistic perspective – tend to look at manuring as an activity aimed at maintaining the balance in the soil and/or feeding the earth, while conventional farmers think about manuring in terms of feeding plants. In relation to these cultural meanings, the biological farmers do not want to use chemical fertilisers and are hesitant about 'precision fertilisation' (e.g. applying manure only in the planting line), while the conventional farmers have no difficulty with this.

Group interests and identity

A second type of social influence associated with the existence of multiple realities is more related to concrete social interests and identity. Wildlife conservationists, for example, often have different goals and interests from the communities of farmers that live in areas well-endowed with wildlife. The farmers may regard elephants as 'a plague' that destroys their crops, while the conservationists look upon elephants as an endangered species, and on agriculture as an activity that seriously hampers natural ecological cycles and habitats. When discussing the appropriateness of particular agricultural or nature conserving practices, these parties are likely to use different evaluation criteria and arguments (as derived from knowledge), and express opposing views of what should happen. We often witness that such different stakeholders *select*, invent and mobilise those arguments and bodies of knowledge that help them to further a particular cause. In other words, they actively and flexibly construct different realities. At least in social terms these different perspectives are 'real'; their existence cannot be denied and in that sense it is irrelevant to try and argue which perspective is more 'true'. But even looking at 'hard scientific evidence', it is often difficult to assess the validity of different perspectives, especially since different parties apply different criteria and values when evaluating knowledge. Wildlife conservationists and farmers may, for example, agree that elephants cause say US$75 of damage per agricultural producer, but they may disagree on the wider meaning of that fact. Farmers may regard it as a reduction to their well-being and may demand compensation, while wildlife conservationists might argue that farmers should not live in a nature reserve in the first place, and that they inflict at least an equal amount of damage upon nature in the reserve.

Self-presentational interests

A third type of social influence that shapes knowledge and reality construction is of a direct interactional nature, and relates to personal 'micro' interests in a relational context. The knowledge people express, support or reject in direct interactions with others is often linked to self-presentational interests (Goffman, 1959). For example, people often want to make a good impression on others, avoid looking stupid, please others, avoid offending particular people and/or maintain good relations with them. Thus, in a group meeting a farmer may pretend to agree with the idea that a particular pesticide is appropriate to a particular problem solely to please the change agent, or so as not to lose face in the eyes of another person present. Likewise, when being evaluated, communication workers may present a rather distorted picture of their achievements or the functioning of their organisation in order to avoid criticism. The concept of self-presentation derives from the sociologist Goffman (1959), who drew parallels between the way actors in a theatre play their roles, and the way people present themselves in everyday life. As illustrated above, people try to create images of themselves or situations in social interaction (e.g. Goffman,1959; Koelen, 1988). In doing so, they try to influence the perceptions of others; in other words, they try to influence the way others construe the world.

We can see that there are various ways in which knowledge is connected to social influence and interests. What people believe or present to be true is closely intertwined with other people's views, and with efforts to pursue certain micro and macro interests and aspirations. Moreover, particular factual beliefs are never isolated, but are always embedded in a wider and often implicit web of beliefs and perceptions, including various types of social and normative evaluations (see Chapter 5). Thus, people may not only have ideas about whether or not, for example, chemical fertilisers work or do not work, but also about whether their use is good or bad, right or wrong, legitimate or illegitimate, just or unjust, etc. When attempting to understand specific aspects of people's knowledge, it is usually important to look for such, often implicit, associated meanings.

6.4   Knowledge and ignorance

For a long time knowledge has been looked upon only as something that *alleviates* and *reduces* ignorance. This mode of thinking stems from a philosophical tradition (ontological realism) which, as Knorr-Cetina (1981a:1) puts it, proposes that 'the world is composed of facts and the goal of knowledge is to provide a literal account of what the world is like'. Associated with this view is the idea (in philosophical terms, the positivist epistemology) that by means of scientific procedures we can make objectively true statements about the world. Following this line of thinking, many people still think that our knowledge of the world only increases and accumulates. This book is not the place for reviewing a major philosophical debate, but it is important to mention that well-known philosophers (e.g. Kuhn, 1970; Feyerabend, 1975) and mathematicians (Gödel, 1962) have argued convincingly that the idea of objectively true knowledge is highly questionable. They have shown that all – even natural science – statements about reality can only be considered true if one makes certain assumptions in advance. These pre-assumptions serve to make scientific observations, but cannot in themselves be proven (unless one makes other pre-assumptions). For example, if one measures nitrogen availability in a specific situation, one draws on a huge body of theory about minerals, forms of availability, soil characteristics, movement of minerals, measurement methods, etc., which one assumes to be correct.

And if one closely scrutinises these theories, one will find that they too rest on pre-assumptions, and so on, and so on. Also, mathematical formulae like Pythagoras' proposition/theorem on equal-sided triangles $(A^2 + B^2 = C^2)$ can only be proven to be true if one makes certain basic assumptions, for example that 1 plus 1 is 2 (which is basically an agreement among mathematicians rather than something that can be proven). In relation to this, we sometimes witness scientific revolutions (Kuhn, 1970) when some basic underlying assumptions are challenged, as happened for example when Newtonian physics was replaced by quantum physics. In such a case, a whole system of interconnected beliefs is turned upside down or shaken up.

The very existence of scientific revolutions, different schools of thought, multiple realities, cultural diversity, etc. shows us that adhering to particular views and explanations not only gives insight about reality, but also *excludes* a range of other possible perspectives. Thus, knowledge fosters not only insight, but also ignorance and 'blindness' (Winograd & Flores, 1986). As Arce and Long (1987:5) put it:

> 'a body of knowledge is constructive in the sense that it is the result of a great number of decisions and selective incorporations of previous ideas, beliefs and images, but at the same time *destructive* [our emphasis] of other possible frames of conceptualisation and understanding. Thus it is not an accumulation of facts but involves ways of construing the world.' (Arce & Long, 1987:5)

As Röling (2002) has argued, bodies of knowledge and perception (he speaks of 'cognitive systems' which include action and aspirations) may be internally *coherent*, but may not (or no longer) *correspond* with the external environment; that is, certain aspects of reality and/or new aspirations are ignored. He gives the example of the still dominant market-oriented agricultural policies that are based on (among others) Cochrane's theory of the agricultural treadmill (see Chapter 3) and which served to achieve specific objectives (e.g. low prices of agricultural products, release of labour for non-agricultural work, etc.). However, Röling argues that this internally coherent set of theories, aspirations and policies is not able to address the challenges of today which, according to Röling, centre primarily on addressing the 'eco-challenge' (see section 1.1.1)[1]. According to Röling the current context requires new theories and conceptions, including new ontological and epistemological positions in science, inspired by constructivism.

Although knowledge and ignorance play a role in almost every situation, we would like to point to some specific expressions of knowledge and ignorance that are important in connection with communicative intervention. These are:

- problem definitions and disciplinary blindness;
- reality reduction on the side of policy institutions;
- classifications;
- self-fulfilling and self-defeating prophecies;
- incomplete images, prejudices and biases.

---

[1] A conceptual reservation one might have about this mode of reasoning is that Röling introduces his own and many others' new aspirations and political objectives, so that one could argue that the 'cognitive system' not only lacks correspondence, but also coherence. For additional reflections see Leeuwis, 2002a.

Problem definitions and disciplinary blindness

Scientific disciplines typically provide a context in which people focus on rather specific issues, using specific languages and concepts. This makes scientists particularly vulnerable to a certain degree of blindness, especially across disciplinary boundaries. Starting from their specific mindset, scientists can easily frame a problematic situation according to their own disciplinary background. An entomologist may, for example, define a particular agricultural problem as resulting from pest infestation. The sociologist may interpret the situation as showing 'a lack of organisational capacity to organise collective crop-protection', while a soil scientist might point to land-degradation as the primary cause for plants becoming more susceptible to pests. In relation to the latter, a legal sociologist might argue that soil degradation results primarily from exploitative land-tenure arrangements, etc. Clearly, the disciplinary perspective one takes has immediate implications with regard to both further scientific endeavours (e.g. the formulation of research questions) and the intervention strategies that might be proposed (e.g. spraying, organisation building, composting or the design and/or imposition of new land-tenure regulations). Integrating different disciplinary perspectives and combating such forms of blindness are important tasks for change agents.

Reality reduction on the side of policy institutions

Politicians and policy-makers too are prone to certain kinds of blindness, which Wagemans (1998, 2002) has called 'reality reduction'. With this term he refers to policy institutions having a tendency to look at society through a formal perspective, and only being receptive to problems and signals that are meaningful through these particular spectacles. Everything that cannot be easily observed or understood within the formal perspective is considered to be irrelevant or non-existent by policy-makers (Wagemans, 2002). Typically in formal modes of thinking, there is a great interest in planning, regulation, quantifiable outcomes, political survival and media attention. And thus, there is a considerable risk that reality is reduced to factors that:

● can be measured quantitatively;
● fit in with the classifications (see below) that have been integrated in policies and regulations;
● are amenable to planning and legal intervention;
● are on the political agenda and/or acquire media attention.

As an example, we refer to the Dutch mineral policies in agriculture in the period 1998–2003. Due to European Union regulations, the Dutch government wanted farmers to reduce their emission of nitrogen per hectare to a specific amount (see for background, the case discussed in section 6.7). To achieve this, the government created 'mineral laws', introduced compulsory 'mineral bookkeeping', reserved special 'mineral funds', funded a range of 'mineral projects', subsidised 'mineral advisory products', and categorised farmers according to their 'mineral status'. However, by focusing on minerals alone, the government tended to overlook some important issues. These included: (a) that farmers had little interest in minerals as a separate learning theme, but were more interested in themes indirectly related to minerals, such as feeding, grassland management and animal health; (b) that farmers were becoming rapidly overwhelmed by the number of forms and regulations which neither they nor the responsible frontline government bureaucracies could handle adequately; and (c) that relations between the ministry and farmers had deteriorated to such an extent that effective policy-implementation was almost impossible (Leeuwis, 2002b). Thus, by reducing the whole situation to a 'mineral problem' and developing 'mineral policies', the government was overlooking circumstances that were relevant to the whole issue. In summary, we can say that effective intervention frequently requires looking beyond formal policy perspectives.

Classifications into target categories

As we have indicated earlier, knowledge serves to reduce uncertainty and complexity and thus provides order to the world. A particularly powerful way of ordering the world is through the use of classifications into different categories. For example, we can reduce complexity in farming by classifying farmers as big, medium size or small. By making classifications we highlight certain differences – in this case between farmers – that we consider relevant to policy-making. At the same time, by using classifications we make other differences 'invisible' in the sense that they are easily overlooked when using a particular classification. For classifying farmers, for example, one could have used a range of other classifications: female headed house-hold versus male headed household; economically viable versus economically non-viable; commercial versus subsistence; rain-fed versus irrigated; or a classification into farming styles (see section 15.2). It is important to reflect on the classifications one uses, and consider carefully what classification is most enlightening and relevant in a particular policy context. If, for example, one wishes to combat soil erosion, none of the above classifications may provide an accurate insight into relevant diversity. In cases where the prevention of soil erosion depends largely on the labour intensive making of contour ridges, it may make sense to create a classification according to the criterion 'labour availability per hectare' or 'degree of slopes in fields'. The resulting categorisation may well cut across all the others mentioned above. More generally, we can say that it is important to have a critical attitude towards pre-existing classifications, as they may make invisible the differences that are relevant to a specific intervention process.

Self-fulfilling and self-defeating prophecies

The idea of a self-fulfilling prophecy derives from Merton (1957) and is somewhat related to the issue of knowledge and ignorance. The basic idea here is that expectations and predictions may become 'true' merely because everybody believes in the prophecy, even if one can question whether it was originally correct. For example, when economists are convinced that small or mixed farmers cannot survive as farmers long-term given the economic climate, communication workers and policy-makers may direct their policies and services in such a way that small or mixed farmers benefit less, so that eventually they do indeed become non-viable due to lack of support systems. Similarly, change agents may communicate with small farmers in such a way that the farmers themselves start to believe that their farms have no future, and quit farming. All of this is no proof that the initial expectations and assumptions about the future of small or mixed farmers were valid, since one could have arrived at a totally different future if different communication for innovation policies had been pursued, aimed, for example, at achieving multi-functional agriculture (see Chapter 1). Thus, 'ignorant' (in the sense of narrow) views can be made more and more 'true' in the long run.

As communicative intervention, and intervention in general, is frequently dealing with future expectations, it is important that communication workers reflect on the validity of their assumptions. However, there is no general answer to the question of whether self-fulfilling prophecies are to be regarded as positive or negative; it depends on the situation, and one's aspirations and ethical considerations (see Chapter 3). In some cases, it may be positive to encourage self-fulfilling mechanisms in order to try and realise something that seems impossible but is nevertheless desirable.

In addition to self-fulfilling prophecies, change agents may also encounter the opposite, the self-defeating prophecy. Here, initially valid expectations become increasingly 'untrue' because people anticipate them. The predicted over-fishing of a lake, for example, may never happen, because it had been expected to occur and in response people organised themselves (e.g. facilitated by a change agent) to prevent it from becoming true. As this example shows, self-defeating prophecies can be quite positive. However, they may also have negative consequences; for example, a new arrangement that is expected to emerge in order to prevent over-fishing, but may not because some people start to frustrate efforts to produce it.

Incomplete images, prejudices and biases

The perceptions that people have are often based on limited or sketchy information. Where gaps in knowledge exist, people frequently extrapolate such information and/or associate it with other images (Van Woerkum, 1997). In doing so, they create a more 'complete' picture, which often means that the validity of different elements is no longer clear. Thus, on the basis of limited experiences, farmers may construct a particular image of a research organisation (ivory tower, non-practical, arrogant, bureaucratic, etc.) that may or may not do justice to the organisation as a whole. Similarly, bombarded by news on television and fund-raising advertisements, many Europeans have come to associate a continent like Africa largely with war, hunger, poverty and heat. Their image of Africa leaves little space for the fact that there are also peaceful regions, booming economies, productive agriculture, wealthy people, joy and happiness, as well as freezing weather conditions.

In the process of creating biased images, several mechanisms can play a role. In perceptions of risk and uncertainty, for example, we have already mentioned that estimations may be biased by the frequency, visibility, magnitude and directness of phenomena (see section 5.2.1). In addition, our images tend to be shaped more by our recent experiences than by things we know from the past, which may result in bias if the experiences diverge considerably. Moreover, when confronted with unstable situations, many people have a tendency to think that 'things were better in the past', essentially because the past is about situations they have already experienced and hence is associated with low uncertainty, while the future is characterised by higher uncertainty – which is something that people tend to feel uncomfortable with regardless of the specifics of the situation.

It is impossible for people to have nuanced and balanced knowledge and perceptions about everything, and in many cases people can afford biased and incomplete views since they relate to distant phenomena. In multi-actor change processes, however, incomplete images about other actors and/or social and natural phenomena frequently play a role, and may need to be exposed and adapted.

Finally, it is important to recognise that, like knowledge, areas of ignorance can be socially constructed (see section 6.3). In relation to the example on disciplinary blindness, scientists belonging to a ministerial plant protection unit may define an agricultural problem as pest infestation because it enables them to attract research funds and save jobs, while they would not be able to claim that money if they looked upon it as a social–organisational problem. Similarly, particular classifications, self-fulfilling prophecies or images can be carefully cultivated and maintained to help further certain interests. In this way, organisations produce and reproduce areas of knowledge and ignorance (see also Chapter 16).

6.5   Epistemic cultures: scientists' versus non-scientists' knowledge

Different cultures and groups of people may not only be characterised by different knowledge and perceptions of the world, but they may also have different ideas as to how new knowledge can be produced and validated; that is, they may have different 'theories of knowing'. In philosophical terms, this is the area of epistemology, and hence we speak of different epistemic communities or epistemic cultures (e.g. Knorr-Cetina, 1981a). In the epistemic culture of many natural scientists, for example, an important role is assigned to controlled experiments, a separation between (one) independent and (several) dependent variables, statistical analysis, randomisation, reproducability, peer reviews, PhD ceremonies, etc. Many natural scientists also have a tendency to 'reduce' complex wholes to their basic constituent parts; the idea is that by focusing on the individual parts (e.g. the different elements of an organism) and the relationships between isolated variables, one can understand the functioning of the complex whole (the organism). As anthropologists have shown, other cultural communities may employ rather different modes of thinking, procedures and rituals for arriving at credible claims, even if some basic similarities exist (Polanyi, 1958; Marwick, 1974). Likewise, farmers tend to use different procedures for arriving at valid knowledge from those used by natural scientists. Although many farmers conduct 'experiments' (e.g. Stolzenbach, 1994), these tend to have rather different characteristics from scientific experiments (for elaboration see section 13.5.8). Moreover, in farmers 'epistemology', things like practical experience, farm comparisons, intuition and discussions with colleagues tend to play an important role.

Misunderstandings about superiority: the local nature of all knowledge

As will be argued in Chapter 8, scientists' and farmers' knowledge can in principle enrich each other and deliver important ingredients for innovation in agriculture. However, this process of enriching is frequently disturbed by the fact that many scientists tend to look at their scientific knowledge as universal, generally applicable and superior to farmers' knowledge. Hence, they regard themselves as experts and others as laymen (see also Chapter 11). Here we touch on a rather unfortunate and unhelpful belief that forms an integral part of the dominant epistemic culture in science. This issue has generated a lot of debate on the usefulness, quality and validity of scientific versus local or indigenous knowledge in farming (e.g. Richards, 1985; Van der Ploeg, 1987, 1999; Röling, 1988; Marglin, 1991; Warren, 1991; Brouwers, 1993; Scoones & Thompson, 1994; Wynne, 1996; Scott, 1998). In this debate some authors have gone to the other extreme, and argued that positivist and reductionistic science inherently produces less relevant knowledge, and that local knowledge is generally superior to scientific knowledge.

In this book we take the position that both extreme positions are essentially flawed. On the one hand, natural scientists should realise that all knowledge is contextual, and in that sense they cannot claim universality and general applicability. If natural science experiments in research facilities show a positive relation between, say, nitrogen levels and rice yields, then scientists should be aware that this relationship is only valid or true within the specific context of an experiment (and as long as the underlying theories, concepts and modes of measurement remain undisputed; see also section 6.4). That is, it is valid within the space of the experimental field and given certain conditions such as a specific (a) amount of care and labour input, (b) type of soil, (c) degree of pest infestation, (d) pest control measures used, (e) planting date, (f) cropping history, (g) climate, (h) crop environment, (i) range of crop interactions, (j) form of crop processing and consumption, etc. Many of the experimental conditions are likely to be different outside the research facility, and hence the generated knowledge cannot be treated as universally valid outside the research station (unless one attempts to reproduce all experimental conditions outside the research facility, which is usually impossible and/or inefficient). In essence, this means that scientific knowledge too is local knowledge: it is created and meaningful in a specific cultural, spatial, technical, climatic and socio-political context (which may or may not coincide with a specific geographical context)[2]. For this reason we do not conform to the often used distinction between scientific and local knowledge in this book, but speak simply of scientists' versus farmers' (local) knowledge. Straightforward application of knowledge that is deemed valid in one locality (e.g. a scientific laboratory and an associated research network), in another locality (e.g. a particular region and its social configuration) is bound to lead to problems and disappointment. Even if eventually aspects of such knowledge may prove valid in another local context, there is always a need for further testing, adaptation, integration and innovation.

---

[2]   Note that we do not use the term 'local' to refer to geographical boundaries only, but to refer to specific networks of social relationships that somehow impinge on how knowledge takes shape.

On the other hand, it does not make much sense to argue that conventional natural science research has nothing to offer to farmers in specific contexts. Much of the existing local farmers' knowledge needs to be renewed, adapted and supplemented because of rapid changes in context (e.g. population growth, migration, climate change, industrialisation, ecological changes, globalisation, degradation, etc.). And farmers' experiments and knowledge do have certain strengths, but also a number of weaknesses (see section 13.5.8), and therefore tend to leave a number of questions unanswered. It has been demonstrated that conventional (positivist and reductionist) laboratory research can at times provide extremely valuable insights into solving farmers' problems (e.g. Van Schoubroeck, 1999; Lee, 2002), even if such insights tend to be 'partial' in that they address only a limited selection of the variables and factors that farmers take into account (see section 13.5.8). In short, conventional applied or fundamental research can be beneficial when it addresses relevant questions and issues. Much of the critique on conventional scientific research boils down to the assessment that it tends to operate in isolation from real-life innovation processes, and generates its own questions rather than addressing the questions and specific problems that societal stakeholders find relevant; hence the frequent plea to make agricultural science more interactive (Röling, 1996; Leeuwis, 1999b). In many ways, however, this is essentially a matter of science organisation, science policy and science politics rather than of epistemology.

The above implies that even if – within the parameters of a well-defined context and conceptual frame – natural scientists can claim to arrive, at least temporarily, at valid or 'objectively true' conclusions, they cannot claim to arrive at *neutral* conclusions. This is because the conclusions arrived at are often directly linked to the research questions that were asked. These questions and problem definitions, of course, are never neutral: they are asked and/or funded by specific stakeholders, for a specific reason, and in connection with specific goals and interests. The director of a fertiliser company, for example, might want to know what combination of fertilisers can best be applied (when, what dosage, etc.) in maize production in a region of Tanzania, while local farmers may be more interested in developing a cropping system that minimises the use of chemical fertiliser. Thus, different stakeholders might ask different questions and set different priorities, and hence are bound to arrive at different conclusions. However, it is clear that in this case the director of the fertiliser company may well be in a much better position (i.e. may have more access to relevant resources) to put into effect his research interests than local farmers are. In that sense, science is intrinsically linked with issues of power and politics.

6.6   Knowledge, power, agency and structure

The above discussions have indicated that knowledge and power are intertwined. To further clarify this relationship it is important to look at the concept of power itself. A classic definition of power is the one provided by Weber (1968:224): 'the capacity of an individual to reach his will, even against the opposition of others'. In view of such a definition, it is not surprising that many people have come to associate power primarily with something 'dirty' or negative. In a more balanced and positive vein, Giddens (1976) has later refined Weber's definition, and speaks of two different forms of power: 'power as transformative capacity' and 'power as domination'. Power as transformative capacity refers to the fact that every human being has the capacity to act and get certain things accomplished. It is:

> 'the capability of the actor to intervene in a series of events so as to alter their course.' (1976:111)

This type of power is also called 'human agency', a notion which:

> 'attributes to the actor (individual or social group) the capacity to process social experience and to devise ways of coping with life, even under the most extreme conditions of coercion. . . . Agency – which we recognise when particular actions make a difference to a pre-existing state of affairs or course of events – is composed of social relations and can only become effective through them.' (Long, 1989:10)

Although human beings can creatively use their agency, and create 'space for maneouvre' (Long, 1989), their capacity to transform is constrained in various ways by their natural and social circumstances (and/or by their perception of these, see Chapter 5). At the same time, these circumstances also provide opportunities. In a given situation the prevailing natural and social conditions may, for example, make it very difficult to grow tea, but provide good circumstances for growing cotton. This combination of constraining and enabling (Giddens, 1984) circumstances and properties we will call 'social structure'[3].

The second type of power relates more to the second part of Webers' definition. Power as domination is described by Giddens as:

> 'the capability to secure outcomes where the realisation of these outcomes depends upon the agency of others. It is in this sense that men have power "over" others: this is power as *domination*.' (1976:111)

This type of power is concerned with effectively mobilising others in order to help achieve certain aims. People can bring in a variety of resources (see section 5.2.4) to make others conform with their aims and interests. The resources can be rather direct or even rude; one can, for example, impose sanctions or use violence to 'discipline' other actors, and make them act in a way that goes against their initial inclinations and desires. But the resources can also be much more subtle, even to the extent that people are hardly aware that they are being influenced by others in order to help them realise certain ends. Private companies, for example, sometimes use very subtle advertising campaigns, in which they try to link their products with certain emotions and images (e.g. fostering the idea that smoking cigarettes is adventurous and an act of freedom), in order to try and make consumers buy their product.

It is important to realise that power as domination is not so much a state of affairs but rather an interactional process. One cannot simply conclude that because some

---

[3]   There is a long-standing and heated debate in the social sciences on how to conceptualise 'structure', and the way action and structure influence and/or determine each other (Giddens, 1984; Long & Van der Ploeg, 1989; Bourdieu, 1990; see also Leeuwis, 1993). However, this debate is beyond the scope of this book.

people have many resources they *automatically* have power over people with fewer resources (see also Latour, 1986). In an interaction context, power is not only exerted by the party with resources, but also to some degree 'handed over' by those with fewer resources (who also play an active role); that is, power can only be exerted if others recognise the resources brought in as meaningful. If, for example, a person does not value money, it will be difficult to bribe him, no matter how much money one possesses.

On the basis of the above we can conceptualise more precisely the ways in which knowledge is related to power.

Knowledge as a source of agency

Availability and access to particular knowledge can enhance or limit a social actor's capacity to exert a particular type of agency. For example, depending on the availability and access to relevant knowledge on crop protection, a farmer may or may not be able to fight a particular disease effectively. In this sense, having access to relevant and valid knowledge is by definition 'empowering'. At the same time, having access to inadequate knowledge can be considered as 'disempowering'. In essence, knowledge can be a *resource* that actors can draw upon in dealing with the world around them. As is the case with other resources (money, land, etc.), it is evident that some categories of actors may have better access to this resource than others.

Knowledge as structure

As we have seen, actors' agency can be constrained and enabled by relatively persistent characteristics, properties and circumstances that actors have to deal with. This may relate to natural conditions (e.g. weather conditions and soil characteristics), but also to social conditions (such as political (in)stability, the functioning of agricultural institutions, etc.) as well as material configurations (e.g. the physical lay-out of a farm, available infrastructure, the dominant technology used in specific realms, etc.).

Some of these structural conditions can be regarded as the result of the previous and/or persisting application of particular knowledge. In the Netherlands, for example, conventional wisdom between 1970 and 1990 was that, for efficiency reasons, agricultural enterprises would have to become more and more specialised; that is, the expectation and policy was that mixed cattle and crop farming would come to an end, and that farmers would focus on either a limited number of crops or on a particular type of animals (cattle, pigs, chicken, dairy, etc.). On the basis of this vision and knowledge, a large number of agricultural institutions (including extension divisions, research institutes, university departments and educational programmes) have become segmented and organised according to the envisaged specialisations; that is, they focus on crop farming, horticulture, dairy farming or pig farming, etc. From the 1990s onwards, however, the interest in mixed farming has returned, in view of environmental concerns, but it has proved far from easy to effectively mobilise research, extension and education on mixed farming, partly because all institutions have been organised along specialised lines.

This example shows that if many actors continue to draw on particular knowledge and classifications (in this case specialised agricultural sectors; see also section 6.4) for a prolonged period, this knowledge is likely to shape institutional and/or other properties (e.g. technology, environmental damage) that cannot be easily changed. In other words, the continued drawing on knowledge can result in relatively persistent structural properties, which constrain or enable human agency. In Science and Technology Studies such structural properties are also referred to as technological regimes (Kemp et al., 2001). This term refers to a coherent set of technologies, rules and institutions in the context of which people are operating at a given time.

Knowledge as a tool and resource for domination

While structural consequences as described in the previous section can be emergent and unintended, they may also be the result of deliberate attempts by some actors to dominate and discipline others. As we have seen, domination is about influencing other people's agency. As knowledge is a vital component of agency, we can say loosely that effective domination implies that an actor is able to impose his or her knowledge on others; that is, an actor manages to make others act according to, or at least in line with, his or her definition of the situation. As we have seen in Chapter 5 (section 5.2.4), the capacity to impose depends to a large extent on actors' ability to mobilise various resources, rewards and sanctions (e.g. money, land, inputs, violence, protection, etc.). These resources can, for example, be used to:

- produce, strengthen or invent 'favourable' knowledge and definitions of reality (e.g. through the funding of research; see also the example of the possibly different research agendas of farmers and fertiliser companies at the end of section 6.5);
- selectively communicate or withhold particular knowledge and information (e.g. withhold or ridicule knowledge on how to reduce fertiliser use);
- push and persuade others to adopt a particular perspective by means of rewards etc. (see section 5.2.4);
- ensure that issues are discussed in a language and terminology unfamiliar to others, which puts them at a disadvantage.

6.7 Practical relevance: the case of farmer experimentation in environmental co-operatives

In this chapter we have discussed issues of knowledge at a fairly abstract level. However, we feel that the insights generated are relevant in practical contexts. We have learned in this chapter that the knowledge and perceptions that stakeholders adhere to in a specific context are likely to be:

- only partly explicit;
- embedded in a wider system of beliefs;
- shaped by action and experience;
- far from accidental or neutral, but shaped by a range of social interests;
- selective, biased and inherently connected to forms of ignorance;
- influenced by different epistemic beliefs and practices;
- tied up with issues of power in various ways;
- in summary: they are *locally contextual* in many respects.

Without some understanding of these kinds of issues, we feel that productive communicative intervention is unlikely to come about. Thus, we conclude this chapter with an illustration of the relevance of our conceptual discussions in a communication for innovation context. At this point we will focus on the relevance for *understanding* a specific situation, as a first step to successful intervention in it.

The case we discuss concerns two groups of dairy farmers in the north of the Netherlands, who have joint activities geared towards improving the sustainability of their farms and who refer to themselves as environmental co-operatives. In describing and analysing the case, we have benefited from project documentation (in particular Atsma et al., 2000) and collaborative research efforts in the context of the AGRINOVIM[4] project (see also Eshuis et al., 2001; Stuiver et al., 2003; Wiskerke & Van der Ploeg, 2003).

An important environmental problem in the Netherlands relates to over-fertilisation of farmland, resulting in pollution of ground water and 'acid rain' due to ammonia emissions in the air. Typically, dairy farmers in the Netherlands have a relatively high stocking density combined with a very high milk production per cow, which can only be realised with the help of imported feed compounds (e.g. soya, cassava) and application of high dosages of chemical fertiliser on grassland (up to 400 kg of nitrogen per hectare). Apart from the chemical fertiliser, dairy farmers apply a large amount of liquid animal manure which is collected in large tanks underneath the slatted floors of the animal sheds where the dairy cows are kept in winter and/or at night. In view of the above, many of the activities (including on-farm experiments) of the environmental co-operatives are geared towards solving the manure problem. In connection with this, they have managed to attract some funds for a 'mineral project', which includes further experimentation and study group activities, together with support for this in terms of logistics, facilitation and scientific back-stopping. In implementing these activities and experiments, the environmental co-operatives have established co-operation and contact with a large number of actors, including governmental bodies, feed industries, the agricultural university and research stations. In a nutshell, the mineral management strategy of the northern farmers consists of various interconnected ingredients (Atsma et al., 2000):

- they strive to significantly reduce the use of chemical fertiliser (especially nitrogen);
- this is thought to be possible only if the quality of animal manure is improved, and if the soil becomes more 'efficient' in making nitrogen available to plants;
- they strive to improve the quality of manure through reducing the amount of protein in animal feed, and by adding compounds (clay or micro-organisms) to the manure through which composting is supposedly enhanced;

---

[4] AGRINOVIM stands for a research project entitled 'The dynamics of AGRicultural Innovation: studies at the interface of NOVelty creation and sociotechnical regIMes'. Discussions with several researchers on this project have helped with writing this case; in particular discussions with Marian Stuiver, Frank Verhoeven, Jan Douwe van der Ploeg, Arie Rip, Jaap van Bruchem, Johan Bouma, Han Wiskerke and Jasper Eshuis. The authors have also benefited indirectly from studies carried out by the VEL & VANLA research team led by Lijbert Brussaard.

- improving the nitrogen 'efficiency' is thought to be possible only if the structure of the soil is improved (i.e. less damaged by heavy machinery), and if soil biology (i.e. life of bacteria, worms and insects) is enhanced;
- in connection with all this, the timing of various activities (e.g. mowing grass, applying fertiliser and manure) is adapted.

However, in pursuing these ideas and corresponding experimentation, the farmers groups have run into various problems related to the theoretical issues raised in this chapter. We discuss some of these in the following sections.

Relating to 'knowledge as structure'

The experimental activities take place in the context of certain *structural properties*, including a dominant *technological regime*, which tend to constrain the activities of the farmers. As most Dutch dairy farmers have slatted floors and feed a lot of protein to their animals, for example, the Dutch government enforces legally that liquid manure is injected *underneath* the grass in order to prevent ammonia emissions. The machinery required for this is heavy and expensive, and so many farmers hire contractors to do the job. However, the northern farmers would like to apply the manure *on top* of the grass, as this requires less heavy machinery and is therefore expected to lead to less destruction of soil structure; and it would also preserve the landscape which consists of relatively small fields surrounded by trees and is therefore not conducive to using large equipment. Moreover, spreading liquid manure on top of the grass can be done by farmers themselves, instead of a contractor, and is thought to lead to a better distribution of minerals from animal manure, which is deemed necessary when application of chemical fertilisers is significantly reduced. Finally, the farmers argue that their liquid manure is of a better quality than average, which considerably reduces ammonia emissions already, and makes manure injection redundant.

Even if environmental co-operatives, after a long fight with the Ministry of Agriculture, were eventually allowed to experiment with top dressing of manure, this is only allowed in the context of *research*, not as regular practice, and the permission needs to be fought for every year. Thus, this example shows how the existing configuration of technology, practices and rules tends to limit the space for experimentation and action. This is also demonstrated by the fact that farmers cannot simply do away with their expensive slatted floored houses and liquid manure, and move to straw bedded sheds and fixed manure. Although this might reduce some of the legal problems and help improve manure quality through composting, it is not feasible due to the farmers' previous investments and wider economic and labour concerns.

Relating to 'tacit knowledge' and 'epistemic cultures'

A second problem for the northern farmers is in their interaction with several scientists (Verhoeven, pers. comm.). From their on-farm experiments, a large majority of farmers have become convinced about a number of points, for example:

- that the existing norm concerning the amount of protein in animal feed is too high, meaning that high milk yields can be maintained even if less protein is fed;
- that animal health tends to improve when less protein and more 'structure' is fed;
- that the compounds added to liquid manure improve manure quality and the ease of handling it;
- that the current norms regarding the quantity and scheduling of chemical nitrogen application are sub-optimal.

On the basis of their experiences and perceptions, the farmers now want to advise their colleagues to follow suit. However, several scientists have argued that none of the conclusions of the farmers are supported by thorough scientific evidence, and that the existing scientific norms should still be regarded as valid. In addition, it was implied that it would be 'bad science' if the scientists who are involved in the on-farm experiments encouraged farmers to spread such advice (Verhoeven, pers. comm.).

What we see here clearly relates to the existence of *tacit knowledge* and different *epistemic communities*. Although the farmers engage in measurement and calculation, and have modelled their experiments partly according to scientific standards (e.g. they have a control group), their on-farm experiments differ considerably from conventional scientific experiments (see also section 13.5.8). For example, there is no random sample of farmers (but only the members of the co-operatives), conditions are not strictly controlled (because every farm is different), and in practice there tend to be several independent variables at the same time (as alternative farming practices usually involve several variables simultaneously). Despite these shortcomings from a natural science perspective, many farmers are convinced that they are on the right track and that their innovations work. Here, not only the many calculations and measurements play a role, but also their everyday handling of animals and equipment, their impressions of the skin of animals (an indication for health), the stories and experiences of other farmers, the colour and texture of grass and soil, their intuition, and the observation of many other details on the various farms. For scientists all this *practical* or *tacit* knowledge has little meaning, and may in fact be regarded as vague and/or as causing biased perceptions. For farmers, however, these direct experiences are utterly convincing. In so far as the scientists argue that farmers should not be 'spreading the news', they seem to express that they regard their own methods and knowledge as superior.

Although we are in no position to judge who is 'right' and 'wrong' with regard to specific details of these disputes, we would – somewhat detached from this case – like to point to some strong and weak points in the reasoning of scientists in situations like this. Scientists may have a point when they argue that farmers' views may be biased by methodological weaknesses or because they might like their experiments to yield positive results and are perhaps all too eager to convince each other that they do so. Also, scientists may rightly caution farmers about giving advice to others when the processes at work are not yet fully understood, and experiences are still based only on a limited number of seasons. However, in making such comments scientists should not disregard the *local* and *socially constructed* character of their own knowledge, and should consider that:

- Their own responses to farmers' research could also be biased in view of personal and/or institutional interests of scientists (see the next paragraph).
- Norms arrived at through conventional scientific procedures may well have weaknesses in that they:
  - (a)   are typically based on research that only takes into account a limited number of variables, and disregard variables that are taken into account in tacit knowledge;
  - (b)   may be based on 'average results' whereby important contextual differences are no longer visible and/or taken into account (e.g. the original research data might have shown a certain amount of contextual or unexplained variation);
  - (c)   may incorporate assumptions about local conditions as well as farmers' goals and aspirations that are not valid in specific contexts (e.g. they may implicitly assume high milk yields as the overriding goal, while farmers grant priority to cost reduction or ecological sustainability).

Apart from this, natural scientists might want to take into account that:

- If large numbers of farmers are convinced about something, this can in itself be a 'hard' social scientific fact, which cannot be reasoned away easily.
- It is often impossible and undesirable for farmers to put off changes in practice until these are officially 'certified' by scientists. In fact, trying to underpin (or refute) the northern farmers' conclusions by conventional academic research could take several years (assuming that funding would be available). It is important to recognise that innovation is not the same as 'science' and has a different rhythm and logic (see Part 3 of this book).

Related to the 'social construction of knowledge'

As we have already hinted, the knowledge disputes between the northern farmers and several scientists may not only be explained in terms of epistemic cultures, but may also relate to various forms of social construction that we have discussed in section 6.3. As we have not done research in this specific context our remarks here are speculative, and serve merely to illustrate the point. The knowledge claims made by scientists and farmers could have been influenced by:

- *Self-presentational interests*: As we have seen, the farmers are challenging scientific norms that have been worked on by scientists for many years. Thus, there could well be personal reputations at stake, which play a role in the way scientists respond to the farmers' conclusions. This is equally true, of course, for the farmers (and the scientists who support them) who might not want to lose face either.

- *Group interests and identity*: In a situation like this there are usually a variety of group interests at stake. As mentioned earlier, the northern farmers may want their experiments to yield positive results, in order to keep attracting funds and to reinforce their credibility as an environmental co-operative. This may lead them to over-emphasise their achievements, a tendency which may be further increased by a feeling that they have not been taken seriously enough by the government and others. Scientists from competing institutes, in turn, may question and play down the results for similar reasons. By questioning the results on scientific grounds they present themselves as 'custodians of the truth', which may help them to mobilise new funds from the government (which does not really know how to handle the situation) to settle the matter once and for all and in a thoroughly scientific way. To some extent, the case presented can also be interpreted as a struggle between competing groups of scientists, i.e. those who strongly believe in on-farm research and a holistic approach, and those who prefer a more reductionist mode of investigation. The scientists' responses to the farmers' conclusions may also have been shaped by the division between those who suggest that more efficient nitrogen use can best be achieved by improving the efficiency of the soil, and those who argue that this efficiency can best be enhanced by improving the efficiency in animals (e.g. by breeding and feeding regimes). The northern mineral project is dominated by people who support the former view, which may have influenced responses by researchers belonging to competing research groups.

Related to 'knowledge and ignorance' and 'knowledge as a resource for domination'

The closing point of the previous paragraph is an illustration of how one problematic situation can be defined differently by different groups of scientists from different disciplines (e.g soil biologists and animal scientists). Social scientists may define the situation in a different way again, for example, as a dominant technological regime that is difficult to overthrow (Wiskerke & Van der Ploeg, 2003). The various views do not necessarily exclude each other, but depending on the resources that the different disciplinary groups have at their disposal, relative areas of knowledge and *blindness* may arise. We can also point to the fact that the existing farming system and technological regime were already characterised by important areas of knowledge and ignorance. Environmental co-operatives started from a situation where there was a lot of knowledge on increasing milk production per cow per hectare, and per labour unit, but relatively little on ways of enhancing manure quality and the interactions between nitrogen processes, soil biology and soil structure. Thus, the northern mineral project can be seen as an attempt to create a different pattern of knowledge and ignorance. The existing pattern was by no means accidental, but the result of prolonged pursuit (by among others the government, agro-industry and knowledge institutions) of a particular model of farming in which environmental issues were not considered, and which coincided with selective research agendas. Therefore, the attempt to open up new areas of knowledge can be regarded as an effort to shift the *power balance* between different modes of farming through the mobilisation of more favourable *knowledge resources*. In other words, through the mineral project northern farmers try to reduce the *dominating* influence of existing knowledge.

A final point in connection with knowledge and ignorance is that the different scientists and farmers who struggle over knowledge claims tend to do so on the basis of rather sketchy information and *images* about each other, including the research practices engaged in. Details about the data and experiments on which the various claims and conclusions are based, for example, are not always mutually transparent, which means that an 'estimate' must be made about their validity. Here all sorts of associations may play a role. One of the natural scientists who co-operates intensively with the northern farmers advocates the consideration of spiritual issues in farming and agricultural research. This makes him quite controversial among agricultural scientists, as it triggers images of 'new age', 'vagueness' and perhaps even 'charlatanry'. It may well be that the way scientists respond to the northern farmers' research results is affected by such images, rather than by the actual research quality, which leads them to take the results less seriously.

Related to 'knowledge as a source of agency'

Despite the fact that the environmental co-operatives have become involved in various disputes and controversies regarding knowledge, the insights they have gained have enhanced their space for maneouvre and hence agency. They are, for example, in a much better position now to meet the criteria set by new government regulations aimed at reducing nitrogen emissions, while other farmers are still facing serious problems in this respect.

Summary

As the case study demonstrates, our deliberations on knowledge and perception give us a better understanding of why and how knowledge disputes emerge. While this case has focused on disputes between scientists and farmers, a similar analysis would be possible for disputes between, for example, farmers and nature conservationists. Change agents typically find themselves in the middle of such complexities, and are supposed to play a positive role stimulating integration between different perspectives. The practical aspects of doing this will be discussed in Parts 3 and 4 of this book.

Questions for discussion

(1) Identify a recent controversy that you have witnessed or been able to follow through the media. Who were the key stakeholders involved, and what knowledge, arguments and information did they bring to bear in order to make their case and/or support their view? Discuss if and how the social influences mentioned in section 6.3 may have affected the knowledge claims presented by stakeholders.

(2) Give an example of a situation in which certain stakeholders purposefully maintained ignorance. Is there a resemblance to the expressions of ignorance mentioned in section 6.4? Why do you think this ignorance was constructed?

(3) We have mentioned three ways in which knowledge may be related to power. Give an example of each of these, building on your own experience.

7 Communication and the construction of meaning

Now that we have explored the concept of knowledge, we can move on to gain a better understanding of 'communication', which is an important process used by people to exchange experiences and ideas, and hence a vital trigger for altering knowledge and perceptions of various kinds (i.e. learning). Hence, communication is a core ingredient of change agents' strategies for inducing change.

7.1   What is communication?

In a simplified form, we can define human communication as the process through which people exchange meanings. Looked at more closely, we can recognise various important ingredients and distinctions.

Communication is about using symbolic signals

To exchange meanings, human beings use a variety of devices: words and language, pictures, drawings, music, Chinese characters, letters of the alphabet, body language, etc. These devices or signals are symbolic, which means that they refer to something else (Oomkes, 1986). The letter combination 'COW' *refers* to an animal we all know, but the letter combination itself is not a cow. Likewise words, body language and music can *refer* to certain feelings we have, but are not feelings themselves. These symbolic signals are what we have earlier called information, i.e. tangible expressions of knowledge, thoughts and feelings (which in a way are also interpretations). The symbolic nature of communication allows for a great deal of interpretative freedom; after all the symbol 'COW' may be associated with totally different animals by different people (e.g. a Holstein-Friesian dairy cow, a Zebu cow, a Charolais meat cow) if no further information is provided.

Communication can be through verbal and intentional 'messages'

Human beings can make deliberate attempts to communicate meanings to others. In such cases, we combine several signals into a *message* (i.e. information). A communication worker, for example, might say: 'I really think it is important that you try this method of biological pest control on your farm,' and he or she can support the statement by emphasising the word 'really' with the help of intonation of the voice, and/or by making gestures. While verbal and written forms of communication are often intentional, many people are less conscious and deliberate about their non-verbal communication (body language). Non-verbal forms of communication include:

- appearance (e.g. wearing a particular style of clothing);
- posture (e.g. standing tall or 'shrinking');
- gestures (e.g. rapid movements of eyes, impatient ticking of fingers);
- spatial position (e.g. standing close to someone or keeping a distance).

From people's body language one can often read whether a person feels at ease or uncertain, is enthusiastic about something or indifferent, is happy about something or disappointed. Many people are unaware that they convey such clear messages/ information about their state of mind, and that others interpret these signals and derive meaning from them. Clearly, intentional and sub-conscious messages can be conveyed simultaneously, so that one can send out several, sometimes contradictory messages at the same time.

Symbolic signals are transferred through channels and media

When we speak to someone, our vocal cords (stimulated by the brain) cause vibrations in the air, which 'transport' the sounds we make to another person's eardrum, which again triggers that person's brain. This combination of brains, vocal cords, sound waves, eardrums, etc. we call a communication *channel*. A variety of such channels exist for the 'transportation' of visual, auditive, tactile and olfactory signals. Communication *media*, then, are composite devices which incorporate several channels at once. Television, for example, is a medium that opens up auditive and visual 'long distance' channels that allow for 'live' (synchronical) or 'delayed' (asynchronical) communication (see below).

We will discuss different media in Chapter 12. It is important here to note the famous statement that 'the medium is the message' (McLuhan, 1964:23), which emphasises that the use of a particular medium may well carry a message in itself. If, for example, a manager chooses to discuss a highly personal issue with a staff member through e-mail or a letter, this says something about the manager or the relationship between manager and staff member.

Apart from the obvious communication media – such as radio, television, posters, internet, etc. (see Chapter 12) – one could argue that objects and man-made artefacts not only play a role as non-verbal signals (see above), but need to be looked upon as 'material communication media/channels' as well. Parallel to what is argued in section 7.2, one could say that man-made objects are often 'encoded' with multi-interpretable 'messages'. A door in a room, for example, may be seen as a medium through which the designer/builder transfers the 'message' that this is the appropriate place to enter or leave. Likewise, agricultural technologies can be seen as embodying designers' knowledge and meanings, constituting 'tangible expressions of knowledge' and hence having much in common with what we have called 'information' in section 6.1. We might even say that they are 'carriers' of information, which comes close to saying that they constitute a communication channel or medium (see also section 20.4.3).

Communication takes place in a historical and relational context

People who communicate with each other often do so in the context of previous communications and experiences. This implies that communicating parties have a relationship with each other; they may or may not (a) know each other well, (b) be aware of each other's identity, (c) like each other, (d) trust each other, (e) be interested in each other or (f) have a conflict with each other, etc. The way communication unfolds in a particular context is shaped partly by the nature of the relationship. When people (e.g. change agents and farmers) have positive experiences of each other, communication may be smooth, whereas people may not even be prepared to listen to each other when there has been a serious conflict.

Communication can be 'interactive' to various degrees

Although all communication is a form of human interaction, and is in that sense 'interactive', some authors make a distinction between different degrees of interactivity (Rafaeli, 1988), i.e. the extent to which different communicative exchanges occur and are connected to each other; or, to put it more simply, the extent to which people make an effort to understand and listen to each other. Rafaeli speaks of 'non-interactive' communication when people exchange messages which do not refer to each other. In other words, people do not react to earlier statements. This could be a monologue, or a lecture without opportunities for posing questions, but also could be a situation were people totally ignore each other's statements, and each follow their own independent line of argument. A second level is called 'reactive' or 'quasi-interactive' communication. This involves patterns of communication where people merely respond to the previous statement of the other person, in much the same way as a press conference where a series of questions are asked and answered. One could also think of the interaction with a coffee machine that first 'asks' whether you want tea, hot chocolate or coffee, and then proceeds after a button is pushed to enquire whether you want your coffee 'expresso' or 'cappucino', with or without milk, etc. Finally, Rafaeli speaks of 'full-interactivity' when later utterances address the *relationship* between previous statements and the utterances preceding these; for example, when someone makes critical remarks in relation to an answer given to an earlier question. Thus, one can speak of a real 'dialogue' between people who listen to each other and connect different statements through time.

Communication can be synchronical or asynchronical

Regardless of the level of interactivity, communicative responses can be immediate to a greater or lesser extent. When people meet face-to-face, or talk to each other on the telephone, they can respond to each other straightaway. We call this synchronical communication, as both parties are involved in the process at the same time. However, there are many forms of communication where responses are 'delayed', because communication takes place in the form of letters, e-mails, articles, etc. Here the communicating parties engage in the process at different times, so we call that asynchronical communication. An important implication is that the context experienced by different parties while communicating may become increasingly different as communication moves from face-to-face, directly mediated by, for example, telephone, to asynchronical. Someone may, for example, write a letter or make a phone call in an atmosphere of intimacy and relaxation, while the letter may be read or the phone call received in a context of hectic work. Thus, the chances of people being on the same 'wavelength' are reduced in non face-to-face situations.

Communication messages have different levels and layers

When rehearsing a play, professional theatre actors soon find out that one sentence can be given different meanings. Depending on the specific meaning that a director wants to highlight, an actor may be encouraged to act out and intonate the famous sentence from Shakespeare's *Hamlet*, 'To be or not to be', as either a question, a command, a plea, an insult, a confession, a surrender or a conclusion. The same is true for much more mundane sentences, and for practically everything we say. Depending on the context (Wittgenstein, 1969) and through the particular mix of signals used, the sentence 'Can you please give me something to eat' can be made to mean: 'Shut up, and bring me my food you stupid idiot!' (insulting, commanding), 'I'll do anything you want, if only you would give me something to eat' (begging, promising) or 'You have misunderstood me; I am not thirsty, but would really like something to eat' (correcting, asking). Thus, messages do not only have a literal meaning, but can also be imbued with more implicit meanings and connotations. To capture this, the language philosopher Austin (1971) argues that language expressions have 'descriptive' and 'performative' dimensions; sentences do not only make statements about the world, but they also *do* things to the world. (For further discussion of this 'speech act theory' see also Searle, 1969; Habermas, 1981). In a similar vein Oomkes (1986; see also Watzlawick et al., 1967) distinguishes between the 'content' and the 'relational' level of messages, i.e. the level of what is said literally and the level of what it *does* to the relationship between the communicating parties. Thus, communication messages can be looked upon as composed of different elements and/or as performing different types of actions (e.g. making truth claims, defining relationships, taking on identity, creating credibility, providing justification, making proposals, regulating interaction, etc.). Several branches in the social sciences delve deeper into this complexity, and have developed their own vocabulary and typology for dealing with it; these include linguistics, 'speech act theory' (Searle, 1969; Habermas, 1981) and discursive psychology (Potter & Wetherell, 1987; Te Molder, 1995).

Communication involves selection processes

It is impossible to tell someone everything one could tell, and neither is it possible to recall everything another person tries to convey. Communicating parties make all sorts of *selections* when sending or interpreting messages. The specific selections made are connected with the processes of social construction discussed in sections 5.2.5 and 6.3. Thus, selections are shaped by people's culture, pre-existing knowledge, goals, aspirations and interests in a specific context. On the basis of these, people may 'frame' messages in a particular way, withhold or emphasise certain information, pay selective attention, remember certain things and forget others, interpret messages in a specific way, select specific areas for further discussion, and/or accept specific interpretations and reject others. Thus, communication is very much a social and 'political' activity.

7.2   Three models of communication

In both communication science and disciplines related to informatics, one can find various conceptual models of communication, which in turn are strongly related to different views on the nature of information (see section 6.1). In this section we will briefly introduce three models, each of which has influenced communication for innovation studies and practice considerably.

The 'objective' or 'transmission' model

The 'transmission' model of communication was followed in the early days of extension studies (see Chapter 2). The idea was that communication could best be understood as information transfer. This model is summarised in Figure 7.1, which shows that there is a 'sender' who composes a 'message' and sends it through a 'channel' to a 'receiver'. In this process, the information (captured in the message) was thought to have a fixed ('objective') meaning. Thus, in principle the receiver would be provided with the same information as the sender had intended to transfer, unless something went wrong in the channel (e.g. radio interference, bad printing, noise, distortion, etc.). Subsequently, sender and receiver can change roles, with the original sender receiving 'feedback' from the original receiver through a similar communication procedure.

However, communication scientists and practitioners soon found that there was something fundamentally wrong with this model. Even if nothing went wrong in the channel, the sender and receiver (e.g. the change agent and farmer) would not end up with the same meanings and information (Berlo, 1960). Considerable discrepancies in the modes of thinking of change agents and farmers continued to exist, no matter how hard change agents tried.

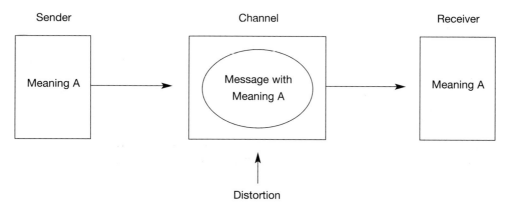

Figure 7.1   The objective or transmission model of communication.

The 'subjective' or 'receiver-oriented' model

In view of the problems mentioned above, the 'transmission' model was refined into a new model, in which important differences in pre-existing knowledge between

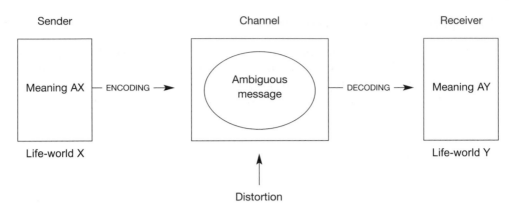

Figure 7.2   The subjective or receiver-oriented model of communication.

sender and receiver are recognised. Based on their personal history and context, the sender and receiver are seen as having a different life-world (see section 6.2) or stock of knowledge (Bosman et al., 1989). When a sender composes (or 'encodes'; Berlo, 1960) a message, he or she tends to draw upon his or her own frame of reference, whereas there is a good chance that the receiver makes use of a totally different stock of knowledge when interpreting ('decoding') the message. Because of this, systematic differences in interpretation of the message (i.e. different meanings) are likely to occur between sender and receiver (Dervin, 1981, 1983) (see Figure 7.2). The same happens again when sender and receiver change roles and give feedback to each other.

In view of this (i.e. that meanings and information are constructed by subjective individuals), it is concluded that effective communication can only take place if a sender makes an effort to *anticipate* the frame of reference of the receiver. Thus, agricultural communication workers would, for example, have to carefully study farmers' perspectives and modes of thinking in order to get their messages attuned and adapted to them. Such empathy would require intensive *interaction* between farmers and communication workers, in which both parties become both senders and receivers in a series of subsequent 'interactive' (see section 7.1) exchanges.

Although this model of communication was a big improvement when compared to the 'transmission' model, some difficulties remained. Despite communication workers' efforts to anticipate their clients' frames of reference and modes of thinking, the latter would in some cases still refuse to accept the meanings proposed by change agents and/or would ignore their advice.

The 'social network' or 'negotiation' model

Deeper analysis of communication processes revealed that the 'subjective' model of communication suffers from two important shortcomings. The first is that it focuses on understanding communication between two *individuals*, but fails to capture the influence of previous and/or more or less simultaneous communications in the wider social network of the sender and receiver. As mentioned in Chapter 5 (section 5.3.3), change agents are not the only ones communicating, which implies that the meanings constructed by a farmer in interaction with 'a change agent may be influenced, directly or indirectly, by others (other farmers, religious leaders, family, local leaders, traders, agents from other organisations, etc.). The second shortcoming is that the 'subjective' model tends to focus on cognitive processes and exchanges only, without taking into account the operation of power, as connected with exchanges of other resources such as money, goods, services, status, etc. (see section 6.6). Thus, the model fails to explain how the construction of meaning is influenced by politics, social relationships, struggle for resources, social interests and aspirations of various kinds (including emotional interests, see section 5.2.1), even if communicative intervention practice shows that such factors play an important role in shaping people's perspectives and meanings. As Giddens puts it:

> 'The communication of meaning in interaction does not take place separately from the operation of relations of power, or outside the context of normative sanctions. All social practices involve these three elements.' (Giddens, 1979:81–82)

Much of the discussion in Chapters 5 and 6 is an attempt to come to grips with these shortcomings, and help us understand how meanings and perceptions come about through 'negotiations' or 'transactions' within a wider social network and context.

It is important to note that, in everyday practice, all three models are still being used, either implictly or explicitly. The 'objective' model of communication, for example, is still being used in mainstream computer science and informatics. It is mainly used for describing communication between computers, where 'information' is indeed regarded as having a fixed quantity (expressed in bits) that is not supposed to change during communication. However, developers of computer-based advisory packages sometimes apply the same model for thinking about the communication between human beings and computers, which causes numerous problems (see Chapters 12 and 13), similar to those in the early days of extension. Moreover, even in regular communication for innovation practice, one can still frequently witness problems of anticipation, even if most development organisations have adopted – at least in words – a client (or receiver)-oriented approach. The 'negotiation' model of communication is relatively recent and must be further developed into appropriate guidelines and tools; it still needs to find its way into many intervention organisations and/or communication divisions.

7.3 Some basic anticipation problems in communicative intervention

On the basis of our discussions on communication and the construction of meaning, we can identify several basic types of problems that may exist in relation to the sending of even one simple intervention message:

- The message does not *reach* the intended audience: This is often connected to inadequate choice of media and methods (see Chapters 12 and 13). Communication workers may, for example, organise meetings or broadcast programmes during hours, or at locations, that are inconvenient for the intended audience.

- The message does not capture *attention*: The message reaches the audience, but the audience does not read or listen seriously and actively. The timing or framing of the message may be wrong: the audience is distracted by other concerns, or gets the feeling, rightly or wrongly, that the message is about something they are not interested in. Also, lack of attention may result from 'message overload' (Cuilenburg, 1983). Finally, it may be that the *relationship* between the communicating parties is such that the audience does not want to listen, for example when the relationship is 'disturbed' by conflicts or lack of trust (see section 5.3.6).

- The message is not *understood* or *reconstructed*: Even if people pay attention, they may not understand the message in the way it was intended, because the audience is not familiar with the language and terminology that is used, and/or does not have the pre-existing knowledge that is assumed. In addition, people may receive additional messages from others in their network that lead them to look at the original message in a specific way, for example when other messages contradict with the original one. In this case, the message does not properly anticipate the life-world and wider network (see sections 5.3.3 and 6.2) of the audience.

- The audience does not *agree* with the message: Even if an audience understands what the sender means to say, they may not agree with the message and may stick to their original views. Here the problem is not so much that people's life-worlds are insufficiently anticipated, but rather that the audience has different aspirations and interests (see section 5.2.1) from those assumed, is subject to social pressures, or doubts the validity, credibility or integrity of the message (see sections 5.2.4 and 5.3.6).

- The audience does not *act* according to the message: People may in principle agree with a message, but still choose to ignore it. This may again be connected to their priorities in terms of aspirations and interests, with judgements regarding self-efficacy or environmental efficacy, and/or with social pressures from the wider social and political environment (see Chapter 5).

- The audience does not *continue to act* according to the message: After some time of following a message or advice, people may discontinue. On the basis of their experiences, they may find that after all it does not suit them and/or the wider environment, or they may find the message less relevant in view of changes in circumstances (see section 5.2.5).

Basically, we see that communicative interventions may – in terms of content and/or mode of communication – fail to anticipate:

- relevant diversity among audiences (see also sections 5.3.4 and 15.2);
- media preferences and media behaviour of audiences, as well as media potential (see Chapter 12);
- the nature and quality of the relationship between the sender and audience;
- audiences' life-world and aspirations of various kinds;
- the messages sent by others in the wider communication network of the audience;
- audiences' perceived environmental efficacy;

- audiences' perceived self-efficacy;
- social pressures experienced by audiences;
- perceived future consequences and developments.

A key task in communicative intervention, then, is to analyse and prevent such anticipation problems; in other words, to arrive at messages that make sense to those they are intended for. In the next chapter we discuss in more detail how communication can be organised in order to contribute effectively to change.

Questions for discussion

(1) We have argued that 'communication is about using symbolic signals'. Discuss whether or not this characteristic is recognised fully in the 'objective' or 'transmission' model of communication.

(2) How would you visualise the 'social network' or 'negotiation' model of communication? You may adapt and/or extend Figure 7.2 in order to capture this model.

(3) Observe and listen to a dialogue between two or more people (e.g. on television, video, radio). Analyse the different levels and layers (e.g. descriptive, performative, content, relational) of meaning that are conveyed through 'messages', and discuss how verbal and non-verbal forms of communication are used to this effect.

PART 3
Innovation as a process of network building, social learning and negotiation

From the previous chapters it has become clear that bringing about change is not easy. It usually requires that different actors and stakeholders change their actions and behavioural patterns, so as to arrive at new interrelations and forms of co-ordination. This in turn requires, among other changes, multiple changes in knowledge and perception. Although we have proposed that communication can be an important strategy for bringing about change, it would be far too naive and simplistic to assume that the mere distribution of communication messages will do the job. What we need is a wider vision on how to organise communication in change processes. In Part 3 of the book we propose that bringing about change requires the organisation of an innovation process (Chapter 8), in which communication is used primarily to facilitate network building, social learning and negotiation (Chapters 8, 9 and 10 respectively). In other words, insights into these processes should provide the basis and organisational principles for using communication in change processes. This way of looking at the relationship between communication and innovation is in line with a more 'interactive' approach to communicative intervention, and can be seen as a response to more 'instrumental' approaches (see Chapter 4). At the same time, insights from negotiation theory in particular suggest that 'instrumental' approaches can still be of use, which leads to a discussion of the balance between the two approaches in Chapter 11.

8 Changing perspectives on innovation

In intervention practice and theory, ideas regarding innovation have changed considerably in association with the shift from instrumental/persuasive models to interactive models of communicative intervention (see Chapter 4). There is a long-standing instrumental tradition in extension studies that looks primarily at the adoption and diffusion of innovations, and the role of communication in these. In this chapter we provide a summary and critical discussion of the 'adoption and diffusion of innovations' perspective (section 8.1) and then move on to discuss a more process-oriented approach (sections 8.2). In Chapters 9 and 10 we focus more specifically on the learning and negotiation dimensions of innovation processes.

8.1   The 'adoption and diffusion of innovations' tradition

Between 1950 and 1970 especially, thousands of studies were conducted across the world which sought to explain why and how people came to adopt, or not, new agricultural technologies and practices (see Havelock, 1973 for an overview). Almost invariably such studies took place in a context where the uptake of particular innovations was deemed too slow. The purpose of the research was frequently to help accelerate the adoption and diffusion of innovations on the basis of the findings. Against this background, typical questions asked by researchers included:

(1)   What are the stages that people go through when considering whether or not to adopt an innovation? Which types and sources of information are important in each stage?
(2)   What are the differences between people who adopt innovations quickly or slowly?
(3)   How do characteristics of innovations, and other factors, affect the rate of adoption?
(4)   How does an innovation diffuse through a society over time? What is the role of communication between different categories of potential users in this process?
(5)   What does this all mean for the role of change agents in stimulating the adoption and diffusion of innovations?

As reflected in this, studies on adoption and diffusion of innovations tended to *start* with a predefined innovation, the uptake of which was regarded as desirable for those being researched. Here, we clearly recognise the instrumental model of communicative intervention discussed in Chapter 4. In section 8.1.1 we discuss some of the main ideas and understandings that derived from this tradition. Subsequently, we elaborate on some important shortcomings in the thinking fostered by the adoption and diffusion of innovations perspective, and explain why communication for innovation scientists have tried to move away from these historical roots (section 8.1.2).

8.1.1   Key conclusions drawn from adoption and diffusion research

Stages in the adoption process

Adoption studies indicated that adoption of innovations is not something that happens overnight, but rather that it is the final step in a sequence of stages. Ideas varied about the precise number, nature and sequence of the stages through which people progressed. However, the most widely used characterisation of stages in connection with the adoption of innovations (as well as the 'acceptance' of policies; see also Van Woerkum et al., 1999) derives from Rogers (1962, 1983). The model built heavily on normative theories about decision-making models (see section 4.1.1) and consisted of the following stages:

(1) Awareness   of the existence of a new innovation or policy measure
(2) Interest   collecting further information about it
(3) Evaluation   reflection on its advantages and disadvantages
(4) Trial   testing innovations/behaviour changes on a small scale
(5) Adoption/acceptance applying innovations/behaviour changes

An important practical conclusion relating to the stimulation of adoption was that people required and searched for different kinds of information during each stage. The information requirements evolved from:

(1) information clarifying the existence of tensions and problems addressed by the innovation or policy measure;
(2) information about the availability of promising solutions;
(3) information about relative advantages and disadvantages of alternative solutions;
(4) feedback information from one's own or other people's practical experiences;
(5) information reinforcing the adoption decision made.

In addition, it appeared that people use different sources of information in connection with different stages of adoption. In countries with a well-developed mass media system (see Chapter 12), farmers usually become aware of innovations through such media. In later stages they tend to prefer interpersonal contact with somebody in whose competence and motivation they have confidence. This person may be a change agent, but for most farmers exchanges of experiences with colleagues are more important. In regions where there are few agricultural mass media, demonstrations often play an important role in the early stages. Dasgupta's overview of 300 studies in India (Dasgupta, 1989) shows that change agents there are mainly influential during the early stages of the adoption process.

It must be noted that in a later edition of his book *Diffusion of Innovations*, Rogers (1995) proposed different stages which are less inspired by normative decision-making theory, and supposedly reflect better what happens in practice:

(1) Knowledge   about the existence of a new innovation or policy measure
(2) Persuasion   shaping attitudes under the influence of others; see section 5.2.4
(3) Decision   adoption or rejection of the innovation or policy measure
(4) Implementation   adapting the innovation and putting it into use
(5) Confirmation   seeking reinforcement from others for decisions made, leading to continuation or discontinuation

Adopter categories and their characteristics

An important finding from adoption research was that innovations are not adopted by everyone at the same time. Particular innovations are used quickly by some and only taken up later by others, while others never adopt them. More importantly, adoption research suggested that there was a *pattern* in the rate at which people adopted innovations, meaning that some would usually adopt early, while others would adopt late. Such conclusions were arrived at through the analysis of adoption indexes which were used as a measure for innovativeness, defined as 'the degree to which an individual is relatively earlier than comparable others in adopting innovations' (Rogers, 1983:22). An adoption index was usually calculated by asking people whether, at a given time, they had adopted any of 10 to 15 innovations recommended by the local extension service. Individuals would receive a point for each one adopted. On the basis of their score, adoption researchers typically classified people into five different categories:

(1) Innovators   2.5%
(2) Early adopters   13.5%
(3) Early majority   34.0%
(4) Late majority   34.0%
(5) Laggards   16.0%

The percentages represent a standardised average of percentages found in different studies, when using one or two standard deviations from the mean as boundaries between categories (under the assumption that distribution of adoption over time is normal in statistical terms). This is illustrated in Figure 8.1.

Many researchers have investigated the relationship between an individual's adoption index and a variety of social characteristics. Such studies have been conducted in highly diverse areas such as agriculture in industrialised and less industrialised countries, education, health services, and consumer behaviour. Remarkably similar

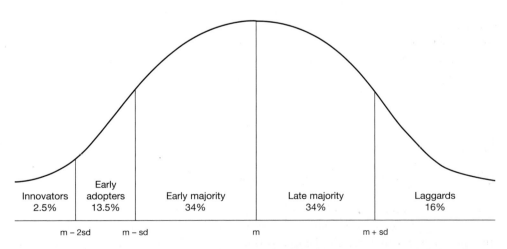

Figure 8.1 Adopter categories and their distribution. Reprinted from Rogers (1983) with permission of the Free Press, a division of Simon & Schuster. © The Free Press.

Table 8.1 Percentage of studies showing a positive relationship between adoption index and selected variables.

Variable	% of studies	Number of studies
Education	74	275
Literacy	63	38
Higher social status	68	402
Larger size units	67	227
Commercial economic orientation	71	28
More favourable attitude to credit	76	25
More favourable attitude to change	75	57
More favourable attitude to education	81	31
Intelligence	100	5
Social participation	73	149
Cosmopolitanism (urban contacts)	76	174
Change agent contact	87	156
Mass media exposure	69	116
Exposure to interpersonal channels	77	60
More active information seeking	86	14
Knowledge of innovations	76	55
Opinion leadership	76	55

Data reproduced from Rogers (1983) with permission of the Free Press, a division of Simon & Schuster. © The Free Press.

results were found in all of these fields. Some of the results are summarised in Table 8.1. Perhaps interestingly, the variable of 'age' is not included in Table 8.1 because, contrary to popular opinion, no relationship was found in about half of the studies, and only one third of the remaining studies suggested that younger people were more innovative than older people.

The types of studies summarised in Table 8.1 were carried out for the purpose of predicting which people could be expected to adopt innovations more quickly or slowly than others. At the same it was expected that such information would help to design more effective communicative intervention strategies. In connection with this, adopter categories were often used, explicitly or implicitly, as *target categories* for communicative intervention, with different media and strategies used for people who supposedly belonged to different adopter categories. When talking to change agents today, one still finds that they tend to distinguish between, on the one hand, 'innovative', 'progressive' or 'vanguard' farmers, and on the other hand, 'laggards', 'conservative' or 'traditional' farmers. This shows how influential adoption and diffusion research has been.

The role of opinion leaders in diffusion

While the term 'adoption' refers to the uptake of innovations by individuals, 'diffusion' relates to the spreading of innovations in a community. Not surprisingly, communication for innovation scientists have always had a great interest in processes of 'diffusion' and 'horizontal exchange' (see also sections 2.2 and 13.3.2). Studies on diffusion suggest that specific people within a community play an important role in stimulating, or preventing, the spreading of innovations. Such people are usually referred to as 'opinion leaders'; that is, people who are influential in shaping opinions of various kinds. It must be noted that different people may be 'opinion leaders' on distinct matters and for different groups of people. Some people in a community may, for example, act as agricultural 'information brokers' with the outside world; that is, they selectively bring in and interpret information from elsewhere, and communicate this, again selectively, to other members of a community. There are others who may serve as 'experiential experts', may be influential in shaping people's norms and values, or may play a role in legitimising or disapproving particular changes. Opinion leadership may or may not coincide with formal leadership roles. Furthermore, diffusion research found that opinion leaders tend:

- to adopt many innovations, but usually are not the first to adopt them;
- to be well educated and enjoy sound financial positions in their communities;
- to lead an active social life and have many contacts outside their immediate surroundings;
- to have a special interest in a particular subject.

On the basis of diffusion research it was suggested that communication workers would be most effective in stimulating diffusion when they targeted their activities at these opinion leaders. The idea was that focusing on relatively 'innovative' or 'progressive' farmers in a particular community, would mean other farmers would benefit almost automatically.

Variables determining and predicting the rate of adoption

An important concern in the adoption and diffusion research tradition was to explain and predict the eventual rate of adoption of particular innovations. Based on an extensive body of literature, Rogers (1983.233) has summarised conclusions in connection with this in the model in Figure 8.2.

We will not elaborate on Figure 8.2 in detail but present it merely as an indication of the overall mode of thinking in this research tradition. Several issues relating to this model are discussed in a different context in other parts of this book. The variables 'relative advantage' and 'compatibility' are closely intertwined with our discussion on 'evaluative frame of reference' in section 5.2.1. The other variables included in category I are discussed as factors that influence learning in section 9.4. With 'type of innovation-decision' Rogers (1983, 1995) refers to the number of people involved in the adoption process, with 'optional' innovation decisions taken by individuals independently of others. 'Collective' innovation decisions require consensus among many people in a system or organisation, allegedly causing adoption to be slower. According to Rogers this is again less so in the case of 'authority' innovation decisions, where only a few relatively powerful individuals can decide on adoption or rejection by a collective.

The issue of 'communication channels' relates to the influence of the media on diffusing innovations, which is something we address in Chapter 12. The category of variables mentioned as 'nature of the social system' is connected largely with social influences and societal characteristics that shape diffusion (e.g. social norms, patterns of interconnectedness in social networks). This is something that relates to

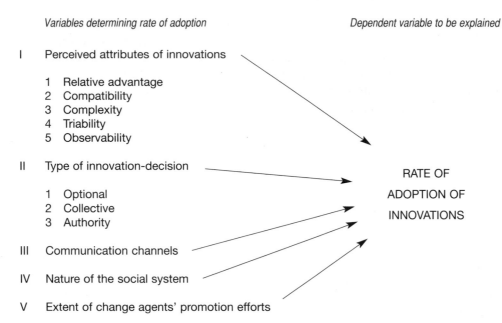

Variables determining rate of adoption   Dependent variable to be explained

I   Perceived attributes of innovations

     1   Relative advantage
     2   Compatibility
     3   Complexity
     4   Triability
     5   Observability

II   Type of innovation-decision   RATE OF

     1   Optional   ADOPTION OF
     2   Collective
     3   Authority   INNOVATIONS

III Communication channels

IV   Nature of the social system

V   Extent of change agents' promotion efforts

Figure 8.2   Model representing variables that influence the rate of adoption. Reprinted from Rogers (1983) with permission of the Free Press, a division of Simon & Schuster. © The Free Press.

our discussions of social construction processes, power, structural properties and efficacy (see sections 5.2.5, 6.3 and 6.6). Finally, in Chapter 16 on organisational issues we discuss issues relating to the efforts and motivation of change agents.

8.1.2   Critical reflections and practical limitations

Over the years the 'adoption and diffusion of innovations perspective' has been criticised theoretically and for the intervention practices it has inspired. In this section we list a number of highly interconnected shortcomings.

Pro-innovation bias

Concern in adoption and diffusion research with the rate of adoption and ways to increase this, implies that the innovations studied are considered worthwhile, and that it would make sense for most farmers to adopt them. This assumption is called the 'pro-innovation bias' (Röling, 1988). In practice, however, many innovations are proposed which do not make sense for many farmers. Conventional adoption and diffusion research tended not to correct for *relevance* when calculating adoption indexes. Thus, people would be classified as 'laggards' on account of their non-adoption of innovations that were not suitable for them in the first place. The pro-innovation bias is further demonstrated by the fact that many discussions on non-adoption were framed in terms of who or what is to 'blame', showing that non-adoption was regarded as negative. The use of the term 'laggard' expressed and/ or reinforced the idea among researchers and change agents that individuals were somehow to blame for non-adoption, due to their assumed 'resistance to change' or 'conservatism'. Although we do not deny that phenomena like conservatism exist and may play a role, many studies indicate that other explanations (e.g. inadequate innovations, structural limitations, conflicting interests, etc.) are at least equally valid. In the adoption and diffusion tradition this latter position was also expressed in terms of 'blame', as the issue was discussed in terms of the 'individual blame' versus the 'system blame' hypothesis (see also Rogers, 1995).

A linear and 'top–down' model of innovation

Innovations studied in this research tradition were usually those proposed by agricultural researchers. It was basically assumed that innovations originate from scientists, are transferred by communication workers and other intermediaries, and are applied by agricultural practitioners. This mode of thinking is called 'the linear model of innovation' (Kline & Rosenberg, 1986) as it draws a straight and one-directional line between science and practice. The model is further characterised by a clear task division between various actors; some actors are supposed to specialise in the *generation* of innovations, others concentrate on their *transfer*, while the farmers' role is merely to *apply* innovations (see Figure 8.3).

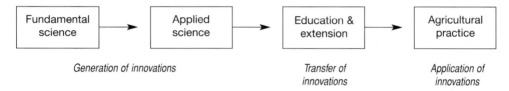

Figure 8.3 The linear model of innovation.

However, when scholars started to analyse in retrospect how successful innovations came about in practice, they soon discovered all sorts of deviations from the linear model. It appeared, for example, that researchers often got 'their' innovative ideas from practitioners, that farmers made significant adaptations to the packages developed by scientists, that many innovations occurred without involvement of scientists, and that the function of communication workers was not so much to transfer knowledge and information from scientists to farmers, but rather the other way round, or even to play a role only in knowledge exchange among farmers (Van den Ban, 1963; Van der Ploeg, 1987; Leeuwis, 1993; Vijverberg, 1997). In view of such findings it was concluded that innovation requires close co-operation in a network of actors, who all contribute to the 'generation' and 'transfer' of knowledge, and to its 'integration' into viable innovations (Engel, 1995). In adoption and diffusion research the active and creative role of farmers in innovation processes has long been overlooked (Röling, 1988, 1994b), as emphasised by the fact that very few, if any, studies have focused on the adoption of *farmers' ideas* by researchers and change agents. Apparently, studying adoption 'the other way round' was not considered worthwhile, which is indicative of the limited value attached to farmers' knowledge and ideas by 'adoption and diffusion of innovations' researchers.

A uni-linear model of farm development

Both the pro-innovation bias and the terminology used in the adopter category classification ('early adopters', 'late majority', 'laggards', etc.) reflect the idea that there is basically one direction in agricultural development which all farmers who want to continue farming should follow sooner or later. We have already discussed (in section 5.3.4) that recent research indicates that there are several viable patterns of developing a farm, even under homogeneous conditions (Leeuwis, 1989, 1993; Van der Ploeg, 1990; see also section 15.2). Each pattern of farm development is characterised by different patterns of innovation.

Blindness, biased perceptions of innovativeness and stigmatisation

It has been shown that, as an expression of uni-linear ideas on farm development, change agents and/or their organisations preferred and favoured *particular* types and *patterns* of innovations (Van der Ploeg, 1987, 1999; Leeuwis, 1989; Roep et al., 1991). This reflects a certain blindness (see also section 6.4) for alternative directions, which is also reflected in adoption indexes. These indexes have been calculated typically on the basis of a list of innovations suggested by extension organisations. However, one can expect these lists to have been rather biased in view of the developmental preferences referred to above, leading to misleading perceptions about 'innovativeness'. A study in Ireland, for example, indicated clearly that large groups of farmers did not adopt the innovations suggested by the extension organisation, and hence were regarded by extensionists as 'laggards', 'stagnant' and 'non-innovative'. However, it appeared that these farmers preferred a *different* pattern of farm development, and had adopted innovations *other* than those promoted by the extensionist. When taking these 'unofficial' innovations into account, they were in fact equally as innovative as the clients of the extension organisation (Leeuwis, 1989). Moreover, had an adoption index been constructed on the basis of these 'unofficial' innovations only, all those regarded by the extension organisation as 'early adopters' would have come out as 'laggards' instead. This example demonstrates the likelihood of adoption indexes leading to inadequate views of the innovativeness of certain groups of farmers, who then become disqualified and stigmatised unjustifiably as 'laggard' and 'stagnant'[1]. The phenomenon that adoption and diffusion research often emphasised incorrectly 'individual blame' explanations of non-adoption (see the discussion on pro-innovation bias) represents another stigmatising tendency in this research tradition.

---

[1]   Another point for discussion is whether negative qualifications and/or connotations are justified at all in connection with people who are *really* 'non-innovative'. Apart from the 'pro-innovation bias' concerning specific innovations referred to earlier, we can speak of a more general bias in communication for innovation studies, including in this book, that innovation as a phenomenon is 'good', or at least 'necessary' and 'inevitable'. However, others may disagree legitimately with our culturally ingrained positive evaluation of change.

Arbitrary and inadequate categorisations for targeting

As we have seen, the classification into adopter categories has often been used to direct and *target* communicative intervention activities. However, it is important to realise that – as a result of the way in which people are assigned to different categories (see section 8.1.1) – members belonging to the same category may have very little in common apart from the fact that they have adopted a similar number of innovations out of a selective list. An adopter category may well include people with a totally different gender, age, farm size, farming style, land tenure position, ethnicity, pest management problem, etc. Thus, in many ways the categories and their statistical cut-off points are rather arbitrary, and it is unlikely that the people belonging to the same category have comparable aspirations or face similar problems, opportunities and constraints (see also Röling, 1988).

Communicative intervention practice has learned that different farmers may need different solutions and a different approach on a similar issue. Thus, for purposes of effectiveness and efficiency, it is important to look for *relevant* differences between people in relation to a broader issue (e.g. pest management), by which to identify relatively homogeneous target categories that require a different kind of service. We will further discuss the issue of target segmentations in Chapter 15. Suffice it to say here that classifications into adopter categories are often of limited practical use for communicative intervention. It is worrying to observe that many change agents still speak and think mainly about 'innovators', 'followers' and 'laggards', because it may hinder the identification of much more relevant and interesting differences between farmers; for example, classifications that centre on livelihood strategies (Hebinck & Ruben, 1998), farming styles (Van der Ploeg, 1990; see section 15.2) or learning styles (Kolb, 1984; see section 9.1).

'Progressive' farmer bias as a self-reinforcing process

The insight that so-called opinion leaders play an important role in diffusion, has been used in practice to justify change agents paying disproportionate attention to such persons. As mentioned earlier, the people regarded as opinion leaders were usually relatively wealthy farmers who had already adopted a relatively large number of the innovations favoured by intervening organisations, and hence were regarded as 'progressive' farmers. Leaving aside for a moment whether diffusion actually occurred, communication workers have frequently been criticised for paying most attention to those who needed it least (Havelock, 1973; Röling, 1988). More precisely, we would argue that communication workers have tended to focus on those farmers and opinion leaders who fitted best with their preferred model of farm development. The fact that opinion leaders show relatively high adoption rates indicates this. In many ways this bias has been a self-reinforcing process (see Röling, 1988). In laymen's terms we could say that change agents, like most human beings, associated themselves most with those who were on the same wavelength, and with whom they got along best. In doing so, they also tended to become more familiar with the problems and issues faced by such farmers, leading to further efforts to cater for their needs, rather than for the needs of others.

The selective and non-automatic nature of diffusion

In view of our conclusion that communicative intervention has often been biased in terms of the innovations proposed and the opinion leaders selected, it is not surprising that diffusion tended to be selective as well. Perhaps more importantly, it has been realised that diffusion is less of an 'automatic' process than assumed earlier. In other words, the fact that people who adopted innovations can be shown to have been influenced by others (e.g. opinion leaders) in retrospect, does not mean that 'opinion leaders' can be expected to actively support diffusion. In fact, they may well have reasons to shield or selectively withhold information from others, especially in an age of competition and commercialisation of knowledge (Oerlemans et al., 1997; Leeuwis, 2000b). Moreover, various forms of social differentiation in communities (e.g. along lines of gender, tribe, political affiliation, social status, religion, etc.) tend to exist. Opinion leaders tend to be influential only in a specific segment of a community, so diffusion across lines of differentiation is far from self-evident.

The selectiveness and non-neutrality of technology

Several of the shortcomings discussed so far centre around the experience that the innovations looked at in adoption and diffusion of innovations research proved to be *selective* for a particular audience (IFAD, 2001); that is, they were only applicable for, and often also communicated to, a specific segment of the farming community, while little service was provided for other, often more vulnerable and/or less influential groups. The fact that technologies are selective is something that cannot be avoided but which can be a positive characteristic. In order to be of use, technologies *need* to be adapted to specific agro-ecological and social environments. The more adapted they are, the more selective they become. Implicit in any technology, therefore, is a social 'code' (Mollinga & Mooij, 1989) that makes it potentially applicable to a limited group. Thus knowledge and technologies are never politically neutral. The problem with much adoption and diffusion of innovation research, then, is that it failed to *recognise* and *explicate* these political dimensions of technology, or even present technology as neutral and potentially beneficial to all.

Innovation as a collective rather than an individual phenomenon

In adoption and diffusion research, the adoption of innovations has been portrayed as relating to an individual. Although it is recognised to some extent that people are influenced by others in taking innovation decisions, and although some special innovations are described as 'collective', there tends to be a greater emphasis on the individual farmer. Adoption rates and categories, for example, are calculated on the basis of individual responses. In Chapters 1 and 2 we have emphasised strongly that virtually *all* innovations require, and are constituted by, changing patterns of co-ordination between interdependent actors, implying that what Rogers calls 'optional' innovation-decisions (see the explanation of Figure 8.2) are an exception. It follows from this that 'adoption' too involves simultaneous and co-ordinated changes by a variety of actors (e.g. male farmers, women farmers, traders, input suppliers, transport companies, etc.), with different actors 'adopting' different interconnected sub-innovations. Conventional adoption and diffusion research, however, does not pay much attention to co-ordination between interdependent actors[2]. It essentially defines adoption of innovation as an individual affair. In our view, this is a severe conceptual inadequacy, which also limits the practical relevance of this body of research.

A one-dimensional view of innovations

In adoption and diffusion research, the innovation is often treated as a single entity. However, from a collective point of view, innovations consist of a variety of new and interdependent practices that may be implemented by a variety of people (see above). We have already mentioned in Chapter 5 that even for a single actor, innovations have a composite nature which includes a variety of technical and social practices at different domains and levels of farming, and at different times (see section 5.1). Furthermore, we have discussed in section 5.3.6 that acceptance and adoption of innovations can be associated with several 'layers', such as 'adoption' of underlying problem definition, 'acceptance' of intervention, and 'acceptance' of manifold consequences and risks. In all, innovations are rather multi-dimensional, and complex interactions can be expected between adoption processes regarding various aspects of 'the innovation'. This complexity is hardly captured in mainstream adoption and diffusion research; but looking into such matters could be important for the understanding of widespread phenomena such as partial adoption, adaptation and 're-invention' of innovations.

A focus on rational decision-making

As we have discussed, the original conception of the adoption process and its stages were heavily influenced by normative models about rational decision-making. As suggested in section 5.3.7, rational decision-making is often a practical impossibility, and experience shows that the way in which people go about altering practices often bears little resemblance to decision-making models. On similar grounds, Rogers (1995) attempts to move away from such models. However, decision-making still figures as an important step in Rogers' current separation between stages. In our view, this remaining concern with decision-making is logically connected with the idea that the adoption of innovations is largely an individual affair. When starting from the assumption that innovation is a collective process, other key processes come to mind when thinking about adoption. These include social learning, conflict management and negotiation (see Chapters 1 and 2).

---

[2] In connection with this, one may wonder whether the assumption of a normally distributed diffusion curve (see Figure 8.1) is always correct. Although several studies seem to support it (e.g. Ryan & Gross, 1943; Griliches, 1957), one would expect more 'stepped' adoption patterns, resulting from different networks of actors adopting more or less at the same time. One cannot exclude the possibility, however, that normal curves are the result of *multiple* network-wise adoptions, or that the 'steps' were at times 'ironed out' by drawing a nice continuous line.

A mechanical view of society and social change

The concern in adoption and diffusion research with pre-defined science-based and/or policy-induced innovations firmly places it in the instrumental tradition (see Chapter 4), which assumed that innovation is something that can be planned and controlled in advance (e.g. Havelock, 1973). Thus, it tends to suffer from similar problems to those mentioned in connection with this tradition (see section 4.2).

Conclusion

The above concerns have made many communication for innovation scientists uneasy about the perspective of adoption and diffusion of innovations, both conceptually and practically. Of course, there is considerable scope for adapting and improving this body of theory and the intervention practices derived from it. One can, for example, pay more attention to:

- the study of interdependencies and the collective dimensions of adoption;
- 'unwrapping' innovations into constituent components, and studying processes of partial adoption and/or 're-packaging';
- studying the adoption of farmer innovations and ideas by researchers, change agents and farmers;
- the construction of less biased or more differentiated adoption indexes;
- the development of more relevant classifications of diversity among farmers;
- the identification and selection of more effective 'opinion leaders' for different segments in communities;
- etc.

However, we see few scholars in our field moving along such paths, essentially because the overall questions asked have changed considerably. In the new questions that are being asked, the idea of 'adoption' itself is no longer a major concern.

8.2   Innovations and processes of innovation design

In line with more interactive and process management-oriented ideas about communicative intervention (see section 4.2), the questions regarding innovation and innovations have changed considerably. Today's questions are geared towards obtaining a better understanding of innovation processes, and include:

(1)   How and why do processes of innovation design unfold? What are the social processes that play a role?
(2)   What characteristics of innovation processes contribute to the emergence of coherent innovations?
(3)   How do conflicts of interest between different stakeholders influence interactive innovation processes?
(4)   What relationships can exist between processes of, on the one hand, social and individual learning, and on the other, negotiation and conflict management?
(5)   What are the roles that facilitators may play in interactive innovation processes?

Inspired by such questions, several insights have come to the fore. In the remainder of section 8.2 we discuss some of these, mainly those deriving from the field of innovation studies itself. The discussion is continued in Chapters 9 and 10 where we discuss learning and negotiation as key components of innovation processes.

8.2.1   The multi-dimensional character of innovations

Innovations are often looked at in a rather isolated and mainly technical way; a new type of plough, for example, would be looked at by many as a distinct 'innovation'. However, in this book we look at innovations in a wider sense. We have already argued in section 5.3.5 that changes 'never come alone', and often include both technical and social–organisational elements. A new plough, for example, is not just a new way of turning the soil, as it is likely to be effective only in conjunction with other agronomic changes (e.g. new maize varieties) as well as new forms of social organisation within the family (e.g. a new way of dividing labour between men and women), the community (e.g. new arrangements regarding exchange of labour and implements), and/or the wider institutional environment (e.g. new arrangements for the provision of inputs, credit, spare parts, etc.). We have already discussed a telling example of the interconnections between the social and technical dimensions of innovation, in section 1.1.2. In any case, we can only speak of a complete innovation if an appropriate mix and balance exists between technical devices and social–organisational arrangements. Thus, we look at an innovation as a *package* of new social and technical arrangements and practices that implies new forms of co-ordination within a *network* of interrelated actors (as well as non-human 'actants'; see section 20.4.3). Phrased differently, we can say that an innovation needs to be understood as a 'novel working whole' (see Chapter 1 and Roep, 2000).

8.2.2   Building effective linkages and networks in an evolutionary process

Of course, innovations defined as above do not emerge instantly. The origin of an innovation can be an original idea, a technical novelty, a policy initiative or a new social arrangement, but in order to become a real innovation such an 'origin' needs to be reworked into a coherent package of technical arrangements and social relationships[3]. Thus, innovation processes need to include deliberate efforts to create effective linkages between technological arrangements, people and social–organisational arrangements. This process of building coherent linkages and networks around a novel idea or technical device is called 'alignment' (Rip, 1995), i.e. bringing different aspects and dimensions of an innovation 'in line' with each other. The importance of alignment activities can hardly be overestimated. The fact that many agricultural, and other, technologies 'fail' (i.e. do not become accepted on a significant scale) can

---

[3]   In a somewhat similar vein, Verkaik et al. (1997) suggests that, in innovation processes, 'knowledge and ideas' need to be translated into 'skills and technologies', and subsequently into real socio-technical 'innovations'.

often be attributed to insufficient, partial or unbalanced alignment. Agricultural scientists, for example, often work mainly on the technical dimensions of the technology 'design', and pay little attention to building effective networks (see also section 17.3) and arrangements with prospective users and/or supporting institutions, which causes their technically advanced products to fall into a social 'vacuum'. Also, in regular industry there are examples of how technically superior products fail because of a lack of network support, while inferior products become widely accepted as a result of the energy that has been spent on network building, etc.

It is relevant to note that the role of networks and networking (see also section 17.3) changes in the course of innovation processes. At an early stage, it is important to mobilise creativity and get new ideas on the table, which in many cases can benefit from establishing linkages with 'outsiders' with whom little contact existed previously, as they may have a 'fresh' outlook on the situation as well as previously unavailable expertise (see also section 11.2.3). Thus, in the earlier stages innovation processes may benefit from widening the network, and from drawing upon what Granovetter (1973) has called 'weak ties' ('weak' in terms of, for example, low frequency of contact, distance, durability; see also section 20.4.1). Later in the process, however, it becomes important to consolidate specific linkages and build an effective support network, i.e. to create 'strong ties'.

In connection with the above, and also in line with the experience that innovations could not be 'planned' (see section 4.2), innovation processes are increasingly looked upon as being *evolutionary* in nature. In essence, the idea is that a variety of innovations and innovation processes compete with each other in a dynamic selection environment, and the 'fittest' (i.e. the best fitting) survives eventually (Bijker et al., 1987; Rotmans et al., 2001). Similar to the idea of alignment, it is proposed that the chances of survival are better when technical and social–organisational arrangements evolve together in a process of 'co-evolution' (Smits, 2000; Geels, 2002). An important implication of this is that, in order to solve problems, it is important to create sufficient variety of solutions that are worked on simultaneously, accepting a certain amount of 'redundancy' (Aarts & Van Woerkum, 2002).

8.2.3   The need for temporary protection in innovation processes

Effective alignment, and the co-evolution of a coherent package of technical and social arrangements, takes time. It requires learning from mistakes, experimentation, careful adaptation, etc. With complex innovations, it is impossible to do everything right immediately. For innovations to become mature, therefore, it is important that a safe space for experimentation and development exists in innovation processes. Such a 'protected space' (Kemp et al., 2001) can take various forms, e.g. special research funds that can be used for making non-profitable investments, compensation funds for farmers who participate in risky on-farm experiments, extra facilities for training and support during pilot projects, etc. However, at a certain point an innovation will have to prove itself without artificial forms of support. It is when support and protection are withdrawn that it becomes clear whether an innovation has been effectively aligned with its environment (it 'survives') or not (it 'vanishes').

8.2.4   Different types of innovations and innovation decisions

Innovations differ in magnitude and scope. A farmer may, for example, optimise his or her farming system by making slight adjustments in the application of chemical fertilisers. We call these '*regular*' innovations; that is, innovations that do not challenge fundamentally the main technological and social–organisational characteristics of the farming system (or, as Kemp et al. 2001 put it, they remain within the rule-set provided by the dominant 'technological regime'; see also section 6.6). Such regular innovations occur almost as an integral part of everyday farming practice. In contrast, '*architectural*' innovations (Abernathy & Clark, 1985) are those which require and incorporate a fundamental reorganisation of social relationships, technical principles and rules[4]. They can be seen as 'overthrowing' the existing regime, and breaking out of the path dependence created by it (see section 5.2.5). Had the above farmer changed over to biological farming, and abandoned chemical fertilisers altogether, we could call it an architectural innovation, as it would have fundamentally altered the logic of farm operations. The distinction between regular and architectural innovations overlaps partly with the separation made in Chapter 5 between practices and decisions at different levels and with different time-scales. Strategic decisions (see section 5.1) are more likely to involve architectural innovations than operational and tactical issues; the same is true of changes that take place at the higher hierarchical levels of the farm.

8.2.5   The problem-driven character of innovation

The energy, motivation and resources to work on an innovation often stem from a wish to solve particular problems. Hayami and Ruttan's (1985) theory of induced innovation suggests, for example, that the direction in which technology development takes place depends on which production factors (labour, land, etc.) are deemed most scarce or 'problematic'. More generally, a 'problem' can be defined as a perceived tension between an existing state of affairs, for example, environmental degradation, or labour scarcity, and a desired state of affairs (less degradation, sufficient labour). As 'problems' are thus inherently connected with human perceptions and aspirations, actors involved in an innovation process are likely to have different goals and problem definitions to begin with. Thus, in connection with innovation processes, it is more accurate to speak of a 'problematic field' than of a 'problem'. The problem fields to which innovations are expected to answer can orginate from various developments:

- *Changed perceptions about reality*: e.g. a given farming system may become increasingly looked upon as causing environmental degradation;
- *Changed human aspirations*: e.g. farmers may increasingly strive to meet consumer concerns, which may render current practices problematical;

---

[4]   Other terms used to refer to these kinds of fundamental and radical innovations are 'system-innovations' or 'transitions' (see for example Geels, 2002).

- *Changed social environment*: e.g. the labour needed to sustain a particular farming system may become increasingly scarce;
- *Changed natural/physical circumstances*: e.g. a farming system may start to yield less optimal results due to ecological change;
- *Changed social opportunities*: e.g. new international trade agreements may create new market opportunities;
- *Changed technical opportunities*: e.g. the availability of computer technology may trigger people to rethink current agricultural practices and technologies, and search for applications of such opportunities in agriculture.

In general, it tends to be easier for people to recognise 'problems' than 'opportunities' (even if the same development can imply both), not least since problems become more easily visible through feedback and experience than opportunities do (see section 9.3). A risk with 'technical opportunities' especially is that they can easily end up in a 'technology push', in which particular *solutions* are already available, for which one then tries to identify or 'invent' a suitable problem. In general, the other types of situation (where one searches for socio-technical solutions in relation to an already experienced problem) tend to more easily generate energy and motivation, basically because they are experienced as less artificial and imposed.

8.2.6   The 'hidden' nature of building blocks for innovation

The conclusion that innovation requires the integration of ideas, knowledge, experience and creativity from a variety of actors (farmers, researchers, service providers, communication workers, etc.) implies that these need to be somehow brought together, mobilised and connected with each other. In Chapters 9 and 10 we will further discuss some aspects of this process, and in Chapter 11 we will focus specifically on the types and areas of knowledge and expertise needed in innovation processes. For now it is important to realise that the relevant 'building blocks' of innovation cannot be gathered together instantly. The sources of innovation are often hidden and invisible (Wiskerke & Van der Ploeg, 2003), largely inside the brains of particular persons who may not even be aware that their experiences and ideas could form relevant building blocks for innovation. In other words, potential building blocks for innovation are often not easily accessible as they are part of people's tacit knowledge (see section 6.2), and relevant people may not yet be part of the network from which the innovation process evolves (see section 8.2.2).

8.2.7   Basic tasks in interactive innovation design processes

In the more linear and top–down tradition of extension studies, it was proposed that processes of adoption (of innovations) or acceptance (of policies) progressed through the stages of awareness, interest, evaluation, trial and adoption/acceptance (see section 8.1.1). However, in the context of interactive processes it is no longer useful to refer to a 'fixed' innovation or policy, as the prime idea is that these are yet to be

---

[5]   Also recall that we have considerably refined the concept of 'acceptance/adoption' in section 5.3.6.

'designed'[5]. Given that innovations consist of a package of new social and technical arrangements, it is clear that 'design' is a multi-faceted process, taking place at different points in time and space, and involving different sets of actors. At the same time, design should not be looked at as a straightforward and controllable process (see section 4.2). Rather, we should think of design as a process of network building, social learning and negotiation that takes place in an evolutionary context (see section 8.2.2). During the process, new social and technical arrangements and practices evolve gradually and through many iterations, with interrelated elements slowly taking on a more concrete and definite form. That is, 'designs' may originally have a rather abstract or vague shape (e.g. an idea expressed in the form of a vision, story, image or drawing) and may crystallise over time into concrete and hopefully coherent technical devices, maps, practices, agreements, contracts, rules, etc. Notwithstanding the changing perspective on innovation, some of the insights and terminology of the 'adoption and diffusion' perspective are still useful in relation to interactive processes. We intend to deal with the following basic tasks in innovation processes:

(1) raising awareness of a problematic situation (see sections 8.2.5 and 9.5);
(2) mobilising interest in a network of stakeholders (see sections 8.1.2 and 9.5);
(3) socio-technical design and redesign under protective conditions (see section 8.2.3), which involves:
    (a) experiential (social) learning and exploration among stakeholders (see Chapter 9 and section 13.5);
    (b) negotiation among stakeholders (see Chapter 10);
(4) gradual lifting of protective conditions (see section 8.2.3);
(5) further socio-technical evolution and redesign or failure (see section 8.2.2).

The term 'task' expresses better than 'stages' that we are talking about fields of activity that are worked on simultaneously and/or that many iterations are likely to occur. In the following chapters we will elaborate on social learning and negotiation as key processes within an interactive approach. In doing so, we will also refine the basic tasks mentioned here. In Chapter 14 we discuss more concrete facilitation matters in relation to interactive/participatory processes. At that point we shall see that it is not only innovations that require 'design' and 'redesign', but also the processes aimed at creating them.

Questions for discussion

(1) Despite all the critique, some conclusions derived from the 'adoption and diffusion of innovations' research (see section 8.1) may still be relevant. Which insights do you still consider relevant in your work?

(2) Both the 'model for understanding human practices' (see Figure 5.1) and the 'model representing variables that influence the rate of adoption' (see Figure 8.2) say something about why people do or do not do things. What are the key differences and similarities between the two models?

(3) The research division of a seed company has developed a new pest-resistant rice variety with the help of genetic engineering techniques. They state that the newly developed variety can be regarded as 'a successful innovation'. When compared to the definition of an 'innovation' used in this book, what is problematic about the statement of the research division?

(4) Efforts to design and establish architectural innovations may at times meet resistance from the existing 'technological regime'. What examples of this can you identify from your own experience?

9 Social and individual learning

As shown by the model for understanding farmers' practices (Chapter 5), cognitive change – in other words, learning[1] – is an integral part of everyday life. We all act, and receive feedback from our environment, which in turn leads us to adapt our cognitions. It is this kind of learning – as distinct from separate educational activities and teaching – that is crucial in the context of adult education (see also Jarvis, 1987; Blum, 1996; Merriam & Caffarella, 1999) and communicative intervention. It may be helpful to recall here that, in change and innovation processes, learning may occur and/or be required on various 'fronts' (see Table 9.1 and Chapter 5).

Moreover, for purposes of arriving at coherent innovations, it is clear that 'individual' learning does not suffice, but that simultaneous learning of interdependent stakeholders is needed; that is, in order to arrive at coherent practices, multiple actors need to develop complementary and/or overlapping (or even fully shared) understandings about the above 'learning fronts' as a basis for effective co-ordinated action. For this, several authors have coined the term 'social learning' (Dunn, 1971; Friedmann, 1984; Röling, 2002; Woodhill, 2002). Röling (2002) describes 'social learning' as a key mechanism for arriving at more desirable futures, and as a 'third way of getting things done' that stands in sharp contrast to the instrumental modes of thinking (see Chapter 4) underlying conventional technological intervention and neo-classical economics (Leeuwis & Pyburn, 2002; Röling, 2002). More specifically, Röling (2002:35) defines social learning as 'a move from multiple to collective or distributed cognition'. In the case of 'collective cognition', coherence is forged primarily through shared perceptions of the 'learning fronts' in Table 9.1, resulting in and from truly 'collective' action[2]. The idea of 'distributed cognition' recognises that stakeholders may well work together and engage in complementary (i.e. coherent) practices while significant differences in perception remain. Here ideas, values and aspirations may be overlapping or mutually supportive, but are not necessarily 'shared'. In our view, 'collective cognition' and 'collective action' are more likely to emerge within stakeholder categories (e.g. among small farmers belonging to an ethnic group), while 'distributed cognition' and 'co-ordinated action' are the best achievable in a multi-stakeholder setting. In many innovation contexts, both tend to be required.

---

[1] A more detailed definition of learning can be derived from Bierly et al. (2000:597). They speak of learning as 'the process of linking, expanding, and improving data, information, knowledge and wisdom' (see section 6.1).

[2] As mentioned in section 6.4, Röling distinguishes between learning at the level of 'coherence' and 'correspondence'. Learning at the level of coherence involves learning about whether or not different 'elements of cognition' (including, for example, perceptions, theories, values, goals and action) are internally consistent with each other (Röling, 2002). Learning at the level of correspondence involves learning about whether or not the 'elements of cognition' are connected in a meaningful way with the wider context (see section 6.4 for an example).

Table 9.1   Broad 'learning fronts' and specific types of cognition that are associated with these, constituting at the same time areas and entry points for cognitive change.

Broad areas of cognitive change or 'learning fronts'	Specific perceptions and types of cognition involved
The functioning of agro-ecological systems	• existing agro-ecological phenomena, categories and variables • associations and causal attributions pertaining to agro-ecological phenomena • beliefs about agro-ecological consequences of human practices • agro-ecological uncertainties, likelihoods and risks
The functioning of social systems	• existing social phenomena, categories and variables • associations and causal attributions pertaining to social phenomena • beliefs about social consequences of human practices • social uncertainties, likelihoods and risks
Human aspirations and tensions in their satisfaction	• existing and salient identities • technical, economic, relational and emotional goals and interests • values, norms and responsibility considerations • problem definitions
The effectiveness of the social environment	• trust in support networks • trust in (inter)community organisation • social-organisational uncertainties, likelihoods and risks
Self-efficacy and confidence	• beliefs about the ability to mobilise resources • beliefs regarding available skills and competencies • beliefs regarding the ability to accommodate risks • trust in the validity of current knowledge and perceptions
The significance and nature of social relations and social pressure	• perspectives of others social actors • desires and expectations from others • resources mobilised by others to influence and persuade • relevance of rewards and sanctions to one's own aspirations • mutual interdependence for realising certain aspirations • quality of relationships with other stakeholders (e.g. in terms of trust, reliability, integrity, sincerity)

As implicit in the above, the adjective 'social' in 'social learning' has several connotations; it refers to:

● the *topics* that need to be learned about, such as other stakeholders' perspectives and interests, the social world, social arrangements (as reflected in various 'learning fronts'; see Table 9.1);

● the *methods* through which social learning may be stimulated; frequently the term 'social learning' carries the methodological connotation of 'learning in a group or platform' (for elaboration see Part 4 of this book);

● the *socio-political nature* of the learning process; here the term 'social' refers to the point that knowledge and perceptions tend to be socially constructed (see section 6.3 and Leeuwis, 2002a), which implies that learning cannot be looked upon as a neutral process.

Each of the above connotations is discussed elsewhere in this book. In this chapter, we present some additional insights into human learning, and indicate what implications they have for practitioners who strive to enhance this process. In doing so, we speak about learning in general, and do not differentiate much between different types of cognition (see Table 9.1) and/or whether it takes place in an individual or group setting.

9.1   A basic model for adult experiential learning

When we mention learning in the context of innovation we are not talking about compulsory classroom situations where teachers try to foster and test learning on a fixed curriculum. In rural settings, we are usually dealing with adults who are involved in farming and/or other livelihood activities, and who are confronted with changing circumstances and problems that require innovation. Here (social) learning is less of a goal in itself, is often more voluntary, and is immediately connected with diverse human interests and changes in professional practice. Because of the immediate relations with practice, Kolb's (1984) model of 'experiential' learning is widely used as a basis for organising communication for innovation (see Figure 9.1). The model describes how people learn through experience. This type of learning is very 'powerful'; it appears that conclusions drawn by people *themselves* on the basis of their *own* experiences tend to have a greater impact than insights formulated by *others* on the basis of experiences that learners cannot identify with. It is also referred to as 'learning by doing' or 'discovery learning'.

The model indicates that learning occurs from a continuous interaction and iteration between *thinking* and *action*: concrete actions result in certain experiences, which are reflected upon (also against the background of relevant non-experiential insights), and subsequently generate cognitive changes, from which new actions can emerge, etc. This model implies that learning can be enhanced by actively supporting the basic steps and translations that take place during learning, and by offering new

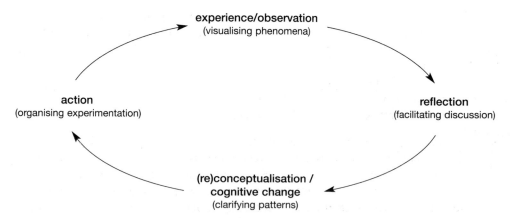

Figure 9.1   The learning cycle with examples in brackets of ways in which the different stages can be supported by communication workers (adapted from Kolb, 1984).

learning opportunities. One can, for example, actively encourage experimentation, widen the range of observations made, stimulate processes of reflection, and assist in drawing conclusions (see also section 13.5.8).

Learning styles

Kolb uses his model not only to describe how experiential learning takes place, but also to indicate that different people learn in different ways. We will not go into detail on the four universal learning styles that Kolb distinguishes (basically because we find them somewhat problematic; see Leeuwis, 1993:287–8). However, Kolb's idea that different people tend to learn in different ways is valuable, as it teaches us that different people may need different forms of support in reaching similar conclusions. Relevant dimensions for diversity in this respect include:

- *Abstract versus concrete*: Some people learn easily with the help of *abstract concepts*, while others learn more effectively through *concrete sensorial experiences*. In relation to the management of minerals in the soil, for example, some Dutch farmers prefer to work with complex calculations and parameters in order to compare and determine nutrient losses and soil conditions over the years. Other farmers find it difficult to relate to such abstract figures, and prefer other indicators. To get an idea about changes in the nutrient status, they may, for example, look at the colour of the grass, study the presence of particular weeds, observe the physical condition of the land, taste the crop, etc. According to their preference, these farmers may value different types of support from communication workers.

- *Diverging versus converging*: Some people tend to jump to conclusions quickly on the basis of certain experiences (i.e. they *converge* easily), whereas others tend to *diverge* into all sorts alternative explanations that require further testing and elaboration, and find it difficult to arrive at solid inferences. Such people run different risks (changing too quickly versus not changing at all), and may want or require different forms of support.

- *Holistic versus reductionistic*: Perhaps partly overlapping with the previous dimension, some learners are more inclined to learn about 'parts' whereas others have a greater affinity with looking at 'wholes' (see also sections 6.5 and 13.5.8). Taking mineral management as an example, some farmers are inclined to focus on the mineral aspects of the farm only, while others immediately forge linkages with other levels and domains (see section 5.1) of farming practice (e.g. Stolzenbach & Leeuwis, 1996; Oerlemans et al., 2002).

- *Individual versus group*: Some people have a clear preference for learning with others, while others are less inclined to involve others in their learning process. Some farmers, for example, like to discuss problems and experiences in a study group or group meeting, while others tend to avoid group sessions and prefer to figure things out by themselves, or through bilateral contacts only. In some instances this seems to be associated with having a *co-operative* versus a *competitive* outlook on the issue at hand (e.g. Leeuwis, 1993).

- *External versus internal motivation*: Depending in part on the issues at hand (as well as on the wider context; see section 9.4), people may experience an 'internal' drive to learn about something, or feel more or less 'forced' by others to engage in it (Stolzenbach & Leeuwis, 1996; Ketelaars & Leeuwis, 2002). In other words, they may have or develop a real interest in a topic and be enthusiastic to learn more about it, or they may learn mainly because they fear negative consequences if they do not (see also section 9.4). Again, communication workers are wise to develop specific strategies and support for the different groups.

9.2   Levels of learning, and the relationship with decision-making

Earlier we made a distiction between regular and architectural innovations, and indicated that this distinction is associated with changes at different hierarchical levels and with different time horizons (see sections 5.1. and 8.2). Parallel to this, we can say that regular innovations (involving, for example, operational decisions) require a different degree of learning from architectural innovations (involving, for example, strategic decisions). In the case of the former, we will – following Argyris and Schön (1996) – talk of 'single loop' learning. This typically involves learning 'how to do things better' within the basic cognitive assumptions and principles (e.g. norms, values, goals) that underlie current practices. When such basic assumptions and principles themselves become the subject of learning (usually in the case of architectural innovation), Argyris and Schön call it 'double loop' learning. This type of learning is much more demanding (and sometimes threatening), because it involves questioning and perhaps letting go of the basic certainties, goals and values that one acted upon previously. If, for example, a farmer who is used to applying fertilisers, pesticides and herbicides in single crops changes over to biological farming, he or she will have to learn how to deal with farmyard manure, intercropping, complex rotations, biological pest control and a new network of people and institutions. Argyris and Schön mention a third level of learning, 'triple loop' learning, which essentially involves learning about learning; in other words, questioning the current methods, techniques and forms of feedback through which learning is organised. A farmer may, for example, realise that he or she always discovers animal health problems too late, and may search for new routines for observation and registration that improve feedback and learning on health conditions.

The relationship between learning and decision-making

In the early years of agricultural extension and extension science, 'decision-making' was *the* central concern. In the earlier editions of this book, extension was even defined as 'the conscious use of communication of information to help people form sound opinions and make good decisions' (see Chapter 2). Not surprisingly, then, a normative procedural model on 'how to make sound decisions' (very similar to the planning model described in section 4.1) formed the backbone of communicative intervention activities. In view of the shortcomings of this model as described earlier, the emphasis has shifted from planning and decision-making to learning and negotiation. It appeared unrealistic, in practical terms, to expect people to adhere strictly to rational decision-making procedures, as this would be very time-consuming given the multi-dimensional character of farming and innovation, and the multiplicity of goals and aspirations involved (see section 5.3.7; Simon, 1976).

Thus, for the purposes of this book it seems more useful and realistic to regard decision-making as the final outcome of longer lasting learning processes with varying degrees of deliberateness and consciousness, involving also what Giddens (1984) has called 'reflexive monitoring of action' (see section 5.2.5). In this view, decisions 'grow' over time, partly unnoticed. When considering the building of a new greenhouse, for example, a horticulturist may take into account all sorts of experiences and stories that he or she has encountered in the last 5 or 10 years when visiting or meeting with other greenhouse horticulturists, even if such encounters themselves were not part of a deliberate, rational, decision-making process. In fact, through time and experience the grower may already have slowly developed a number of specific ideas regarding the optimal lay-out of the greenhouse, the materials and glass to be used, the type of heating system to be installed, etc. When the moment of taking final decisions arrives, the most important ideas regarding the design of the greenhouse may have crystallised long ago, while the only decision remaining may be who will build the greenhouse against what price.

In our view, the best way of enhancing and supporting decision-making, then, is to stimulate and encourage continuous experiential learning. On the basis of such regular learning, people can identify which issues and problems need to be tackled, and can gradually collect the necessary insights and experiences to inform and shape conclusions that, in retrospect, may be called 'decisions'. However, all of this does not imply that normative decision-making models are useless. They may help structure one's thoughts and reach sensible actions, particularly in areas or situations (e.g. unexpected crises) where limited experiential learning has taken place, while at the same time urgent action is needed.

Central and peripheral routes of learning

In relation to attitude change (i.e. changes in one's frame of reference, see section 5.2.1) Petty and Cacioppo (1986) speak of a 'central' and a 'peripheral' route towards attitude change. In terms of learning, we could say that these two routes are characterised by a different 'depth' of learning[3]. Petty and Cacioppo speak of the 'central route', when people change their perceptions and aspirations on the basis of careful elaboration of arguments and counter-arguments (i.e. in much the same way as suggested by rational decision-making models). However, they emphasise that people may also form or change attitudes in a much more peripheral way, in which arguments are not carefully scrutinised. Here, people rely on triggers and cues that accompany the use of arguments, such as the number of arguments for and against

---

[3]   Note that Petty and Cacioppo (1986) do not link these routes to different modes of learning in interactive innovation processes. On the contrary, they discuss the issue in the context of explaining the effectiveness of persuasive efforts. Thus, the distinction between central and peripheral routes of *learning* represents a different framing of their theory which remains our responsibility.

that are presented, the attractiveness of presentation, the perceived credibility of the source, the opinions of significant others, the number of people who seem to agree, etc. (Petty & Cacioppo, 1986; O'Keefe, 1990). Emotional appeals and affective associations too can be considered as peripheral triggers (Batra & Ray, 1986; Petty & Cacioppo, 1986; Louw & Midden, 1991). It can be 'rational' and effective to change perceptions, aspirations and practices on such grounds, as it may help to deal with complex issues and situations, without spending too much energy on elaboration (Van Meegeren, 1997). However, studies by Petty and Cacioppo (1986) and others (Verplanken, 1989) suggest that changes arrived at through the central route tend to be more robust than changes induced through peripheral triggers; that is, in the former case perceptions and aspirations are less prone to alter in the face of newly presented counter-arguments, are more likely to be accompanied by changes in practice, and tend to last longer. For change agents these insights are important, because they imply that striving for durable changes requires the stimulation of more active (i.e. central) forms of learning. This is in line with the earlier emphasised importance of 'discovery learning' (see section 9.1).

9.3   The centrality of relevant feedback

As we have seen, *feedback* plays an important role in shaping human practices (see section 5.2.5). This is basically because it is a crucial mechanism in human learning. Feedback is information we get about the outcomes, characteristics and/or consequences of our actions, and it helps us to evaluate these. Such information can come from different sources, can be varied in nature (depending on the area of learning involved), and can vary in quality of precision, reliability, validity, etc. In particular, when feedback is somehow 'disturbing' it can trigger learning processes (see section 5.2.5). Almost anything can be feedback in relation to something. A few examples may help to clarify this:

(1)   In schools teachers often give grades to students, which are supposed to be indicative of students' knowledge and abilities in certain areas; i.e. grades are meant to give feedback on how students are performing.
(2)   A farmer may make a rough visual estimate or a very precise measurement of maize yields per hectare, and compare that information with similar estimates/ measurements from previous years and/or other farmers in order to assess whether certain changes in farming practice were positive or negative.
(3)   A communication worker may organise a meeting and observe that several people fell asleep during that meeting, and infer that something was wrong with the topic of the meeting and/or the way it was administered.
(4)   A person may be told by a colleague that other people in the organisation find the way he operates offensive and non-tactical, and conclude that it might be a good idea to work on his social skills.
(5)   A farmer may accidentally get to talk to an urban dweller, and conclude from the discussion that consumers have totally different perceptions regarding food quality and safety from those of farmers, and that there is a need to bridge the gap.

Of course, the feedback we get is not always optimal or complete, and also the conclusions we draw from it can be misguided. In relation to the above examples, it may well be that:

(Example 1)   Students' grades are affected by the bad quality of the teachers' exams.
(Example 2)   Yield estimates are too rough to be meaningful and/or yield measurements are affected by unaccounted harvesting for own consumption.
(Example 3)   The attendees at the meeting fell asleep because they attended a party the previous night.
(Example 4)   The colleague was not genuine and/or was expressing his own rather than other people's views.
(Example 5)   The urban dweller was totally unrepresentative of consumers in general.

It frequently happens that people simply do *not* get feedback on certain matters. On a personal level it can be threatening for others to give feedback, for example because they do not know how a person will respond and do not want to put good relations at risk. For similar reasons people in powerful positions may receive inadequate feedback on various matters, simply because subordinates do not want to bring 'bad news'. Thus, leaders may lose touch with things happening around them, which may have serious consequences as it may result in totally inadequate policy decisions.

Stimulating and contributing to learning, then, is almost synonymous with organising and providing good quality feedback. In particular, the provision of new forms of feedback can be very stimulating as it can make things tangible, manageable and debatable that were not tangible, manageable and debatable before. For example, when rice farmers in Indonesia started to count 'good bugs' and 'bad bugs' in order to get precise feedback on the balance between pest infestation and the presence of natural enemies, this dramatically improved their capacity and enthusiasm for using biological pest control and reducing the use of pesticides (Van de Fliert, 1993).

The type of feedback provided by communication workers may vary considerably according to the communication for innovation strategies involved (see section 2.2, Table 2.1). In the context of efforts to facilitate conflict resolution or interactive policy design, for example, the exploration of the various stakeholder perspectives may in itself already provide important feedback to the participants. Similarly, in the context of farm management communication, powerful strategies and methods that communication workers can use to organise feedback for farmer learning include:

- *measuring* things that have not been measured before (e.g. the example of counting bugs);
- making arrangements through which farmers can *compare* their own farm operations and results with those of *other farmers*;
- making arrangements through which farmers can *compare* their operations and results in one season with those of *other seasons or years*;
- supporting the design and evaluation of *on-farm experiments*;
- *visualising* agro-ecological processes that are difficult to observe.

These and other methods and strategies will be discussed in more detail in Chapter 13. At this point, it is perhaps significant to note that giving feedback may or may not include forms of systematic *measurement*. When considering the *quality* of measurement-based forms of feedback, questions may be raised about its:

- *reliability*, e.g: How consistent are the measurement procedures/tools used in farm comparisons? Are the various measurements really comparable?
- *internal validity*, e.g: Do the measurement procedures/tools really measure what we claim to measure?
- *external validity*, e.g: Is the feedback provided relevant in view of the problems and issues that people are confronted with?
- *sincerity*, e.g: Are those giving feedback sincere in their intentions or trying to manipulate?

In principle, similar issues can be raised with regard to forms of feedback that do *not* originate from systematic measurement, for example, the feedback people give and receive in the course of everyday practice, or the claims made and/or the perspectives portrayed during negotiation processes.

Positive (supportive) and negative (confrontational) feedback

In the literature, a distinction is often made between positive and negative feedback. Positive (or supportive) feedback is information that indicates one is on the right track: a compliment by someone, a series of measurements that shows progression in solving a problem, or even a financial reward indicating, for example, that the government appreciates someone's achievements. In contrast, negative feedback is information that indicates the existence of a problem and/or that one is not doing well. Others call this type of feedback *confrontational* feedback (Heymann, 1999). Social psychological research indicates that positive rewards often have more impact on people than negative sanctions; this is because people tend to close themselves off from 'bad news' or try to reason it away (Martijn, 1995; Van Meegeren, 1997; see also section 9.4). On the other hand, confrontational feedback has the potential to cause mental tension, which is an important pre-condition for learning and problematising current practices and routines (see section 5.2.5). This implies that it can be very effective, especially when people are already open to learning, or when the feedback is so strong that it can hardly be denied or ignored.

In pursuing change, communication workers would be wise to look for a careful balance between positive and confrontational forms of feedback (see also section 12.2.3). Some ground rules for giving feedback on issues relating to group dynamics are discussed in section 14.2.3.

9.4   Factors that may affect learning (pre-conditions and obstacles)

We live in a dynamic society and ecosystem, which implies that there are many situations in which human learning is required. Despite the need for learning, however, groups and/or individuals are often not inclined to learn, or only start learning when problems have become immense. This can even happen in situations where confrontational feedback is readily available. In this section we point to a number of factors and processes that help us understand better why learning takes place or not, which adds to the idea that learning requires a certain amount of feedback. In general terms, the question of why people learn or not can be understood with the help of our model for understanding farmers' practices (Chapter 5), as 'learning' can be considered as 'a practice' as well. In other words, here too factors like frame of reference, social pressure and self-efficacy are important. We will highlight and translate some issues that are specifically important in relation to learning.

First, it is important to recognise that learning takes effort, energy and time. This means that learning can be considered 'a scarce resource'. In other words, people are selective in their investments in learning. A factor often mentioned is *'motivation to learn'*. This variable expresses that in order to take on a particular learning challenge, people must be motivated to do so. What interrelated factors and processes may influence people's motivation to learn?

The relative importance/seriousness of an experienced problem

Learning requires first of all that people experience a problem (see section 8.2.5), which means that in their frame of reference (see section 5.2.1) there must be a tension between their aspirations and their perception of reality. Depending on the *priority* of the aspirations involved, and the perceived *magnitude* of the tension between the desired state-of-affairs and the current state-of-affairs, people may define a problem as relatively important and serious, or not. In principle, people can be expected to select the more serious problems for learning, provided that they have some confidence in the possibilities of solving the problem (see below). However, serious problems may also be ignored when they are somehow experienced as highly threatening (Hruschka, 1994).

Direct involvement with a problem

A slightly different issue is whether or not people are personally affected by the consequences of a problem. People may regard a problem, e.g. poverty, as serious and important, but may not experience the consequences personally. When people are personally involved with an issue (e.g. a pest management problem) in the sense that their immediate aspirations are threatened, they may be more inclined to learn (e.g. about pest control). In such situations we refer to high 'personal relevance' (Verplanken, 1989) or 'outcome-relevant involvement' (Johnson & Eagly, 1989). However, it is important to recognise that even then, people's eagerness to learn may be *restricted* to certain topics, and may in fact actively *exclude* other topics. If, for example, farmers are affected by pest infestations from a nearby nature reserve, they may want to learn about pest control (or about strategies to get rid of the reserve) but it does not mean that they are necessarily interested in learning about the importance of nature conservation. Thus, high involvement may simultaneously enhance *and* obstruct learning.

In our view the same is true for forms of involvement other than direct 'personal relevance'. Johnson and Eagly (1989) speak, for example, about 'value relevant involvement' when wider values are involved instead of concrete interests. A person may, on the basis of religious values and convictions, be strongly against or in favour of the use of contraceptives, without any direct personal involvement with the issue. In connection with this, he or she may be eager to learn about certain things (e.g. how to prevent extra-marital sex) and not about others (e.g. how to encourage people to use condoms).

Urgency

When people feel there is an urgent need to solve a problem they are often more motivated to engage in learning than when learning can easily be postponed. Urgency can be more or less inherent in a problem, but it can also be created artificially. The occurrence of a particular pest, for example, may well require immediate action and learning if one wishes to prevent substantial yield losses. In less urgent cases (e.g. solving erosion problems) urgency may be deliberately created by governments or NGOs who wish to subsidise erosion prevention activities and infrastructures, for example by issuing a clear deadline for submitting community plans to combat erosion.

Self-efficacy and environmental efficacy

To invest in learning, people must have some confidence that they can solve the problem; that is, they must trust their own capacities with regard to problem-solving and/or have the idea that they will be supported effectively by others in finding and implementing solutions (see Chapter 5). Whenever such confidence is lacking (e.g. because of negative experiences in the past), learning is less likely to occur. In the context of multi-actor situations, an important aspect of efficacy is that of *mutually experienced interdependence*. If an actor does not feel interdependent on others for solving a problematic situation, or if the impression exists that others do not reciprocate his or her feelings of interdependence, willingness to engage in social learning may be reduced (see also section 10.3.2).

Complexity, observability and triability

In connection with efficacy, the complexity of problems may also indirectly affect people's motivation to learn. If people feel that problems are highly complex technically or socially, their perceived self-efficacy and environmental efficacy may be reduced. Complexity, then, is clearly related to the *level of learning* required (see section 9.2). In addition, some problem areas can be more easily learned about than others. Here two aspects play a role: observability and triability (Rogers, 1983)[4].

---

[4]   The terms 'complexity', 'observability' and 'triability' stem from Rogers (1983) who uses them in a slightly different way. In Rogers' view they are characteristics of innovations (along with 'relative advantage' and 'compatibility') that help to explain the adoption of innovations (see section 8.1; Figure 8.2). We prefer to look at them as characteristics of learning areas that help to understand why learning occurs easily, or not.

In some areas of learning, the processes involved can be easily *observed* with the help of the human senses. In the technical sphere it is probably easier for many to learn about how the soil responds physically to different mechanical treatments, than about how fungi respond to different chemical treatments. Similarly, in the social sphere it is easier to learn about how to organise a stimulating group meeting on a particular topic, than it is to learn about how to facilitate negotiation processes among stakeholders. A group meeting is an overseeable event where one can receive direct feedback on the way people feel about it, whereas negotiation processes involve numerous interactions over time, many of which take place behind the scenes, and on which it may be less easy to get clear feedback due to complex interdependencies. Tangibility and observability can sometimes be enhanced considerably by the provision of creative forms of *feedback* (see section 9.3). With the help of certain measurement tools or computer animations, for example, it may be possible to get a better understanding of the way fungi behave and respond to chemicals.

Triabilility can also facilitate or hinder learning. Triability refers to the extent to which learning can be supported through small-scale experiments. Small trials allow people to optimise new practices and technologies before applying them on a bigger scale, and thus reduce the risks of large-scale failures. However, some devices or treatments are difficult to incorporate in a small-scale learning trial: it is not so easy to experiment on a small scale with a modern combine harvester, a completely different rotation system, a new irrigation scheme, or new land tenure arrangements. This 'inflexibility' may slow down learning processes in particular areas. In some cases triability can be enhanced by simulations, including computer simulations. For example, it can be possible for farmers to 'experiment' with irrigated agriculture by providing small water tanks before an irrigation system is ready. Or a computer simulation model can calculate the benefits, costs and risks of using a totally different rotation system (see, for example, section 13.5.10). In essence, issues like observability and triability relate to whether or not relevant *feedback* for learning (see section 9.3) can be easily organised.

Clarity about the nature of a problem

In some cases it is unclear and/or uncertain whether or not a problem exists, and how serious and/or urgent it is. There have been debates, for example, on whether or not global warming exists, and if so, whether it is caused primarily by human energy use or by variations in solar activity (Calder, 1997), and whether it should be seen as a blessing or a threat. Thus, an important component of clarity is whether or not different stakeholders are in *agreement* about the nature and seriousness of the problem. If people are confronted with *contradictory information and arguments* in relation to a problem they may become confused and discouraged from dealing with it.

Perceived social consequences and risks associated with accepting alternative cognitions

At several points in this book we have argued that knowledge is not neutral, but closely intertwined with social interests. The novel cognitions that people encounter in a learning process, therefore, may be experienced as either threatening or rewarding. They are threatening when people feel that accepting the alternative views may jeopardise their macro or micro interests in a specific context. For example, in a learning process farmers may be reluctant to accept the idea (including the supporting evidence) that their farming practices cause serious environmental damage, as the very acceptance of that idea may have significant practical, political and/or economic consequences. Here we see that new cognitions may conflict (i.e. they are dissonant) with previously existing cognitions (e.g. farmers cause far less pollution than chemical industries) and aspirations (e.g. to maximise profits with the help of high levels of fertiliser). Social psychological research indicates that people are less inclined to accept radically different ideas, and find it easier to incorporate those that are less conflictive with existing perspectives (Sherif et al., 1965)[5]. The studies by Sherif et al. point out that the range of alternative cognitions that people are ready to accept tends to correlate negatively with their personal involvement, which suggests that those who are most involved (i.e. those whose direct interests are on the line) tend to hold closest to their original views (i.e. be least ready to learn). This is partly in line with our earlier statement that personal involvement may not only obstruct but also enhance people's readiness to learn.

Similarly, Festinger's cognitive dissonance theory (Festinger, 1957) describes what people do when they are confronted with cognitions that conflict with each other or with existing cognitions: they try to reduce 'dissonance' by rejecting and denying unfavourable cognitions and/or by downplaying the importance of the aspirations affected. A heavy smoker may, for example, deny that smoking significantly increases the chances of getting lung cancer, because accepting that relationship would seriously jeopardise his peace of mind. Similarly, he may convince himself that 'health is not an important issue' and/or invent and search for all sorts of arguments that do not conflict (i.e. are consonant) with his heavy smoking behaviour, such as the ideas that 'smoking reduces stress' and that 'smoking prevents one from gaining weight'. In contrast, someone who has just taken the difficult step to stop smoking, may look for additional arguments that support and reinforce that decision (e.g. that 'smoking is very expensive' and that 'women find smokers unattractive') (see Zimbardo & Leippe, 1991; Martijn & Koelen, 1999).

All of this indicates that learning may be constrained by a wish to prevent the possibly negative consequences of accepting new ideas, or a wish to avoid uncertainties that arise from them. The opposite may also occur. When people feel that novel cognitions can be rewarding, learning processes may speed up considerably. If, for example, a new director in a development organisation strongly believes in the future of the internet as a communication medium (see section 12.3), communication workers

---

[5] In connection with this, Sherif et al. (1965) talk of a person's 'latitude of acceptance' (the range of alternative ideas that are still accepted because they remain relatively close to one's original views and attitudes), 'latitude of non-commitment' (the range of ideas to which a person holds a neutral position) and 'latitude of rejection' (the range of ideas that a person rejects because they conflict most with one's original views).

may be stimulated to learn about it, especially if they expect that this may strengthen their position in the organisation.

Social and organisational space

The last example already hints at the importance of the social and organisational environment in which learning takes place. Individual learning takes place in the context of a social environment (e.g. an interest group, sub-culture, community or organisation) in which new ideas may or may not be appreciated. The points discussed in the previous section about the avoidance of possible negative consequences by individuals, applies equally to social groups. If people are part of a group in which the leaders and/or the majority of people see certain new ideas and views as threatening to the interests of the group, individual actors who are open to these ideas and see positive dimensions may be discouraged from expressing and further developing these views. If the situation is the other way round, learning may be accelerated.

We can distinguish between situations in which organisations, cultures and communities act in an 'open' or 'closed' way. People often speak about 'open' or 'closed' groups as fixed categories, but that does not do justice to the fact that organisations, cultures or communities can be 'open' to one thing and 'closed' to others. Nevertheless, many organisations and people (and especially those who have something to lose) tend to more easily recognise the immediate threats and risks connected with new ideas and developments, rather than the opportunities and possible gains in the future. In other words, people and organisations often have a tendency to emphasise the threats, become 'closed', discourage learning, and act in a 'defensive' mode towards their environment. The phenomenon that organisations are 'inward looking' and continue to interpret the world around them through 'fixed' perspectives and schemes of interpretation (see section 6.1) is called 'autopoiesis' (self-referentiality) (see also sections 16.2.5 and 16.4). It is argued by many that, in an age of rapid change, autopoiesis can be risky and unproductive. If organisations and groups do not learn and adapt, their survival is at risk. Thus, much attention is paid to strategies that enhance organisational learning, so that organisations become 'learning organisations' (Senge, 1993).

We deal in more detail with learning within and between organisations and networks of actors in Chapter 16.

Resources and safe space for experimentation

Closely related to organisational space and triability is the need for access to resources and a safe space for experimentation (or 'protected space', see section 8.2.3) in innovation processes. Experiential learning usually requires energy, time, forms of equipment and infrastructure. Thus, even if an eagerness to learn exists, learning may be constrained by a lack of resources. Farmers' learning in the area of mineral management, for example, may be slowed by a lack of reliable measurement devices. Similarly, learning by members of an organisation may be frustrated when no time is allocated for it in their job description. Resources may also be important for resolving constraints on triability. Some issues can only be experimented with (i.e. become triable) when sufficient practical resources exist, and/or when the risks involved can be neutralised by means of, for example, financial compensations.

Stress and trauma

We have seen that learning can be stimulated by dissatisfaction with existing situations and outside pressures. In such cases problems and tensions work productively. However, in certain situations people may face so many problems and tensions that they cannot deal with them. People become overwhelmed by the different pressures, can no longer distinguish between priorities, and 'break down' and/or become apathetic to their environment. Similar states of mental trauma can result from various forms of abuse and violence. It usually takes a long time for people to recover.

9.5   Aspects of learning

When considering both the insights on learning presented so far and the tasks in interactive innovation processes outlined in section 8.2.7, we can identify the following aspects of learning that participants in a social learning process must go through:

- becoming aware;
- becoming interested/mobilised;
- becoming involved in active experiential (social) learning (in the context of negotiation);
- establishing adapted practices and routines.

The third and fourth dimensions may involve learning on numerous interconnected topics and issues (see Table 9.1 and Chapter 5). We refer to 'aspects of learning' rather than 'stages' since the order in which 'awareness', 'interest' and 'active experiential learning' occur may vary. Farmers may, for example, become 'aware' and 'interested' in a particular seed variety only *after* they have had to use it due to unavailability of their preferred seed. It is important for change agents to realise that the learning of different stakeholders and groups of people may, at a given time, have encompassed different aspects of learning, and that for each aspect different types of information can be relevant (see also section 8.1.1).

To become aware of a problematic situation (aspect 1 above) people require adequate information and feedback on the nature, importance, magnitude and seriousness of a problematic field (see section 9.4). For becoming interested and mobilised (aspect 2) other issues may be more relevant, for example, information regarding personal consequences, opportunities and threats, urgency and the possibility of effectively contributing to problem-solving (efficacy issues). When people become actively involved in experiential learning and negotiation (aspect 3), different matters become important, such as information on organisational and technical solutions, and the perspectives and positions of other stakeholders. For the establishment of new socio-technical routines and practices (aspect 4), stakeholders may require feedback on the effectiveness of their practices, as well as information on whether or not other stakeholders follow the agreements and arrangements made.

Questions for discussion

(1) Discuss the key differences between the learning that teachers usually try to foster in classroom situations and the kind of learning discussed in this chapter.

(2) Consider the issue of global warming. Which factors may inhibit and/or encourage learning on this issue by ordinary citizens? What strategies could one employ to overcome obstacles for learning?

(3) It has been argued in this chapter that high involvement in a certain issue or problem may both obstruct *and* stimulate learning. Discuss why this is so.

(4) We have argued that new cognitions can become accepted through a central or a peripheral route. Discuss examples from your own experience in which the peripheral route was dominant.

10   Negotiation within interactive processes

We have emphasised at various points that negotiation is important within innovation processes. In this chapter we explain why, and provide some insights from negotiation theory that are relevant to facilitators of interactive innovation processes. Practical guidelines on negotiation are discussed later in Chapter 14, when we discuss process management.

10.1   Why look at negotiation?

In Chapters 1 and 2 we have already emphasised both that the innovations needed in today's agriculture typically require co-ordinated action, and, more generally, that innovation needs to be understood theoretically as a collective process (see also section 8.2). Also, we have indicated that wherever different actors and stakeholders are involved in processes of meaningful change (e.g. architectural innovation), conflicts are likely to emerge, since such changes may have consequences that affect the values and interests of many stakeholders. In the conventional literature on participation and interactive processes, it is assumed that such conflicts can be overcome by social learning and decision-making (see Chapter 14). Typically, it is suggested that *all* relevant stakeholders should be involved in the process, and that conflicts of interests between these stakeholders can be resolved through the development of a shared understanding of a situation, as a result of learning and improved communication (e.g. Pretty & Chambers, 1994; Röling, 1994a; Thrupp et al., 1994; Engel, 1995; Maarleveld & Dangbégnon, 1999). The idea is that participatory processes almost 'automatically' become *consensus* and co-operation-oriented (provided that sufficient attention is given to social learning), and that consensus is a pre-condition for development and innovation.

A range of studies, however, has shown that innovation processes organised along such lines often produce disappointing results for those who initiate or participate in them (Eyben & Ladbury, 1995; Leeuwis, 1995; Mosse, 1995; Dangbégnon, 1998; Khamis, 1998; Wagemans & Boerma, 1998; Zuñiga Valerin, 1998; Amankwah, 2000; Mutimukuru, 2000). On closer investigation, one finds that such difficulties result from an inability to either resolve or use productively conflicts of interests that tend to emerge during the innovation process (regardless of their specific purpose, i.e. resource management, community development, technology design).

10.1.1   Conflict-related frictions in innovation processes: six cases

Below, we analyse some examples of frictions in innovation processes, and argue that they are related to the limited capacity of conventional participatory methods to deal practically with conflict situations.

*Case 1: Conflict, unproductive consensus and fruitful competition
(Leeuwis, 1995)*

A group of glasshouse horticulturists in the Netherlands manages to arrange a national subsidy in order to build a computer programme for electronic data exchange and enterprise comparisons. A condition for getting the subsidy is that the programme must cater for the needs of all fruit–vegetable growers. Thus, a large range of growers, representing different crops and regions, becomes involved in the design process. However, the information needs of different growers vary considerably, the only common denominator being that they are all to some degree interested in exchanging glasshouse climatic data. However, the growers are much more interested in exchanging a range of production data, which unfortunately differ considerably from crop to crop. As a compromise, it is agreed to focus on exchanging climatic data. However, due to unhappiness with the compromise, a struggle over money, the slowness of the process and lack of progress, a number of conflicts emerge and several growers drop out. Among them is a small group of cucumber growers who start a competing initiative. In a short time this group develops a programme that caters very well for cucumber growers' needs, and which is later adapted to cater for the needs of other growers as well. For a few years, the two computer packages compete on the market. Eventually the two packages are integrated within a third initiative, which builds on the lessons learned in both projects.

What we see in this case is essentially that the diversity of interests among the stakeholders – the result of an inclusive approach in the selection of participants – hampers the opportunity to reach a productive consensus in the context of the available experience, budget and technological opportunities. In fact, it is only when a range of growers is excluded, causing two rival groups of growers to compete, that learning is accelerated and an appropriate technology is developed, which eventually caters for the needs of all.

*Case 2: Reinforcing conflict through public decision-making
(Zuñiga Valerin, 1998)*

Zuñiga Valerin describes how in the San Bosco community in Costa Rica, ranking techniques (see section 13.5.6) are used during group meetings to identify the priority of general and agricultural community problems. She also describes how institutional interests and mandates, as well as conflicts within the community, have a bearing on which problems are dealt with, and how. Typically, decisions on activities are voted for by hand-raising during subsequent group meetings. Within the community this leads to considerable tensions, as the San Bosco community consists of different sub-communities, divided by a river and, to some extent, by family lines. Due to differences in infrastructure, the two sub-communities have different needs and priorities, which are defended by the respective community leaders. Both leaders engage in lobbying to influence the outcome of the voting during group meetings, which puts various community members in a difficult position. The responsible field worker knows about the political struggles but admits he does not know how to deal with it, other than by encouraging everyone to express their view, and eventually following the majority vote. As a result, one sub-community tends to feel discriminated against in the participatory process.

In this case we see an emphasis on 'democratic' decision-making by means of ranking during public group meetings. Even while the different actors tend to be aware of each other's views and positions, the open group meetings provide limited opportunities for the kind of 'give and take' that is necessary for settling the disputes. This is not surprising. The presence of an audience is not exactly a conducive environment for leaders to give in or make deals; rather it is an opportunity to show strength and reinforce one's position as a leader.

Case 3: Lack of manoeuvring space, institutional mandate and capacity to manage conflict (sources: interviews with Ministry of Agriculture staff, Manicaland, Zimbabwe; Mutimukuru, 2000; Mutimukuru & Leeuwis, 2003)

In the context of a wider land-use planning effort, the inhabitants of the Chitora ward in Manicaland, Zimbabwe, agree during a village meeting that they are interested in establishing a communal grazing scheme. However, an important reason to agree is that planning regulations define such a scheme as compulsory in any land-use plan; without it the plan (including a much desired road and a bridge) cannot be realised. Subsequently, the Ministry of Agriculture develops a plan (in a rather top–down mode), and manages to find a donor willing to provide the necessary funding for the grazing scheme. All parties agree that the donor will provide the material for fences, while the community will deliver the necessary labour for constructing the fences. However, the plan implies that some farmers, including a traditional leader, will have to move their stone-built homes to other areas. The donor refuses to pay compensation to those who have to leave. Also, members of neighbouring communities object to the scheme, as it will limit their access to grazing lands and water. Over time, competence conflicts emerge between and within traditional authority structures and the newly introduced Grazing Scheme Management Committee. These and several other conflicts are never resolved during subsequent village meetings initiated occasionally by Ministry of Agriculture staff. At first, some community members start to build the fences in the context of a 'food for work' drought relief programme, but as soon as this initially unanticipated programme ends, nobody turns up. Several attempts to revive and adapt the project during additional village meetings prove short-lived (Mutimukuru, 2000), and eventually the grazing scheme is given up by the donor and the Ministry of Agriculture and never completed; this despite the fact that over time at least part of the community develops a genuine interest in having a grazing scheme.

This case demonstrates the risks of developing plans in isolation from relevant stakeholders. The Ministry of Agriculture at first operates in a rather top–down mode. This is encouraged by formal land-use planning guidelines and procedures, which also at a later stage limit the space for settling the problems that emerge. It is interesting to note, however, that quite similar conflicts to the ones described emerge in more 'bottom–up' community initiatives and plans in Zimbabwe (e.g. Shambare, 2000), which suggests that meaningful plans are likely to be accompanied by conflict regardless of their history and origin. Furthermore, this case resembles the previous one in that the participatory methods applied for community participation (plenary village meetings) do not allow for a process of give and take, so shared agreements on what should happen to resolve the conflicts never emerge. An additional dimension here is that the main driving forces behind the project lack the capacity, willingness and/or mandate to take responsibility for solving the conflicts. No credible mediator emerges from the local community (partly because the local leaders themselves are involved in the conflicts), while the Ministry of Agriculture and the donor tend to perceive their roles primarily as 'giving technical advice' and 'providing funds', and are not willing and/or able to organise an intensive conflict management process. Thus, a vacuum exists at the level of process leadership.

Case 4: Neglecting lingering conflicts around technical solutions (Khamis, 1998)

Khamis describes how through a group ranking exercise, a farmer research group in Zanzibar identifies reduced soil fertility as the problem with the highest priority. The group agrees to conduct on-farm experiments with agro-forestry and green manuring. In both cases, trees are to be used in combination with cassava. The nature of the experiments is such that it is a long time before farmers can conclude that the technologies work, to the detriment of enthusiasm within the research group (Khamis, 1998). However, even after the technologies have proved themselves technically, it appears that relatively few farmers are willing to use them. This appears to be related to the land tenure system in Zanzibar. Before 1964 most of the land in Zanzibar was owned by Arabs. After the revolution the land was officially confiscated and redistributed by the government. Informally, however, the Islamic laws and habits of heritage and land tenure are still respected (Khamis, 1998; Wipfler et al., 1998). Thus, much land is considered to have been borrowed from neighbours or now absentee Arab landlords and their local representatives. Some of the respected rules prohibit the permanent and/or temporary planting of trees on borrowed land. Moreover, this dual legal system causes a degree of land insecurity, which discourages farmers from making semi-permanent investments in soil-fertility. In fact, the planting of trees only increases insecurity, as sanctions may have to be faced.

This case-study shows how particular technical solutions can only be applied if new social arrangements are made between various actors with diverging interests, to support the technology (see also Leeuwis & Remmers, 1999). In this case, the use of agro-forestry and green manuring on borrowed land would require a new arrangement between formal and informal landowners, their local representatives, and farmers, concerning the use of trees. However, this need (or 'hidden conflict') was not seriously tackled within the context of the participatory project, as it was felt to be outside the project's scope and mandate.

Case 5: Conflict management without proper follow-up (Bolding et al., 1996)

Bolding et al. (1996) describe a situation where farmers in Zimbabwe's Nyanyadzi irrigation scheme have a conflict over water rights with upstream farmers who abstract water through self-made weirs. The Nyanyadzi farmers regard the upstream farmers as squatters who illegally reduce their water supply, and they repeatedly organise raids in which they destroy the weirs of the upstream farmers. The upstream farmers simply reconstruct them time and time again, and increasingly manage to get formally recognised water rights. A district administrator acts as a mediator. Through bilateral and group meetings with community representatives he eventually manages to arrange a water-sharing arrangement between the farmers in the Nyanyadzi scheme and several upstream communities of water users. However, the solution proves short-lived. It appears that, technically speaking, the solution does not work well, and that there are no viable organisational mechanisms to control and enforce the deal. Moreover, it becomes clear in retrospect that to a considerable degree the water shortages in Nyanyadzi are related to water extraction by farmers on tributaries to the Nyanyadzi river, who do not take part in the deal (Bolding, pers. comm.).

We see here a conflict reach a climax, after which a deal is secured under the supervision of a mediator. However, it appears that no adequate follow-up arrangements have been incorporated in the agreement on how to enforce it, and on how to deal with unforeseen developments, unanticipated interdependencies and newly emerging insights.

Case 6: A lack of government pressure and authority in conflict management (Dangbégnon, 1998)

Dangbégnon describes the dynamics of long-standing conflict between groups of fishermen who derive their livelihood from fishing in Aheme lake, Benin. The groups of fishermen differ in their lineage, cultural identity, geographic location, the marine-ecological zones of the lake in which they fish, the fishing techniques they use, and the governmental jurisdiction under which they reside. Over time, different initiatives emerge which are aimed at bringing about a more sustainable management of the severely over-fished lake. Most of these initiatives include the establishment of platforms (see also section 2.2.1), which typically exclude one or more of the major stakeholders, and are geared towards preserving the interests of particular coalitions. In this context, the conflict sharpens and continues. Eventually, the national government becomes involved and organises a 'Journée de Réflection' in which the various communities, groups, organisations and administrative units are invited to participate. During the meeting, the government more or less imposes certain measures (including the banning of several fishing techniques), and encourages the stakeholders to suggest alternatives. Also, a monitoring committee (Comité de Suivi) is established to control and enforce the new regulations. Initially, a number of semi-permanent 'Xha' and 'Akaja' fishing constructions are removed. However, a number of fishermen from various communities illegally reconstruct them later. Due to different policies in different jurisdictions – associated with electoral concerns – some fishermen face sanctions and others do not. Eventually, this leads to a total breakdown of the 'agreement'.

Here we see that, due mainly to the administrative complexity of the situation, different groups of fishermen tend to have plenty of room to frustrate joint regulation. The central government is weak and has very limited means of consistently enforcing regulations, or of putting pressure on the stakeholders to work seriously towards an agreement.

Types of friction

The above case-studies indicate that conflicts tend to accompany three broad types of friction during innovation processes:

(1) Problems in securing agreements or compromises on change and innovation. Here a variety of circumstances can play a role:
   — interventionists lack the time, skills, mandate and/or status to help resolve emerging conflicts;
   — methods and procedures used are not suitable for making deals (e.g. ranking or voting);
   — interests among participants are too diverse;
   — participants are not seriously interested in reaching agreement;
   — failure to address unequal distribution of resources;
   — representatives get into conflict with their constituents.

(2) Difficulties in maintaining such agreements or compromises after they have been secured. This may, for example, be because:
   — none of the parties is really happy with the agreement;
   — stakeholders do not trust others to do what they have promised;
   — there are insufficient arrangements for monitoring and controlling implementation of agreements;
   — there are no follow-up arrangements for dealing with new insights and circumstances;
   — the wider institutional environment does not support the agreement.

(3) Failure to tackle the most significant problems in the first place. For example:
   — emerging conflicts in the process are not recognised by (or remain hidden from) interventionists;
   — underlying conflicts are defined as outside the scope and mandate of the intervention.

10.1.2 Towards a better language to deal with conflict

The case-studies above suggest clearly – and in line with findings by others (e.g. Long & Long, 1992; Arce, 1993; Nelson & Wright, 1995) – that conflict is never far away from development and innovation, even where participatory methods and modes of thinking play a role. In some cases, conflict gives rise to participatory intervention, while in other cases conflicts emerge during the innovation process and/or affect them negatively or positively. Moreover, the cases present little evidence that the stakeholders' lack of knowledge and understanding of each other's perspectives (i.e. sub-optimal learning and communication) is either the central cause of frictions or the key obstacle to overcoming them. Rather, they suggest that stakeholders are often unable and/or unwilling to take other actors' viewpoints and interests seriously. This is associated with a variety of circumstances, such as the inclusive selection of participants, the superficial decision-making methods used, the lack of institutional space provided, insufficient leadership in conflict management, the non-existence of follow-up arrangements, and the non-availability of external pressures.

One could argue that the above case-studies are simply examples of bad participatory practice (Jiggins, pers. comm.), and that the frictions identified could have been prevented by better ex-ante situational analysis, the offering of give and take opportunities behind the scenes in community decision-making, the provision of responsible leadership in conflict resolution, proper follow-up arrangements and adequate use of external pressures. However, although the cases have many shortcomings from the perspective of participation literature, this would still be too easy an argument. The point is that the types of 'improvements' advocated here are hardly mentioned in either mainstream discussions on participation, or in popular training manuals for facilitators of interactive processes (Pretty et al., 1995; Van Veldhuizen et al., 1997).

Our conclusion from this is that we need to find a better language for anticipating and making use of the dynamics of conflict in interactive processes (for a more conceptual discussion see section 14.1). As a solution, we propose that the organisation of interactive routes should be inspired by theories on negotiation and conflict management. As we will see, learning and communication remain important concerns within such a negotiation approach to participation. However, effective social learning is unlikely to happen if it is not embedded in a well-managed negotiation process. At the same time, effective negotiation is impossible without a properly facilitated social learning process.

10.2   Distributive and integrative negotiations

Typically, negotiation processes can be sub-divided into two broad categories: distributive and integrative (Pruitt & Carnevale, 1993; Aarts, 1998). Many negotiation processes can be described as 'distributive' in nature. In such cases the various stakeholders tend to hold on to their own perceptions and positions (i.e. little learning occurs), and basically use negotiations to divide the cake (or the pain). In a struggle over land use, for example, wildlife conservationists and farmers can simply 'distribute' the land; that is, they can agree to allocate some areas specifically to wildlife and others to farming, with the party with the strongest power position getting the largest share. The gains of one party represent the losses of another. According to Aarts (1998) such compromises tend to be relatively unstable, since the source of conflict remains intact.

Other negotiation processes can be called 'integrative'. In such processes the stakeholders develop new and at least partly shared problem definitions and cognitions on the basis of a creative social learning process (see Chapter 9), resulting in the identification of so-called win–win solutions. Farmers and wildlife conservationists may, for example, rephrase the question 'farming or wildlife?' into 'how to make farming communities benefit from wildlife conservation?'. In connection with this they may come to set up joint tourist facilities or a value-added production chain for 'wildlife-friendly food products', from which both wildlife conservationists and farmers benefit. Obviously the latter type of negotiation is of greater interest for innovation and problem-solving in interactive processes.

10.2.1   Facilitation tasks in integrative negotiations

Building on a variety of negotiation literature (Susskind & Cruikshank, 1987; Van der Veen & Glasbergen, 1992; Huguenin, 1994) a number of tasks have been identified by Van Meegeren and Leeuwis (1999) that need special attention when facilitating integrative negotiations (Box 10.1).

Some of the tasks mentioned in Box 10.1 are especially important during the early stages of a negotiation process, whereas others become important while the process proceeds. However, all tasks remain relevant throughout the process as many iterations are likely to occur. Thus, the overview of tasks should not be interpreted as

Box 10.1   Tasks in integrative negotiation processes (adapted from Van Meegeren & Leeuwis, 1999). Copyright Wageningen University.

Task 1: Preparing the process
- preliminary exploratory analysis of conflicts, problems, social (including power) relations, practices, etc. in historical perspective;
- selecting participants;
- securing participation by stakeholders;
- establishing relations with the wider policy environment.

Task 2: Reaching and maintaining process agreements
- creating an agreed-upon code of conduct and provisional agenda;
- preliminary establishment of an overall objective/terms of reference;
- provisional distribution of facilitation tasks;
- definition of the role of external facilitators and other outsiders;
- maintaining process agreements;
- securing new process agreements as the process unfolds.

Task 3: Joint exploration and situation analysis (social learning A)
- supporting group formation and group dynamics;
- exchanging perspectives, interests, goals;
- further analysis of conflicts, problems and interrelations;
- integration of visions into new problem definitions;
- preliminary identification of alternative solutions and win-win strategies;
- identification of knowledge conflicts and gaps in insight.

Task 4: Joint fact-finding and uncertainty reduction (social learning B)
- developing and implementing action-plans to fill knowledge gaps and/or to build commonly agreed upon knowledge and trust.

Task 5: Forging agreement
- supporting manoeuvre: clarifying positions and claims, use of pressure to secure concessions, create and resolve impasses;
- soliciting proposals and counter-proposals;
- securing an agreement on a coherent package of measures and action plans.

Task 6: Communication of representatives with constituencies
- transferring the learning process;
- 'ratification' of agreement by constituencies.

Task 7: Co-ordinated action
- implementing the agreements made;
- monitoring implementation;
- creating contexts of renegotiation.

'stages' or 'phases'. From the viewpoint of ownership, all parties involved bear responsibility for these tasks. However, early in an interactive process the initiators have a special responsibility. Specific persons may be attracted whose job it is to facilitate and monitor the process. At the same time, it should not be automatically assumed that external facilitators are the ones in the best position to facilitate the various tasks. In fact, there may well be tasks that they cannot perform well or legitimately, and which are best delegated or left to others (see Chapter 11). Similarly, stakeholders themselves can take on certain tasks, sometimes without even being aware of it. In addition, it may be extremely difficult to identify one person who is acceptable as a facilitator to all parties involved, especially in complex institutional settings where many independent institutions have their own agenda. Thus, the precise role of an external facilitator is something that may transpire as the process unfolds, and can in itself be a point of negotiation as well. We will return to the role of outsiders in section 11.2.

For each of the tasks mentioned in this section a number of concrete recommendations can be made. These are discussed in more detail in Chapter 14.

10.2.2   The status of facilitation tasks and guidelines

The proposed tasks (and the guidelines associated with these; see Chapter 14) are based on the idea that a participatory process can be organised along the lines of a negotiation process, in which special attention is paid to the facilitation of social learning. In many cases there will be multiple *parallel* learning and negotiation trajectories that take place at more or less the same time. Within negotiations there are usually a number of issues at stake, and new paths can continually be added, for example by means of joint fact-finding. Negotiations can also take place on various fronts: between the different parties, but also between representatives and their constituencies, or between the negotiation partners and third parties. The complexities involved demand flexible and creative facilitation. The interactions within negotiation processes can have all kinds of unintended consequences and can also be influenced by chance or unanticipated events and developments in the external environment. Within integrative negotiations especially, all kinds of new directions can open up, making adjustments of the chosen trajectory essential. The planning scope for interactive processes is therefore limited. It would therefore be a mistake to approach the proposed tasks and guidelines (see Chapter 14) as a step-by-step plan that has to be closely followed in order to achieve success.

10.2.3   Layers or types of conflict

When facilitating conflict management through negotiations, it is important to recognise that several layers or types of conflict exist. These include:

● conflicting *cultural values*: In a conflict about sustainability, for example, some groups feel that mankind has been given the responsibility to protect and look after nature, while others feel that nature is there to be exploited by man. Similarly, in a conflict about agricultural policy, farmers may attach a totally

different value to farming (i.e. farming is a unique way of life) from that of policy-makers (i.e. farming is one out of many possible occupations).
- conflicts about *norms*: Even if groups subscribe to similar values, they may disagree about the rules for conduct. As Te Velde et al. (2002) have shown, for example, Dutch farmers and consumers agree that animal welfare is an important issue, but they do not agree about norms and rules on how to treat animals.
- conflicts about *resources* and diverging *interests*: People in a society and/or a community are characterised by many differences, from which different interests may emerge. Farmers may, for example, want land for farming, while urban dwellers want space for recreation. Similarly, the men in a village may want a project to build a new road, while women may prefer better healthcare facilities or a new well. Often such conflicts involve struggle about the use of and access to scarce resources such as land, water, subsidies, jobs, economic opportunities, etc. (see also section 5.2.4).
- conflicts about *power* and *influence*: A special 'resource' that people may struggle over is power and influence over other people. People may fight about formal leadership positions and other resources that are valuable for exerting influence in a specific context (see section 5.2.4).
- conflicts about *knowledge*: Regularly, we witness conflicts about what is 'true' and what is 'not true'. We have already discussed an example of this in section 6.7. Often, such disputes are connected with other types or layers of conflict discussed above, in that different arguments and knowledge claims are used as 'weapons' in the struggle about, for example, resources, norms and values. Farmers may, for example, present 'evidence' indicating that pigs do not suffer in intensive piggeries, while consumer groups provide counter-arguments.

As already hinted above, many conflicts involve several of the dimensions discussed at the same time. However, this is often not immediately clear. A conflict may, for example, seem to centre initially around contradictory economic interests, while on closer investigation it may transpire that different value systems are involved as well. Clarifying the precise nature of conflicts is important during the facilitation of integrative negotiations (belonging to Task 3 in Box 10.1).

10.3   Pre-conditions for integrative negotiation

An important question that needs to be addressed is under what circumstances and in what types of conflict situations an integrative negotiation approach towards interactive processes may be helpful. Negotiation theory suggests that three fundamental conditions must be met before serious negotiations can take place (Mastenbroek, 1997; Aarts, 1998):

(1)   There must be a divergence of interests.
(2)   Stakeholders must feel mutually interdependent in solving a problematic situation.
(3)   The key players must be able to communicate with each other.

To this we can add a more pragmatic fourth condition:

(4)   There must be institutional space for using innovative negotiation results.

10.3.1   Divergence of interests

In the preceding paragraphs it has been shown that conflicts of interest are likely to emerge wherever meaningful change is attempted. These conflicts can differ greatly in terms of the vitality of interests involved, historical roots, the number of actors involved, complexity, cultural setting, salience, intensity, etc. Each conflict situation will, depending on its specific characteristics, require a tailor-made mode of organising and facilitating the negotiation process. In any case, suffice it to say that, in the context of rural development, the criterion 'divergence of interests' will rarely form an obstacle for adopting a negotiation approach.

10.3.2   Mutual interdependence

If key stakeholders do not feel they need each other in order to arrive at an acceptable solution to their respective problems, a negotiation approach does not make sense. It often happens that one actor (e.g. a farming community in West Africa) feels interdependent on another (e.g. the European agro-industry) in solving what they perceive to be a problem (e.g. market distortion), while the other party does not recognise that problem as their concern, and thus feels no pressure to talk seriously, let alone negotiate, about this issue. Similarly, resource-rich stakeholders may feel that they can defend their interests best through means other than negotiation, for example by litigation, lobbying, violence, etc. Here it is important to recognise that conflicts are dynamic, and often evolve along a particular pattern. In the early stages of conflicts stakeholders tend to explore and follow their opportunities to win the battle with the means they have available. During such an exploration of 'Best Alternatives To Negotiated Agreement' (BATNA: Fisher & Ury, 1981) conflicts tend to reach a climax, and the relationships between the opposing parties usually deteriorate. In cases where both parties have considerable resources at their disposal, however, stakeholders tend to find out eventually that fighting each other does not lead to a satisfactory solution for any of them, and they start to realise that the only way forward is to restore relationships and negotiate a solution (Aarts, 1998).

An important lesson can be derived from this in that an inclusive participatory approach (i.e. one that brings all relevant stakeholders together) only makes sense during the final stages in a conflict cycle. In cases where the key stakeholders do not feel interdependent, interventions may more usefully focus on enhancing feelings/ perceptions of mutual interdependence. This may be achieved, for example, by strengthening the position of particular (coalitions of) actors in relation to other stakeholders, or with the help of conventional policy instruments such as new regulations, 'carrots' and 'sticks' and/or strategic communication campaigns (see Chapter 4). We can conclude that there are many situations in which an inclusive participatory negotiation trajectory is not (yet) an option, as key actors do not feel mutually interdependent. However, policy-makers and interventionists can employ various strategies in changing this situation. This may include setting up negotiations between a sub-set of relevant actors who do already feel interdependent. Thus, by focusing attention on feelings of mutual interdependence, a negotiation approach helps to identify appropriate *boundaries* for interactive trajectories (see also Leeuwis, 1995). A related implication is that one cannot start an interactive trajectory instantly; to assess promising boundaries of participatory activities a thorough understanding of the socio-historical context is required. If such understanding is lacking, various modes of exploratory research and conflict assessment can be used before decisions are made about the eventual set-up of participatory intervention (Leeuwis, 1993; ODA, 1995).

10.3.3   Ability to communicate

The ability of relevant stakeholders to communicate with each other (e.g. in a negotiation 'platform') can be hampered in various ways. Most obviously, physical distance may form an obstacle. In our globalised society, relevant stakeholders (e.g. in relation to the food market) may be spread across the world, which seriously hampers the forms of communication required in a negotiation process. Effective communication may also be affected significantly by institutional and organisational difficulties; that is, stakeholders must somehow be *organised* in order to be represented, strike deals with and allow effective communication between representatives and their constituencies during the negotiation process. In the case of the international food market this again poses problems, for example with respect to the position of consumers of agricultural products. Finally, deep cultural and historical divides between relevant stakeholders may make communication unthinkable. This may, for example, be the case in conflicts between different casts, ethnic groups or religious communities. In such situations direct negotiations can be impossible.

  If any of these barriers exist, an interactive (negotiation) approach is not a real option. Other forms of intervention may be more appropriate, for example, those aimed at containing conflict and/or reducing negative consequences. At the same time, efforts can be directed towards creating better conditions for communication and negotiation. Such efforts may well take years, or even decades, but can pay off eventually.

10.3.4   Institutional space for using innovative negotiation results

If there is little hope from the outset that innovative negotiation results can be implemented in practice, an attempt to organise an integrative negotiation process may be a waste of time. Often legal, political or bureaucratic concerns play a role here. In relation to problems and conflicts regarding water use in a specific catchment area, for example, it may well be known in advance that a government is not ready to change certain water laws that, according to stakeholders, are part of the problem. Similarly, it may be clear from the start that any innovative solution would require coherent action from several ministries (e.g. Ministry of Agriculture, Ministry of Economic Affairs, Ministry of Land and Water, Ministry of Transport, etc.), and that the prospects of such co-operation are extremely slim, given existing competency struggles and animosities between these government bureaucracies.

Of course, such problems may be ameliorated by effective alignment, or by incorporation of such actors as participants in the negotiation process. However, the latter my be in violation of previously mentioned conditions, such as a feeling of mutual interdependence. Therefore, when crucial institutions surrounding an integrative negotiation process are expected to be unwilling or unable to incorporate innovative solutions agreed upon by stakeholders, one may have to fight, lobby and campaign first in order to increase the institutional space for innovation.

In summary, it is difficult to arrive at a clear-cut typology of situations in which a negotiation approach towards participation may be helpful. This is mainly because of the large number of variables that can be used to construct such a typology, and because conflicts are dynamic: their characteristics change over time. However, it is possible to identify some fundamental conditions that must be met before one can adopt a participatory (negotiation) approach. Thus, each conflict situation will have to be carefully analysed in terms of these conditions before an appropriate strategy can be decided on (see Task 1 in Box 10.1).

10.3.5   An implication: the 'political' dimension of facilitation

As we have seen, starting from a negotiation model redefines the tasks of a facilitator in an interactive process (see section 10.2.1). Although the facilitation of communication and social learning remains an important point (i.e. in Tasks 3 and 4 in Box 10.1), these activities are now embedded within a wider negotiation setting which accompanies new tasks and tools. It is important to realise that this has far-reaching implications for the overall role of the facilitator. In conventional participation literature the facilitator tends to be portrayed as a fairly *neutral* figure whose prime concern is to enhance communication and learning. However, in a context of negotiation it becomes more evident that a facilitator needs to have an active strategy, resources and a power base in order to forge sustainable agreements (see also Chapter 14). For example, a facilitator may have to strategically select participants and exclude others, put pressure on certain stakeholders, and/or impose sanctions if actors do not follow the agreed code of conduct, etc. In applying such strategies, the prime 'political' interest of the facilitator should be to ensure a fair process with productive outcomes. Negotation processes in which facilitators pursue their own non-process-related interests and/or side with the economic or cultural interests of one particular stakeholder, are unlikely to succeed. In order to defend process interests, it is clear that – in addition to the power base provided by agreed-upon rules of conduct – a facilitator needs a certain amount of status, credibility, charisma, influence and trustworthiness in order to be successful. Thus, the selection of facilitators may require a lot more attention than in the past. The facilitation of negotiations is not a task that can easily be left to relatively junior project staff who have attended one or two communication courses. Apart from a certain amount of 'leverage', a facilitator will have to possess the necessary insights and capabilities for social interactions, the shaping of negotiations and the organisation of learning processes.

Questions for discussion

(1) Identify an unsuccessful attempt to induce innovation, that you are familiar with. Discuss whether or not, and in what way, conflict and/or lack of
conflict management played a role in this situation.

(2) One rarely finds explicit reference to the significance of conflict and conflict
management in project documents and proposals for development intervention. Discuss possible reasons for this. What are likely consequences?

(3) A well-known conflict is that between Israelis and Palestinians. Discuss
what layers and types of conflicts are involved. Which layer is, according to
you, most significant? Discuss whether or not important pre-conditions for
productive negotiations are fulfilled in this case.

11 The role of outsiders and different intervention approaches

This final chapter of Part 3 pulls together some of the insights we have gained on the connections between innovation, social learning and negotiation. Building on the discussions in Chapters 8, 9 and 10, we explore the relations between interactive and instrumental/persuasive models of communication in innovation processes (section 11.1). Similarly, we draw more precise conclusions on the role and importance of 'outsiders', 'insiders', and different areas of knowledge in such contexts (section 11.2).

11.1 The relationship between instrumental/persuasive and interactive models: alternation and sequencing

In Chapter 4 we emphasised a number of problems in relation to the instrumental model of communicative intervention, and indicated the need for a more interactive approach towards innovation. However, when discussing learning and negotiation as the key processes within such an interactive approach, we have seen that there may be situations in which productive learning and negotiation are unlikely to occur (see especially sections 9.4. and 10.3). Hence, it is important to draw some explicit conclusions on the usefulness of interactive and instrumental approaches, and the interrelations between the two.

The discussions in sections 9.4 and 10.3 suggest that it makes sense to use instrumental/persuasive strategies and interactive approaches alternately. This can take various forms.

Instrumental/persuasive strategies **preceding** *interactive trajectories*

It may make sense to use instrumental/persuasive strategies to help create the conditions for learning and negotiation in an interactive trajectory. In relation to negotiation (see section 10.3) this may involve strategic campaigns to:

- enhance feelings of interdependence among relevant stakeholders;
- improve the possibilities of communication through encouraging organisation development among less organised stakeholders;
- create institutional space for innovative solutions/interactive processes;

and also to:

- create awareness about the significance of diverging interests with respect to a problematic field.

Here the idea is essentially to make stakeholders aware of a potential conflict of interest that has not surfaced so far. For example, interventionists may try to make a community aware of their conviction that a nearby chemical plant produces serious environmental pollution and health hazards. This is in order to eventually mobilise a community to defend their interests, and thus bring closer negotiations between, for example, the chemical plant, the community and environmental organisations. In many ways this latter point coincides with stimulating a readiness for learning, or overcoming obstacles for learning (see section 9.4). In relation to this, such an instrumental campaign may well be geared towards:

● fostering certain ideas about the seriousness, urgency and personal relevance of problems;
● influencing stakeholders' perception of self-efficacy and environmental efficacy in solving the problem;
● improving the clarity and observability of the problem.

Instrumental/persuasive strategies **following** *interactive trajectories*

Once an interactive trajectory has resulted in innovative socio-technical solutions (e.g. to combat a particular pest), interventionists and/or policy-makers may wish to encourage the application of these (or similar) solutions by a wider audience. Basically, this situation resembles that of a purely top–down planning process, where the idea was to persuade as many people as possible to accept or adopt a given policy or solution. And although one may expect that interactively developed solutions are better adapted, and have a better chance of being applied, than solutions that are developed top–down, a conventional instrumental/persuasive campaign may still face many problems. This is because the targeted audiences have not experienced the learning and negotiation processes on the basis of which the solutions were designed, and which formed the social foundations for their adoption among the participants in the interactive process. There are many indications that one cannot just 'transfer' solutions that were developed in an interactive mode (e.g. Röling & Van de Fliert, 1994; Van Schoubroeck & Leeuwis, 1999). At the same time, it will often be practically impossible and undesirable to organise and facilitate interactive processes in *all* the contexts where similar problems exist to the one tackled earlier. Thus, a useful strategy may be to try and 'transfer', in a condensed form, the underlying learning and negotiation processes, rather than to 'transfer' only the solutions that were eventually arrived at, or to 'repeat' fully the interactive process. This means that attention must be paid to communicating the learning and negotiation experiences from elsewhere (e.g. by bringing different communities in contact with each other), which may help to shorten the process in a new situation. Also, this may well be combined with more conventional persuasive approaches than were applied during the original interactive process. In any case, however, the 'scaling-up' of interactively developed innovations is likely to require a certain amount of 'redesign', i.e. the adaptation of social–organisational and technical arrangements to fit a different context.

*Instrumental/persuasive strategies **during** interactive trajectories*

As we have argued earlier, interactive innovation processes tend to be complex and multi-faceted, and include essentially multiple learning and negotiation processes at the same time (see section 10.2.2). Since these multiple processes are unlikely to be synchronised, it can be expected that the types of alternation described above may occur simultaneously. Thus, it becomes somewhat artificial to use categories like 'preceding', 'following' and 'during'. This separation may be relevant analytically, but in everyday practice interactive and instrumental/persuasive strategies will have to be used alongside each other. Moreover, we have seen in section 10.3.5 that facilitation of interactive processes may require a variety of deliberate strategies from the facilitator in order to keep all parties on board, to enforce agreed-upon codes of conduct, to make sure that weaker parties are heard, to enhance learning, etc. This means that a facilitator may have to apply persuasive communicative strategies during the negotiation process as an integral part of facilitation.

We can conclude that the combined use of instrumental/persuasive and interactive models of communicative intervention is necessary and legitimate in the context of innovation processes. However, this partial 'rehabilitation' of the instrumental/ persuasive model should not be accompanied by a restoration of the associated assumptions about the controllability, predictability and 'explainability' of human action and society at large. Although one can attempt to facilitate and enhance social learning and negotiation processes, with the help of persuasive strategies, among others, their dynamics and outcomes remain messy and unpredictable.

11.2   The role and expertise of 'insiders' and 'outsiders'

Now that we have discussed in some detail how we look at innovation and change processes, we are in a better position to assess what expertise may be needed in such contexts, and who may be expected to contribute it. In connection with intervention, we are especially interested in the possible contributions by relative 'outsiders' (e.g. scientists or external facilitators) and 'insiders' (e.g. farmers, community members and other stakeholders) in a rural innovation context.

11.2.1   Different areas of knowledge and competence needed in intervention

In Chapters 5 and 9 we identified different areas of perception and cognition that are relevant in innovation, in that they constitute areas for learning and cognitive change. Thus, it has become clear that what is conventionally termed 'knowledge' (see Chapter 6) is only one of many types of perception that is relevant. Nevertheless, knowledge remains an important issue as it may be a trigger for other kinds of cognitive change, and also since innovation processes are expected to benefit from the integration of different kinds of knowledge, originating from different stakeholders and localities (see section 8.1.2). In connection with this we have seen in section 6.5 that there have been heated debates on the qualities of 'scientific' versus 'local' knowledge. We have argued that this separation is somewhat misleading, since *all* knowledge is *local* in nature; i.e. it is specific to a particular locality (be it a scientific laboratory or a farmer's field) and cannot be regarded as automatically valid outside such a context. Therefore, we are especially interested in the possible contributions from different actors (including scientists) who bring in different kinds of 'local' knowledge. In the debates on the value of 'local'/'indigenous' versus 'scientific' knowledge, the question 'knowledge about what?' has hardly been discussed explicitly or comprehensively. Simplifying our discussions in previous chapters, we can say that in innovation processes it is important to generate and/or access the following kinds of 'knowledge':

Issue-related knowledge: Here one can think about knowledge regarding the characteristics and functioning of biotic, abiotic and social systems, as well as the problems and dilemmas that different stakeholders experience in connection with this. Furthermore, a crucial area of issue-related knowledge includes the variety of, sometimes hidden and implicit, ideas and experiences with regard to alternative social and technical solutions and/or the criteria that these should meet. The same is true for background knowledge regarding the social, cultural, political and economic context, including the history of intervention and earlier problem-solving efforts. In essence, then, issue-related knowledge is about:

● diverging problem definitions;
● potential solutions;
● context.

Knowledge about people and networks: It is important to have an insight into who the relevant stakeholders, are from the viewpoint of alignment, inside and outside the immediate innovation context, and what their perspectives, interests and values may be. Knowledge of the nature of social relations, for example, interdependencies, hidden and open conflicts, co-operation and coalitions between stakeholders in the past and present cannot be ignored either. Moreover, it is essential to mobilise knowledge of the special qualities and capacities of different actors, in view of their potential contribution to conflict management, learning and/or the provision of access to certain networks and expertise. In essence, then, this type of knowledge is about:

● relevant stakeholders;
● the social relations between them;
● availability of special qualities.

Social process knowledge: This involves knowledge of how network building, social learning and negotiation processes can be shaped in a specific innovation context. This includes, for example, insight into local preferences and obstacles with regard to learning and conflict resolution, as well as insight into possible ways of overcoming difficulties. Thus, *contextually adapted* knowledge and skill is needed regarding:

● network building;
● social learning;
● negotiation.

It is important to recall that the specific knowledge and competence needed in these areas is usually not readily available and/or static; it must be discovered, invented, generated, made explicit, searched for and adapted as the innovation process proceeds.

11.2.2   From 'experts' and 'laymen' to 'outsiders' and 'insiders'

When we look at the areas of knowledge and competence needed in an innovation process, we can draw several conclusions.

First, we can conclude that the knowledge needed is *diverse*; an innovation process will have to involve many people with different capacities and areas of expertise. For this reason alone, it is already a gross simplification to speak of 'experts' and 'laymen' in an innovation process. Someone who can be regarded as an 'expert' in one of the relevant knowledge domains, can at the same time be a 'layman' in other areas. A natural scientist may, for example, have expertise on biological pest control in rice, but be totally ignorant about ecological conditions in a region, the social organisation of pest control, conflict management arrangements in a community, etc. Another problem with the separation between 'experts' and 'laymen' is that the terminology reinforces the idea that *within* a particular knowledge domain there are people with 'more valid' and 'less valid' knowledge. There are many cases, however, in which both scientists *and* local farmers have sensible (and in fact complementary) knowledge and expertise on, for example, matters of biological pest control in a specific context. Scientists may, for example, know more about which insects are predators of a particular pest, while farmers may have more insight into the local ecological niches were such predators can be found and/or kept. Both forms of expertise may be important in an innovation process. Given these imprecisions, we shall avoid the use of the separation between 'experts' and 'laymen'.

Second, the above listing of relevant areas of knowledge makes clear that much of the knowledge needed is related to the specific situation in which innovation is required, and to the stakeholders who are supposed to practise and use the eventual innovative solutions. This implies that a great deal of the relevant knowledge will have to be *provided from the 'inside'* (i.e. by societal stakeholders) rather than from the 'outside' (i.e. by external scientists or facilitators). Of course, knowledge from the 'outside' may remain very useful and enriching, but it is unlikely to be the dominant force in an innovation process.

11.2.3   The role and contribution of external facilitators and scientists

Following our interest in communicative intervention in rural settings, we will close this chapter by discussing the specific role of two types of outsiders whom intervention organisations may mobilise in order to contribute to innovation processes: process facilitators and agricultural scientists. In practice, both types of outsiders may be united in one person, e.g. a communication worker with training in both agricultural sciences and communicative process management. Nevertheless, it is important to distinguish between the two functions.

The role of process facilitators/communication workers

Process facilitators or process managers are communication specialists with a special responsibility to guide and induce innovation processes. Although many managers of development organisations may not think about their staff in terms of 'process facilitators', we feel that nowadays this should be regarded as an essential function of change agents. In our view, they need to play a role in:

- *Overseeing/monitoring learning and negotiation processes*: We have already seen that a number of tasks are important during social learning and negotiation processes (see section 10.2.1). It is important here to repeat that it is not necessarily an external facilitator's job or mandate to conduct these tasks him or herself. Nevertheless, an important function of an external facilitator can be to actively monitor and oversee the dynamics of learning and negotiation, and look at which tasks are fulfilled by whom, how, and with what consequences.

- *Intervening in learning and negotiation processes*: It may transpire from process monitoring that certain tasks are overlooked, or conducted in a way that is not very satisfactory or productive. Here an external facilitator may want to intervene in the process in order to enhance learning and negotiation and/or to defend process interests (see section 10.3.5). In doing so he or she can make use of various persuasive and interactive communicative strategies (see also section 11.1 and Chapter 14). Of course, an external facilitator may try to prevent the process from going astray in advance by drawing up a detailed process design early in the process (see Task 2 in Box 10.1), to be agreed by participants. However, it is important to recall that learning and negotiation processes are capricious by definition, which means that any advance process design will have to be flexibly adapted. This implies that detailed plans with a long time-scale are not likely to be very helpful (see Chapter 14).

- *Collecting and connecting relevant knowledge and actors*: As we have seen, innovation processes require the collection and integration of inputs from different areas of expertise (see section 11.1), and the bringing together of a variety of actors, including actors from outside the immediate innovation context (see sections 8.2.2 and 11.2.2). All stakeholders have responsibilities in this respect. However, external facilitators are sometimes in a better position to access specific networks of actors (e.g. donors, scientists or policy-makers) than stakeholders. Similarly, it would be extremely useful if facilitators had special capacities and skills in *eliciting* knowledge from different stakeholders, and in *translating* it in such a way that other stakeholders can relate to it.

In many cases, playing a role as outlined above does not require a formalised position as a process facilitator. There is already much that communication workers can do when they define their own role in these terms, and have adequate knowledge and skills. It is important to note here that process facilitation should ideally be about *serving others* to make progress (see also section 14.2.2), which requires modesty rather than a wish to always be in the spotlight. More specific guidelines in connection with these basic roles are explained in Chapter 14.

The role of scientists

A key conclusion to be drawn here is that, contrary to what many scientists believe, innovation is not primarily about 'doing scientific research' (see also Leeuwis & Remmers, 1999). Science can be rather strong at *analysing* what happened in the past, but is weak in composing, or *synthesising*, the future (Remmers, 1998:321 onwards). Innovation is essentially *synthesis*, research is essentially *analysis*, but doing research and gathering data can include interactions between researchers and stakeholders that imply learning moments for both. Thus, scientific insight and investigation can play an important role in social learning processes and joint fact-finding within a context of negotiation (Van Meegeren & Leeuwis, 1999). But innovation processes are not likely to be successful if they are scientist owned and/or initiated (Broerse & Bunders, 1999; Leeuwis, 1999a). In a learning and negotiation process, knowledge generated in various locations (e.g. research stations and farmers' fields), by different stakeholders (e.g. researchers and farmers), for dissimilar purposes (e.g. assessing the 'truth' and promoting stakeholder interests) and through different procedures of validation (e.g. scientific method and farmer experience) must be creatively articulated and integrated. In innovation processes, then, scientists can be seen as *resource persons* who can play four basic roles during social learning and negotiation processes:

- *Help explicate implicit assumptions, knowledge claims and questions*: Discussions among stakeholders usually contain a range of implicit knowledge claims, assumptions and questions. Frequently, progress in social learning and negotiation processes is hampered when these remain implicit and do not become a point of explicit discussion and reflection. Such explication is far from easy and can never be complete. Nevertheless, not only process facilitators, but also scientists from different disciplines, can play a useful role in this respect. One may expect scientists to have a special sensitivity for the assumptions, knowledge claims and questions that are hidden in what stakeholders say or do not say about their specific field of expertise. Hence, dialogue between stakeholders and scientists may contribute to making explicit what was implicit previously, and may result simultaneously in a coherent set of relevant natural and social science questions (see section 19.2).
- *Joint fact-finding and uncertainty reduction*: Research can play a role in joint fact-finding geared towards answering shared questions and reducing uncertainties that affect the innovation process (see section 14.2.4). The purpose of this type of natural and/or social science research is not only to provide answers, but also to build confidence, trust and shared perspectives among stakeholders by working together on an issue (Van Meegeren & Leeuwis, 1999). Depending on the questions addressed, such research may involve on-farm research, laboratory research by scientists, computer simulations, etc., as long as it remains part of a commonly agreed upon – and preferably iterative (see Vereijken, 1997) – procedure. In the context of such research, scientists also need 'free space' to follow their own intuitions (see Van Schoubroeck & Leeuwis, 1999).
- *Feedback*: Results from research can serve as *feedback* in order to induce learning, i.e. through the creation of new problem definitions (see section 9.3). Such

feedback from natural and/or social scientists may be provided by research data on the existing situation, but may also arise from comparison with totally different situations (including laboratories) or computer-based projections about the future (see section 13.5.10; Röling, 1999; Rossing et al., 1999). This can also include comparison with radically new technological and organisational solutions. These latter kinds of feedback may serve to enlarge the space within which solutions are searched for. Given that scientists' questions, concerns and conclusions are never neutral (see section 6.5), it is important that, when giving feedback, scientists are transparent and explicit about the implicit dimensions (e.g. underlying aspirations and assumptions) of the knowledge and insights they bring (Alrøe & Kristensen, 2002). Such transparency does not imply that scientists become 'politicians'. On the contrary. When scientists are clear about their underlying aspirations and values, it becomes evident that clashes between interests cannot be resolved by scientists; instead it is the task of societal stakeholders, administrators and politicians to value and appreciate the insights put forward and to make choices.

- *Process monitoring*: Research can play a role in monitoring the social dynamics of the learning and negotiation process itself, in order to inform its organisation and further facilitation. How are relations between stakeholders developing? Which new developments, questions, wishes and problems emerge? How do these affect progress, and what can be done about it?

The above view of the role of scientists is consistent with what the philosophers Funtowicz and Ravetz (1993) have called '*post-normal*' science. They argue that in situations where uncertainty is high and where different values and interests are at stake, applied scientists cannot resort usefully to 'normal' strategies of 'puzzle solving' and/or professional consultancy. Rather, they need to play an active role in societal discussions and innovation processes. It is important to realise here that playing a role as outlined above requires modes of operation by scientists different from those currently dominant. It requires, for example, (a) intensive co-operation between stakeholders, change agents and researchers; (b) cross-disciplinary co-operation among scientists (as the solving of problems may well involve integration of insights from various disciplines); (c) greater emphasis on on-farm experimentation (see section 13.5.8); and (d) new procedures for setting research agendas, etc. (see also Vereijken, 1997; Bouma, 1999; Van Schoubroeck & Leeuwis, 1999). Contrary to critiques that such new modes of working do not allow for 'real science', we feel that they by no means imply a devaluation of scientific endeavour (see also section 19.3). It is true that 'interactive science' (Röling, 1996) may require changes in the type of research questions asked, their origin, and/or the objects of research, but we prefer to look at these as new academic (conceptual, methodological and epistemological) challenges.

However, many universities and research institutes are not well equipped for 'interactive' and 'cross-disciplinary' science (Röling, 1996). They often employ scientists who follow the linear model of innovation (see section 8.1.2), and are often more 'research' than 'innovation' oriented due to prevailing reward structures and funding arrangements. We discuss these organisational dimensions and obstacles in more detail in Chapter 19.

Questions for discussion

(1) Identify a relatively successful attempt to induce change and innovation that you are familiar with. Discuss if and how 'instrumental' and 'interactive' moments and strategies have contributed to intervention becoming relatively effective.

(2) We have argued that facilitation tasks are not necessarily performed by a single facilitator. What are the pros and cons of dividing facilitation tasks among several people? In your view, which facilitation tasks can and/or should be left to societal stakeholders and/or other institutions (e.g. legal professionals and political bodies) rather than communicative interventionists?

(3) Consider a group of scientists that you are familiar with, and discuss to what extent they are able and willing to perform the roles we have proposed for them in innovation processes. Do you feel we have left out any important roles?

PART 4
Media, methods and process management

Now that we have discussed (in Parts 2 and 3) the basic underlying concepts and theories that are relevant to communicative intervention, we will in Part 4 discuss more practical issues. As communication is an important ingredient in any communication for innovation strategy, we first discuss the practical potential of some basic communication forms and media (Chapter 12). In later chapters we focus more on specific methods (Chapter 13) and the management of interactive processes (Chapter 14). This part of the book concludes with a chapter on communication planning, focusing mainly on individual activities in the context of wider processes that, in our view, are not themselves amenable to planning (Chapter 15).

12 The potential of basic communication forms and media

In Chapter 7 we described communication *media* as devices that help to combine different communication channels for the 'transportation' of textual, visual, auditive, tactile and/or olfactory signals. We mentioned the example of television, which is basically a medium that opens up auditive and visual 'long distance' channels. Nowadays, there are a large number of such media, which can be roughly divided into three main classes:

- conventional mass media;
- interpersonal 'media';
- hybrid media.

Conventional mass media are, for example, newspapers, farm journals, leaflets, radio and television; their basic characteristic is that a sender can reach many people with such media while remaining at a distance, and without the possibility of engaging in direct interaction with the audience. With interpersonal media a more direct (in the sense of synchronical, see Chapter 7) exchange between the communicating parties can take place; that is, media in which sender and receiver can easily change roles. The telephone is an example. However, most *interpersonal communication* takes place *without* artificial media (i.e. without technological devices), as it involves the physical presence of people. Basic forms of such face-to-face communication are group meetings and meetings between two people. Following the rapid advances in computer and telecommunication technology, we have since the early 1990s witnessed the upsurge of new hybrid media, which combine the potential offered by mass media and interpersonal communication. The internet and CD-ROM technologies, for example, are media which can potentially reach large audiences, but also allow a certain degree of interactivity between the receiver and the sender.

Currently, these diverse media are more and more combined into new packages, so that the boundaries between the categories of media are becoming more vague. For example, the telephone and internet are increasingly used to interact with the audience during radio and television programmes, which results in 'interactive radio' and 'interactive television'. However, despite these tendencies it remains useful to take a separate look at the basic categories of mediated and non-mediated communication.

Before doing so, we would like to emphasise that it is important to refer to media 'potentials' rather than media 'characteristics' (as one often finds in media literature). It is a mistake to talk about media as if they have fixed qualities and characteristics, because the specific context affects whether such 'characteristics' are actually realised. For example, people often speak about the internet as an 'interactive' or 'fast' medium. However, whether or not communication through the internet is 'fast'

Table 12.1 Different functional qualities (in most contexts) related to basic communication media and forms.

	Conventional mass media	Basic forms of interpersonal communication	Hybrid media
Possibility to deliver tailor made messages (in view of differential potential for interactivity)	–	+	0
Potential to attract attention	+	0	–
Potential to support active learning and decision-making	–	+	0
Possibility to develop and use relationships of trust and mutual involvement	–	+	–
Costs involved per person reached	–	+	0
Potential to reach large audiences	+	–	0 / –*

+ = relatively high (in most contexts)
0 = average
– = relatively low
* When access to computers is high, the potential to reach large audiences is medium; when access to computers is low, the potential to reach large audiences is low.

or 'interactive' depends in many ways on the people involved. If a receiver responds to an e-mail only after three weeks, communication is not fast and hardly interactive, and if the response does not address the issue raised in the original sender's e-mail, 'interactivity' is even further reduced (see Chapter 7). In other words, qualifications like 'speed' and 'interactivity' are characteristics of human interaction in a specific context, and not characteristics of the media. While specific media can or cannot offer the potential to communicate fast or interactively, it depends on people whether these potentialities are realised. Keeping the above in mind, the three categories of media tend to have a number of functional (enabling or constraining) qualities in a context of communicative intervention (see Table 12.1). More detailed discussions follow in the next sections.

12.1   Conventional mass media

Box 12.1 gives an overview of conventional mass media, according to their dominant communication channel(s).

Box 12.1 Overview of conventional mass-media according to dominant communication channel(s) (adapted from Van Woerkum, 1994)

Mainly textual	Mainly auditive	Mainly visual	Combinations (e.g. audio-visual)
newspaper	radio	poster	television
farm journal	speech	drawings/pictures	(video)film
flyer	songs	animation	exhibitions
brochure	story telling	slide show	theatre or drama
advertisement	cassette		
book/manual	audio CD		

12.1.1   The way mass media work

Mass media, and particularly radio, television and newspapers, have the image of being very powerful. Not surprisingly, therefore, one of the first things that new authoritarian regimes do is to make sure they get control of the mass media. The idea here is that if you control the mass media, you can selectively influence the way large masses of people think and see reality, and can prevent others from presenting a different picture. Also, mass media are increasingly used as an integral strategy in military warfare, where they are used, for example, to create certain images about the military situation, to mobilise public and political support and/or to demoralise the enemy. Although we would hope that the use of mass media for communicative intervention purposes in rural settings is of a higher ethical standard than its manipulative use in warfare (see Chapter 3), this comparison allows us to point to a number of factors which tend to limit the 'power' of mass media. In a situation where one can receive numerous television channels (in western countries often more than 50), and an even larger number of national, regional and local newspapers and radio stations, it is not only extremely difficult to control the media, but also to reach large audiences. Apart from this media fragmentation, the 'power' of mass media is further reduced by the fact that audiences are not passive recipients into which certain messages and opinions can be 'injected' (see Stappers et al., 1997). People actively construct meanings in interaction with their social network (see Chapters 6 and 7), and efforts to install certain views and opinions may therefore be unsuccessful or even counterproductive.

It has been argued that the influence of mass media is mainly *indirect* (Klapper, 1960; Oomkes, 1986; Van Woerkum, 1999); that is, mass media serve primarily to put things on the 'agenda' for further discussion among an audience. Thus, they can have an influence on *what* people talk about, but *how* people talk about it is something that depends to a large extent on the context in which messages are received, including the network of people to which actors belong and talk to. For example, the CNN reports on Iraq's occupation of Kuwait in 1990 and the Gulf war that followed in 1991 have been discussed across the globe, but were received differently in different parts of the world and/or in different cultural communities within regions. Some considered them to be factual information, while others looked at them as military propaganda; for some they induced feelings of enthusiasm and support, while they made others extremely worried and sad. This may be an extreme example, but in essence this is the way mass media work. Similarly, a radio campaign in Zimbabwe aimed at encouraging farmers to make contour ridges may be positively discussed by irrigation farmers who are faced with siltation, but framed negatively as 'a continuation of exploitative colonial policies' by dryland farmers who are supposed to deliver the heavy work in the face of labour shortages (e.g. Shambare, 2000).

It is no small achievement to put things on the public agenda, even if one cannot predict the eventual outcomes of the discussions that follow. This is why mass media remain an important concern in communicative intervention. However, the extent to which mass media set the public agenda is not something that can be fully controlled. In some cases, mass media attention may lead to media *hypes*; here media exposure of an issue leads to intensified public concern, which again leads to more media attention, etc. (see Van Ginniken, 1999). As a result of such a self-reinforcing process, it seems that eventually everybody talks about the issue. Studies of such hypes (e.g. Van Ginnikken, 1999) show that mass media coverage is in itself not a sufficient condition for a hype to emerge. A message or issue must somehow fall on fertile ground, which depends largely on accidental or coincidental combinations of circumstances and happenings. In Europe, for example, the issue of food safety became a 'hot topic' in a situation where (a) health had become an increasingly important general concern for people; (b) a number of food scares and scandals (e.g. around the use of polluted animal feeds) followed each other in a short period; (c) trust in governments and food-industries had eroded; and (d) several contagious animal diseases occurred on a large scale. Diseases and food scandals, of course, had happened and been reported on before, but had never had so much impact. In this instance, however, new issues started to 'resonate' quickly, and became part of intensive public debate. Many issues that rural change agents are interested in may never assume the proportion of a 'mini hype', or only after having been raised for the twentieth time. As one cannot plan for such things to happen, Aarts (1998) argues that the timing and framing of mass media messages are important issues to consider when the purpose is to set public agendas, as is a certain amount of repetition and redundancy of messages in order to 'give coincidence a chance' (see also McQuail, 1994; Aarts & Van Woerkum, 1999).

12.1.2   Functional qualities in relation to communicative intervention

In a context of communicative intervention, mass media potentially have a number of functional (enabling or constraining) qualities (derived largely from Van Woerkum, 1999; see also Windahl et al., 1992; McQuail, 1994):

- *Large audiences*: With mass media one can potentially reach a relatively large audience, either directly or indirectly. This quality depends partly on levels of media fragmentation, media access (i.e. the percentage of people with access to, for example, television) and the media preferences of specific audiences (some groups in society, for example, hardly ever listen to radio, even if they have access to it).

- *Mobilising interest*: Mass media can be especially helpful for mobilising attention and getting issues on the public agenda, and for keeping them there by repetition and the broadcasting of reminders (e.g. Singhal & Rogers, 1999). In other words, mass media offer relatively good opportunities to help create awareness of, and/or interest in, certain issues (the first and second aspect of learning; see section 9.5).

- *Audience skills*: For written mass media it is evident that audiences need to be able to read. In some situations this can pose an important limitation to its use (Hoffmann, 2000). Skills for understanding messages communicated through auditive and visual mass media are often more widely available.

- *Non-specificity*: Conventional mass media do not allow much synchronical interaction between sender and receiver, and therefore it is impossible to respond in a tailor-made fashion to the needs, concerns and questions of individuals. This does not exclude the possibility, however, that mass media are directed at a specific audience (e.g. a specialist journal for tomato growers). In any case, mass media may deal with issues only globally and out of context. They have limited potential to play a role in active experiential learning and opinion formation (the third aspect of learning; see section 9.5).

- *Low relational support*: Through mass media it is difficult to establish relationships of trust with audiences and/or to express personal involvement. This may hamper the credibility of messages, and considerably reduces the possibility of stimulating and motivating people to deal with certain issues.

- *Reinforcing existing views*: Mass media tend to be much stronger in reinforcing already existing views and opinions, than in fostering significant opinion change (Klapper, 1960). However, in relation to issues that people have no strong views about yet, mass media can have an impact (directly or indirectly). They can also have an impact if they offer solutions that people are already looking for.

- *Little insight into audience*: Mass media tend to give senders little control over and/or insight into whether or not people receive messages, and the way audiences process, interpret and use them. This makes it difficult to adapt messages and communication strategies when necessary.

- *Speed/actuality*: Mass media (especially radio) can be very fast in communicating news and actualities to relatively large audiences. At the same time other forms of mass media require considerable preparation time (for example making a regular television programme, videofilm or exhibition) and hence are far less quick.

- *Differential time flexibility*: The time flexibility of many mass media can be high, meaning that the time of receiving can be adjusted to the preferences of audiences. Written mass media can often be consulted whenever people choose (provided that distribution is adequate, which is often not the case), while the timing of radio and television broadcasting can, in principle, be tuned to the working schedule of farmers. However, radio and television often have other audiences (e.g. city dwellers) who pose different demands (e.g. amusement) which may compete with farmers in terms of timing. The needs of these other audiences often prevail in view of commercial considerations.

- *Differential spatial flexibility*: The spatial flexibility of written mass media is often high; they can be distributed and read on many different locations. For mass media that involve electronic equipment (television, videofilm, slide show) this flexibility is reduced, as this equipment tends to be tied to specific places.

- *Low cost*: If the audiences reached are sufficiently large, mass media can be a relatively cheap method of communication per member of the audience reached. However, there are considerable variations in costs for different mass media, and therefore in the size of the audience needed to justify these spendings.

- *High storage capacity*: Several mass media (especially printed ones) offer the opportunity for people to 'store' the message, and receive it again whenever they wish.

- *Status associations*: Some mass media (e.g. newspapers, television channels) tend to have a particular social status as a source of information; some may be regarded as 'credible' while others are not. Selecting particular media may thus help to bring about positive or negative associations with a message.

- *External skills*: The use of several mass media (e.g. television, videofilm) requires special skills and expensive equipment that may not be available within intervention organisations. Hence, outsiders (graphic designers, script writers, cameramen, video editors, etc.) need to be involved.

- *Dependence on others*: In order to make use of certain mass media (e.g. television, radio and newspapers) other than in the form of advertisements, organisations for communicative intervention may depend largely on the willingness of others (i.e. editorial committees) to co-operate. In many cases this willingness may depend on whether or not issues can be constructed as 'news'. According to McQuail (1994), western media tend to regard events as 'news' if they rank sufficiently high on the following criteria: involvement of prominent people; level of 'drama'; connections with earlier news; unexpectedness; possibility to evaluate (the consequences of) events as negative. Not surprisingly, therefore, western media for the general public mainly report on agricultural issues when there is a crisis or natural disaster.

There are some differences between different mass media in connection with the functional qualities mentioned above. Although these differences depend on the specific context, we have summarised some of them in Table 12.2.

12.1.3   Basic guidelines for presenting messages through written mass media

In connection with written mass media in particular, research findings suggest some basic guidelines for producing comprehensible leaflets, brochures and technical documentation (see Langer et al., 1981):

- *Use simple (i.e. commonly understood) language*: It is important to tune in to the common terminology and vernacular that audiences use when speaking about an issue. New technical terms should be explained in short and simple sentences. Abstract language and 'jargon' should be avoided.
- *Structure and arrange arguments clearly*: Ideas should be presented in a logical order, clearly distinguishing between main and side issues. Presentation must be clear, with the central theme remaining visible so that the whole message can be reviewed easily. Careful use of layout and typography helps to separate key points or sections of the message.
- *Make main points briefly*: Arguments should be restricted to the main issues and clearly directed towards achieving stated goals without unnecessary use of words. Mass media are not appropriate for conveying all sorts of nuances and complexities.
- *Make writing stimulating to read*: The style should be interesting, inspiring, exciting, personal and sufficiently diverse to maintain the reader's interest.

Table 12.2 Gradual differences between different mass media in relation to functional qualities in most contexts.

	Radio	Television/ video	Leaflet/ brochure	Article in newspaper/ farm journal	Poster
Likelihood of stimulating debate in context of reception	+	+	−	0	0
Ease to draw attention/mobilise interest	+	+	−	0	+
Capacity to go in depth, and support active learning	0	0	+	0	−
Potential to function as a reminder	0	−	+	0	+
Ease of storing messages	−	−	+	+	0
Speed/actuality (short production time)	+	−	0	0	+
Possibility to receive at different times (time flexibility)	−	−	+	0	+
Possibility to receive at different locations (spatial flexibility)	0	−	+	0	+
Likely costs for intervention organisation (sender)	0	+	0	−	−
Likely costs for receiver	0	+	−	0	−
Freedom/independence of change agents in determining content	0	−	+	0	+

− = relatively low in comparison with other mass media (in most contexts)
0 = comparable to other mass media (in most contexts)
+ = relatively high in comparison with other mass media (in most contexts)

Such guidelines probably have a wider relevance. The basic points (i.e. use common language, develop a clear structure, be brief, make it vivid and stimulating) also make sense when one develops, for example, a speech on a particular agricultural topic. However, for other non-written mass media (e.g. theatre, songs, series of pictures, films, etc.) other and/or additional guidelines and skills apply (e.g. Fraser & Restrepo-Estrada, 1998; Hoffmann, 2000). Without going into detail here, we can say that these latter forms of communication make use of more indirect symbols than words, and that it is therefore all the more important to check whether such symbols have the intended meaning in a given context and culture. In a case reported by Hakutangwi (pers. comm.), for example, farmers were shown a picture which tried to convey that feeding cows well, and watching over their health, would improve milk yields. The picture showed a big fat cow, with a big udder and many bottles underneath. The picture was interpreted as 'fat cows need a lot of medicine' since the bottles were associated with medicine rather than with milk.

In any case it is important to *pre-test* mass media products such as leaflets, pictures, theatre productions and posters; that is, it is important to get feedback on the quality and clarity of such products from people who belong to the eventual audience (see also Chapter 15 on communication planning).

12.2   Interpersonal communication

In this book, we refer to interpersonal communication when synchronical exchange between the communicating parties takes place. In most cases, the communicating parties not only interact at the same time, but also at the same place, so we can refer to face-to-face or non-mediated communication. The basic forms of interpersonal communication are group meetings/discussions and bilateral meetings/discussions (including telephone conversations).

12.2.1   Functional qualities in relation to communicative intervention

The basic forms of interpersonal communication potentially have a number of functional (enabling or constraining) qualities in a communicative intervention context:

- *Limited audiences*: In principle the potential to reach large audiences is limited. However, if mass media penetration is low or if mass media are extremely fragmented, group meetings may effectively reach more people, directly or indirectly, than mass media.
- *In-depth dialogue*: Group and bilateral meetings allow for interactive (synchronical) discussion between communication workers and audiences, which means that all parties involved can bring relevant experiences and ideas, and receive appropriate feedback. This means that issues can be dealt with in depth and considering various aspirations (see section 5.2.1), particularly when groups are small. Thus, these kinds of meetings can also discover questions and issues that play a role in a problematic context, but which are not immediately clear at the outset (i.e. they are suited to revealing 'the question behind the question').
- *Active learning and opinion change*: In connection with the previous point, some interpersonal forms of communication are particularly suited to supporting active learning and opinion formation/change in connection with problem-solving.
- *Audience skill*: Interpersonal communication may require considerable abilities and social skills on the part of audiences/participants, especially when dialogue is attempted in a group setting. People need to be able and willing to articulate their thoughts, listen to others and respond constructively.
- *Tailor-made*: Bilateral meetings or small group meetings make it possible to deal with the specific problems, concerns and circumstances of individuals.
- *Pre-assumed interest*: Taking part in a group meeting or bilateral meeting means a relatively large investment of time for audiences. Before getting involved, therefore, people usually need to be interested in the subject concerned. Although interpersonal communication can help to reinforce awareness and interest, its potential for mobilising attention and interest on a significant scale is limited.
- *Collective issues*: Group meetings in particular allow change agents to tackle problems that can only be resolved by collective action and/or simultaneous behaviour change. Such meetings also make it possible to capitalise on group processes and group pressure in furthering change.

- *Insight into audience responses*: When using forms of interpersonal communication, change agents can in principle get feedback on the way people process and interpret messages, and can adjust and adapt messages when needed.
- *Time flexibility within limits*: Group and bilateral meetings allow for some flexibility in time; the timing of meetings can be adjusted to suit different groups or individuals. However, as the preferences of different sub-groups (e.g. male and female farmers) can differ significantly, it may be difficult to schedule meetings at times suitable for all. In any case, interpersonal communication is less flexible in terms of time than, for example, written mass media.
- *High spatial flexibility*: The spatial flexibility of interpersonal forms of communication can be high; that is, meetings can take place in diverse environments, such as farmers' homesteads, farmers' fields, offices, markets, community centres, etc. This makes it possible to adjust the meeting place to the preferences of audiences and/or the requirements of the issue at hand. For technical agricultural issues in particular, there can be many advantages to meeting in farmers' fields (Van Schoubroeck, 1999; see also section 13.3.2).
- *High relational support*: Through interpersonal communication it is possible for change agents to develop relationships of trust with particular audiences, and/or to express personal involvement with people and issues. This may improve one's credibility as a source of information, and help to stimulate people to tackle certain issues or overcome difficulties.
- *Interpersonal skills*: The use of interpersonal communication pre-supposes that change agents have interpersonal skills (see section 12.2.3). In principle, such skills should be available in any organisation aiming to stimulate change.
- *High costs*: The cost per person reached is usually relatively high when compared to mass media. However, interpersonal communication has a much greater potential for fostering active (individual and collective) learning and opinion forming, than mass media, so these extra costs may well be justified.
- *Low storage capacity*: Group meeting and bilateral meetings cannot be easily 'stored'. One could make a recording (e.g. on cassette) of a meeting, but this is rarely done, partly because it is time-consuming and troublesome to re-listen to a meeting, and because it is difficult in advance to determine whether the meeting is worth recording. However, if audiences are literate, they can make notes at a meeting to remind them of the main points, or they can be provided with a handout giving these points.

From the above we can conclude that group meetings and bilateral meetings have similar qualities, and are essential if the purpose is to deliver tailor-made advice and/or to stimulate active learning and opinion forming. Group meetings are especially useful in contexts where:

- Farmers can, and want to, learn from each other's experiences (e.g. when many farmers have similar problems);
- Problem-solving requires collective action among farmers and/or negotiation and co-ordinated action among a variety of stakeholders;
- Funding opportunities and/or the number of clients do not allow for bilateral meetings with all farmers.

In contrast, bilateral meetings can be particularly useful when:

- There is a need to tackle a specific and unique problem;
- When problems are too sensitive and emotionally laden to discuss in a group;
- When problem-solving has no collective dimensions;
- When funding arrangements and/or the number of clients allow for bilateral meetings.

12.2.2   Basic modes of administrating group and bilateral meetings

There are many different types and purposes of group and bilateral meetings (see also section 12.2.1). However, there are some basic modes and principles by which meetings can be shaped (see Windahl et al., 1992; Heymann, 1999; Van Woerkum et al., 1999):

- lecturing;
- diagnosis-prescription;
- counselling;
- participatory;
- persuasive.

- *Lecturing*: An often used mode of organising group meetings is to give a lecture about a particular topic, and allow for questions and/or discussions (in plenary or in small groups) afterwards. Here the change agent is in a rather dominant position, while the audience remains relatively passive. In some cases such meetings can be functional, and even asked for by farmers or their organisations. Lectures in general should not last longer than 20 minutes, as it is very difficult to maintain attention beyond that point. If possible it is important to explore the specific interests and needs of the audience in advance. Reserving sufficient time for discussion and debate is crucial for processing and clarifying the contents of the lecture, and for the stimulation of active learning.

- *Diagnosis-prescription*: The diagnosis–prescription mode of conducting a bilateral or group meeting resembles the method that medical doctors often use to communicate with patients. They ask the patient a series of questions, the sense of which the patient may not understand. On the basis of the answers given, the doctor diagnoses the nature of the disease and/or the likely causes of the problem and gives a prescription or advice. Change agents who have had only technical training are often inclined to use this mode. In such meetings, the change agent is clearly in charge, while the client remains rather passive and dependent. Since the client has little insight into the knowledge that is being used to structure the meeting and ask questions, it is difficult for him or her to contribute actively. In addition, little learning takes place at the end for the client as he or she merely receives a 'prescription'. This mode of conducting a meeting is only useful if all parties involved feel comfortable with it and agree that the change agent is the one with the appropriate expertise. In some cases, farmers may just want advice, without having to think much and learn actively. In a study by Stolzenbach and Leeuwis (1996), for example, it appeared that some farmers preferred this mode of operation in connection with government enforced environmental protection and mineral management measures. They did not like the government policy, and hence were not prepared to invest much in learning.

- *Counselling*: Counselling is a method that has been developed for psychotherapeutic purposes (e.g. Rogers, 1962). Here the client does most of the work. The client determines the topics discussed, while the communication worker mainly listens to the client, tries to order the information and asks questions, but refrains largely from giving his own views. The idea is that clients can discover and explore their own aspirations (including emotions, see section 5.2.1) and the problems associated with these, and find possible solutions themselves. In agriculture, this mode of operation may be useful when problems have an emotional dimension (e.g. in the case of farm succession, family problems and/or migration) and/or when people are unhappy about their situation but find it difficult to define why (e.g. when people lose enthusiasm for farming and do not know what to do about it).

- *Persuasive*: Frequently, change agents want to convince their audiences of the relevance of certain problem definitions, goals, practices, technologies, policies, etc. In practice, this typically results in stressing the adverse consequences of existing practices and views, and promoting alternative ones. Meetings then become easily focused on presenting arguments in favour of one thing, and against the other. The credibility and ethics of such a one-sided approach is questionable (see section 3.4). It may well be that painting a more balanced picture – i.e. also presenting the negative consequences of the suggested solutions, and conversely the positive dimensions of current practices – is eventually more effective, since it tends to contribute to central rather than peripheral forms of learning (see section 9.2; Martijn & Koelen, 1999). As well as aiming to convince people, persuasive strategies can also be used to try and remove obstacles to learning (see section 9.4). When discussing persuasive and interactive models of communication (see Chapter 4), we have already seen that persuasive modes of communication can be functional when *linked* to more interactive strategies (i.e. preceding, during or following interactive processes, see Chapter 11). When communication workers try to convince people of ideas and practices that have been developed elsewhere (see section 6.5) and without the involvement of those who are supposed to apply them, the chances are high that crucial issues and interests are overlooked, meaning that the presented solutions cannot be realistically applied. Persuasive styles of admininistrating meetings may be useful, especially to create awareness of problems of which people are unaware.

- *Participatory*: In a meeting run along participatory lines, both the communication worker and the clients play an active role in discussions towards problem-solving. Both parties bring relevant problem definitions, expertise, experiences, networks, etc., and try to arrive at an integration. The significance and importance of this mode of operation have already been outlined in Chapter 4. In many instances, the participatory mode is the preferred option.

It is important to note that communication workers cannot uniliterally decide about or control the way in which a meeting evolves, as this results from an interplay between different people. Nevertheless, it is important that they try to adapt their own way of operating to the context and preferences of clients, and do not let their own personal preferences and default inclinations prevail (see also section 13.3.1). However, in some cases it may be necessary to use directive (i.e. persuasive and diagnosis–prescription) styles even against the wishes of clients. This may, for example, be the case in matters of serious risk and/or when the interests of others are on the line. A communication worker, for example, may observe that a farmer's cattle are showing early signals of a deadly contagious disease, without the farmer noticing the problem or wanting to hear about it (see section 9.4). Here the communication worker has little option other than to resort to directive intervention at first. However, to really solve the problem and/or prevent it from happening again, other modes of discussion will be needed at a later stage. Thus, different styles of administrating meetings may complement each other at different times. At a particular time, communication workers may consider the following criteria when thinking about their style of operating:

- the clients' awareness of different aspects of a problematic situation;
- the consequences that a lack of attention may have for others;
- the extent to which farmers' knowledge and information is needed to further analyse problems and find appropriate solutions;
- the preferences of the clients.

Only when awareness and attention is low, and/or when farmers' knowledge can be expected to be irrelevant (which is seldomly the case), may it be worth working in a directive mode initially.

12.2.3   Skills needed for facilitating interpersonal communication

To facilitate active learning (see also Chapter 9) in bilateral and/or group meetings, communication workers generally need a number of specific skills. According to Heymann (1999) these include:

- *Empathy*: Communication workers need to be able to see things from the perspective of their clients. In other words, they must have a capacity to enter, at least to a degree, the 'life-world' (see Chapter 6) of their audiences, and connect with their modes of reasoning, aspirations, norms, values, feelings and context. This capacity is called empathy. We are not implying that a communication worker needs to agree and identify totally with a client's perspective, but he or she must at least be interested in it, try to understand it, take it seriously, respect it, and be willing to accept it as a point of departure.

- *Monitoring and active listening*: It is important for a communication worker to register and analyse the dynamics in a meeting; just listening and responding to verbal statements is not enough. One also needs to try and understand *why* certain statements are made (and others not) and register and interpret nonverbal signals (e.g. silences, facial expressions) in order to get an idea of emotions that play a role. On the basis of such 'active listening' one can intervene in the process if necessary, and try to make it more productive.

- *Self-reflection*: To play a positive role in meetings, communication workers must have a certain degree of self-understanding. They need to be aware of their own biases, goals, perspectives and feelings during a meeting, and to be able to 'correct' or ignore them if necessary (i.e. if following their own inclinations is likely to hamper the process).

- *Giving confrontational feedback*: We have already seen in Chapter 9 that confrontational feedback can be an important stimulant for learning. Therefore, providing good quality feedback (see Chapter 9) is an important task for change agents. Such feedback points to inconsistencies or tensions between people's aspirations and achievements, between what people say and what people do, between what people believe and what others believe, between dissonant cognitions and/or incompatible practices, etc. As receiving confrontational feedback can be threatening to people, change agents need tact and should avoid making people feel rejected or less valued as a person. A useful strategy can be to make sure that there is always a balance between positive and confrontational feedback (see sections 9.3 and 14.2.3).

- *Posing questions and probing*: Posing questions can be an important means of directing discussions in a meeting. Communication workers must be able to use questions strategically, as a means of focusing on or shifting to specific issues and/or persons. Closed questions are questions that can only be answered with 'yes' or 'no', and can be useful if a communication worker wants to make a point without 'losing control' in a meeting. Open questions allow a more elaborate answer, and can help to explore and clarify the experiences and views of participants, and also to place these in a new perspective. Another distinction is between direct and indirect questions. Direct questions can hardly be ignored, and can be embarrassing to people if they are not prepared to give an answer (e.g. 'How many cows do you have?'). In contrast, indirect questions ('It would be interesting to know how many cows people have around here') leave more space for people to tell whatever they feel is appropriate. To get to the heart of matters, it is often necessary to keep asking new questions in connection with already given answers; this is called probing.

- *Activating discovery*: According to the principles of experiential learning (see Chapter 9), it is important that change agents give space to participants to draw their own conclusions, and play an active role in setting the learning agenda and/or in administrating meetings. This means not only that a change agent may at times need to 'withdraw', but also that appropriate learning tools (e.g. demonstrations and experiments; see sections 13.5.8 and 13.7.1) need to be used.

- *Maintaining structure*: For people to feel comfortable in a group meeting, it is usually important that people are clear at an early stage about its purpose, agenda, duration, method, etc. (see also Chapter 15 on communication planning). In giving direction and structure to a meeting it can also be useful to think about it as having a beginning (getting to know each other, building trust, exploration, agenda setting), a middle (dealing with problems and issues) and an end (drawing final conclusions, deciding on follow-up). Each of these 'stages' needs sufficient time and attention, depending partly on the history of the group or contact. To maintain direction and structure during a meeting it is important that communication workers regularly provide summaries and/or conclusions as a means of closing a particular topic and/or providing a new start. It can also be important to invite participants to summarise and formulate conclusions, as a means of getting feedback from the audience.

- *Managing group dynamics*: It is important to recognise that all sorts of dynamics (including conflicts; see Chapter 10) can emerge during group meetings. According to Pretty et al. (1995), groups tend to go through four stages: forming, storming, norming and performing. In the 'forming' stage there tends to be little trust among participants, and people are searching for direction, purpose and identity. During the 'storming' stage differences between people (in terms of interests, goals, style, opinion, etc.) become much more visible, resulting in all sorts of tensions. Such frictions may be settled during the 'norming' stage, through creative solutions and acceptance of differences. Finally, a group may start to 'perform' and become productive, among other means by making optimal use of the different qualities of the group members (e.g. leadership skills, analytical skills, capacity to create a positive atmosphere, enthusiasm and energy, networking capacities, practical skills, etc.) (for more details see Pretty et al., 1995). In dynamic situations, the stages described by Pretty et al. may best be looked at as a cycle that is in continuous motion. Communication workers need to be able to facilitate processes of group development (see also section 14.2.3), using the skills mentioned above as well as insights on learning, negotiation and process management elaborated on in Chapters 9, 10 and 14.

12.3   Hybrid media: the internet

New media have emerged that tend to combine the functional properties of mass media and of interpersonal communication, in that they can potentially reach large numbers of people in many different locations, but at the same time support a level of interactivity that is higher than with conventional mass media. Most of these hybrid media are based on computer technology. This is why they are often referred to as information and communication technologies (ICT) or 'new media'. The costs involved in using such technologies are declining rapidly (Shapiro & Varian, 1999; World Bank, 1998), which means that their potential for communicative intervention is increasing. In the early 1990s there were a large variety of 'new media', and for each of these the user needed to have special software and hardware in order to use them. Examples are CD-ROM (Compact Disc – Read Only Memory), CD-i (Compact Disc – interactive), videotex, expertsystems, electronic conferencing systems, etc. Since the early 1990s most of these systems have become accessible and/or integrated into one electronic platform: the internet. The user deals with a relatively limited amount of hardware and software compared to the past, even though behind the screens, literally, all sorts of different techniques remain to play a role. For simplicity, therefore, we have chosen to deal in this section with the internet and its different functions and opportunities. We will refer to 'internet applications' to indicate that there can be large functional differences between, for example, different *websites* on the internet, even if they resemble each other technically.

12.3.1   Modalities of the internet

The internet can be sub-divided into five basic modalities, which can be flexibly used and combined to construct internet applications for communicative intervention purposes:

- *World wide web*: Many organisations and individuals nowadays have a *website*. This is essentially an advanced multi-channel (textual, auditive, visual) brochure (or book), that can be 'opened' at a specific electronic address, i.e. a computer that is connected to a worldwide computer network. The 'brochure' can contain just text pages, but can also include animations, pictures, video clips, voices, sounds or music. To open different 'pages' of the brochure, users 'click' (with the computer mouse) on specific words in the text or on special 'buttons' and through such 'links' can jump to and fro through the content. Such links also exist between websites, and so it is possible, for example, to directly 'jump' (surf) from the website of a company in India to the website of a related company in the USA, provided that the designer of the Indian website included such a link. As we discuss later, all sorts of search facilities exist on the world wide web, which allow users to look for websites on specific topics.

- *Electronic mail*: In addition to websites there are electronic mailboxes. Individuals and organisations can use these mailboxes to send and receive electronic messages. Specific mail is sent from one address to one or more others, as with the postal service, but at much higher speed and with much lower variable costs.

- *Newsgroups*: Newsgroups can be seen as thematic mailboxes that can be accessed publicly. Their are thousands of such groups, each with their own specific theme, such as 'traditional South African music' or 'indigenous knowledge'. Everybody who is interested can open such a central mailbox and look at the messages that others have posted. They can also reply to such messages, so that worldwide exchanges take place on a specific topic. There are also more restricted electronic platforms for exchange and discussion, to which users subscribe.

- *Chatrooms*: Electronic mail and newsgroups work in an asynchronous fashion; the communicating parties need not be active on the internet at the same time. Chatrooms (or Internet Relay Chat, IRC), however, are synchronical. Those who open a specific chatroom or channel communicate with each other directly online. When a participant types a line or sentence, it appears simultaneously on the computer screens of everyone who is logged on. Then participants across the world can react to it so that a group conversation emerges.

- *File transfer*: Through the internet one can transfer not only messages, but also electronic packages (files). Possibilities for file transfer (containing text, music, pictures, etc.) are often incorporated into other modalities. It is possible, for example, to send files along with electronic mail, to collect (download) files from websites, or to send files (e.g. pictures) to people that one is chatting with.

12.3.2   Functional qualities of hybrid media

Due to their hybrid nature it is not surprising that hybrid media resemble basic forms of interpersonal communication in some respects, and mass media in others. In some functional qualities they have an intermediate potential. We have summarised this in Table 12.3.

Table 12.3   Potential functional qualities of hybrid media in comparison with those of mass media and basic forms of interpersonal communication, from the perspective of communicative intervention.

Potential functional quality	Resemblance to mass media or interpersonal forms	Clarification
Audience reached	like mass media	• potentially a worldwide audience (if access exists; this varies greatly across countries and social strata*) • less indirect spin-offs than, for example, television as messages are often not received in groups
Mobilising attention	less than other media	• hybrid media tend to be less visible in everyday life • difficult to find relevant information due to overload
Specificity/ tailor made/ active learning	in between	• some degree of specificity is possible in case of (prestructured) man–computer interaction • less intensive exchange in case of man–man interaction
Relational support	like mass media	• difficult to establish relationships of trust due to limited social presence
Insight into audience	in between	• audiences can respond to messages through e-mail • users' way of using hybrid media can be registered
Speed/ actuality	faster than mass media	• news and actualities are often available on the internet before they are broadcast by radio/television • websites/programmes can be centrally updated, and be immediately available to everyone
Time flexibility	like written mass media	• internet can be consulted whenever it suits the user
Spatial flexibility	often like audio-visual mass media	• equipment, electricity and network access are often tied to locations (but mobile equipment is spreading)
Storage capacity	like written mass media	• all messages received can be stored on a computer or printed, and accessed again if needed
External/ internal skills	like mass media	• a simple website can be built easily by internal staff, but advanced applications require special skills
Dependence on others	much less than mass media	• with the internet everybody has in principle his or her own broadcasting station and editorial board
Costs per person	mostly in between	• development and maintenance costs of hybrid media can be rather high

* See World Development Report 1998/1999 (World Bank, 1998)

As well as some resemblance to other media, hybrid media have some fairly unique capabilities:

- *Modelling and processing*: Computers are typically suited to performing complex calculations rapidly, and/or making selections from large quantities of information (using, for example, search criteria). This offers various possibilities. Scientists can, for example, represent their knowledge and views about a complex problem in a computer model that can test or calculate certain solutions and/or make diagnoses or projections about the future. With a simulation model of the way wheat grows and a database of the rainfall distribution in different months in the past 30 years, it may be possible to predict the relationship between date of sowing and crop yield for different varieties (Hamilton, 1990). In fact such computer models (distributed, for example, through the internet or CD-ROMs) can be seen as a complex mode of communication between the builders of the model and those who use it (Leeuwis, 1993). With the help of hybrid media it is thus possible to make use of a wider range of modes of reasoning and expertise than might otherwise be possible; after all, it would usually not be feasible to bring a group of international scientists to a specific problem, whereas it may be possible to access their ideas by using a computer model built by them (see sections 13.3.1 and 13.5.10 for a more detailed discussion of risks and opportunities of modelling).

- *Multi-channel communication*: The internet's capacity to combine textual, auditive, visual and modelling modes of communication can considerably enrich the way messages are communicated. One can, for example, clarify written descriptions (e.g. what a particular bacterial disease looks like) or instructions (e.g. how to overhaul a tractor engine) with the help of pictures, video fragments and/or animations.

- *Anonymity*: Communication through hybrid media can be much more anonymous than through other forms of communication (e.g. interpersonal). When interacting with other people through hybrid media it is quite easy to hide one's identity, or even to take on a fake or imagined identity (Turkle, 1995). In many cases this may be detrimental to the quality of the interaction, but in some cases it can also be an advantage. Communication on sensitive issues (e.g. sexuality, disease, political controversy) can sometimes be much easier if there is a degree of anonymity.

12.3.3   Internet applications for communicative intervention

The internet has a wide range of applications, many of which are related to communicative intervention (in various societal domains, including agriculture and resource management). Although these may use similar modalities of the internet, they can serve very different purposes (as is the case with conventional media; see also Chapter 15 on communication planning). In the early days of hybrid media it was expected by some that these media would, in industrialised societies, eventually make agricultural communication workers redundant, as the workers' function would be

Table 12.4 Different types of internet/ICT applications for communicative intervention.

	Search and access	Memory and feedback	Advisory	Self-help and request	Public debate
Underlying idea	• providing efficient access to information	• securing adequate feedback for learning	• providing tailor-made advice	• exchanging similar experiences elsewhere	• inventory and/or evaluation of opinions
Communication for innovation strategy/function	• information provision	• exploration and awareness raising	• advisory communication	• supporting horizontal exchange	• generating (policy) innovations
Key means used in the application	• search and selection procedures	• registration, manipulation and representation of information	• calculation, optimisation, simulation and reasoning	• facilities for networking	• facilities for networking and debate structuration
Source of information	• information providers	• end-users	• end-users and 'experts'	• unknown receiver	• information providers and participants
Main role of communication worker in utilisation	• supplying and updating information • creating weblinks	• discussion partner in interpretative process	• discussion partner • co-user • adapting advice	• referring to relevant others • information supplier	• supplying information • facilitating the debate
Typical problems (to be prevented or corrected by communication specialists in the development or maintenance process)	• search procedures do not match the logic and knowledge of users • information overload • ageing of information • biases in information found	• the feedback given does not connect well to dynamic learning interests	• the validity of the underlying model is dubious • complexity and black-box character cause interpretation problems	• difficult to find an appropriate electronic community	• quality of the debate is limited • difficult to reach agreement • biased composition of participants

largely taken over by computers and computer models (IBM, 1988). More recently, it is recognised that this will not be the case. On the contrary, communication workers are increasingly regarded as a critical factor for the successful use and introduction of hybrid media (Nitsch, 1990; Klink, 1991). For example, they have an important role to play in the development and maintenance of internet applications, they can help to improve the quality of debates on the internet, and they can assist users in the finding, selection, processing and interpretation of information. Thus, communication specialists remain important, even if their role may change in situations where the internet becomes a common medium for communication.

In the sphere of communicative intervention, the most significant internet applications are:

- Search and access applications;
- Memory and feedback applications;
- Advisory applications;
- Self-help and request applications;
- Public debate applications.

In Table 12.4 we have summarised some relevant aspects of these types of applications. They will be discussed in more detail in Chapter 13 where we discuss a wider range of specific methods that are relevant to different forms and types of communicative intervention.

12.4   Media access and audience selectivity

When considering the use of media it is important to realise that media choices can have 'political' implications in the sense that they are to the benefit of some and to the disadvantage of others. This is because each medium assumes the availability of certain resources by audiences, for example:

- equipment (e.g. radio, television, computer hardware and software, etc.);
- skills (e.g. literacy, verbal skills, writing skills, computer skills);
- knowledge (e.g. to adequately interpret information about hybrid media available on the internet);
- time (e.g. time to spare on income generation, in order to attend a meeting);
- facilities (e.g. transport to a meeting, childcare);
- permission (e.g. consent from a spouse or religious leader to attend a meeting);
- money (e.g. to pay for internet access or subscription to a journal).

Depending on the availability of such resources, different categories of people may have more or less access to specific media. In other words, by choosing a particular medium one is likely to 'select' a specific audience. With group meetings, for example, one 'selects' those who have time, and during the interaction the quick thinkers and the verbally strong can easily dominate discussions. In contrast, an asynchronical debate on the internet may allow for more egalitarian participation by those who have no time to attend, need time to think and/or find it easier to write down their ideas instead of presenting them orally (provided that they are computer literate) (see Leeuwis, 1999c). These kinds of media selectivity cannot be avoided, but they may be ameliorated with special techniques. To stimulate more 'equal' participation in a group meeting one can, for example, split it up into smaller groups for a time, or give all participants cards on which to write their views as an input to plenary debate (see section 13.5.3). Similarly, in the case of an internet debate one may provide special facilities (training, public internet connections) for people without computers. In any case, it is important for communication workers to reflect on the significance of unequal media access in a particular context, and to take action to avoid negative consequences.

12.5   Media mixes

Most communicative intervention activities and/or programmes will need to use different media simultaneously or in succession. Media combinations are often necessary because (a) different sub-audiences have different media preferences and/or media access; and (b) because different media have different potential qualities. For example, it will often make sense to use interpersonal communication or hybrid media to stimulate active learning and opinion forming about an issue, but to raise attention about an issue in the first place, and/or to announce that certain activities take place, some kind of mass medium may have to be employed (for more detail see Chapter 15 on communication planning).

Questions for discussion

(1)   Identify a recent media hype. How and why did it take on the proportions of a hype? Do you think it is feasible to deliberately 'orchestrate' such a hype?

(2)   Observe a facilitated group discussion (real life or on video/television). Evaluate the performance of the facilitator in terms of the skills mentioned in section 12.2.3.

(3)   Imagine that you are preparing group meetings for youngsters in the context of an AIDS prevention programme. Assuming that many youngsters have access to the internet, what are the pros and cons of organising the 'group meetings' on the internet as opposed to organising 'real' group meetings?

---

13   Communication for innovation methods

Although it is important to understand in general terms the potential of different communication media (see Chapter 12), in practice communication workers need not immediately worry about basic media. This is because specific ways of using and combining media have been developed, which – depending on their scope and horizon – we call methods or methodologies. Often, methods are a more relevant entry point than media when thinking about ways of organising communication in a specific intervention context. For each communication for innovation strategy or function, change agents can make use of various methods that have proved to be relevant in such a context. In this chapter, we first clarify the relationships between approaches, strategies, media and methods for communicative intervention. We then proceed to identify and discuss several key methods and/or methodological strategies in relation to each of the communication for innovation strategies and functions that were identified in Chapter 2.

13.1   Clarifying the terminology used

By now we have introduced a number of terms to distinguish between different forms of communicative intervention. It is at this point important to clearly position them in relation to each other. We can depict the interrelation in terms of interrelated 'hierarchies', with the lower levels forming the 'building blocks' of the higher levels:

- *Strategies (or services)*: Different communication for innovation strategies can be distinguished on the basis of their *wider intervention purpose*, which again relates closely to the *assumed nature of the problematic situation*. In Chapter 2 (see Table 2.1) we have differentiated between six basic strategies or services:
  (1)   advisory communication;
  (2)   supporting horizontal knowledge exchange;
  (3)   generation of policy and/or technological innovations;
  (4)   conflict management;
  (5)   supporting organisation development and capacity building;
  (6)   persuasive transfer of policy and/or technological innovations.

- *Functions*: Within each strategy, different communication for innovation *functions* may be relevant. Basically, these functions relate to different *communicative sub-goals* that are deemed useful as part of the wider strategy. We have distinguished four such functions (see Table 2.2):
  (1)   raising awareness and consciousness;
  (2)   exploration of views and issues;
  (3)   information provision;
  (4)   training.

Box 13.1 The relationship between communication for innovation strategies, functions and approaches.

Level 1	communication for innovation strategies/services
Level 2	can involve several functions
Level 3	strategies and functions can be identified and/or approached instrumentally or interactively
Levels 4 to 7 (see Table 13.2)	strategies, functions and approaches can be implemented with the use of various methodologies, methods, tools and media

- *Approaches*: With the term *approach* we refer to the basic *planning philosophy* that is being adopted by communicative interventionists. Communication for innovation strategies and functions can be initiated and/or organised on the basis of an *instrumental* or an *interactive* mindset, in a context that allows or does not allow for an interactive approach (see Chapter 4).

The above is summarised in Box 13.1. Activities with regard to strategies, functions and approaches can, in turn, involve various media, methods, tools and methodologies, and these have a similar hierarchy:

- *Methodologies*: For each communication for innovation strategy or function, several *methodologies* are usually available. Methodologies are basically pre-defined *series of steps, procedures and activities*, with each step involving the use of one or several methods (see below). Methodologies are often known under a particular label or acronym. Examples of methodologies relevant to communicative intervention in rural settings are:
  - Farmer Field Schools (FFS; Van de Fliert, 1993; Scarborough et al., 1997);
  - Participatory Rural Appraisal (PRA; Chambers, 1994a);
  - Participatory Technology Development (PTD; Jiggins & De Zeeuw, 1992);
  - Rapid Appraisal of Agricultural Knowledge Systems (RAAKS; Engel & Salomon, 1997).

- *Methods*: Methods can be seen as a particular mode of using media and media combinations within the context of a confined *activity*. A method can (but need not) be an element in a methodology. Examples of methods include:
  - a farm visit;
  - a workshop;
  - a group discussion (as an element of, for example, FFS);
  - priority ranking (as an element of, for example, PRA).

- *Tools and techniques*: Tools and techniques are particular ways of operating a method. Whether something is considered a method or a tool is often debatable; the boundaries are not so sharp. A ranking exercise, for example, can involve drawing a matrix in the sand and using pebbles or stones as counters, or it can be conducted on a sheet of paper using stickers or markers. Similarly, a farm visit in which a farmer's problem is discussed can be conducted in various modes (see section 12.2.2):

- diagnosis–prescription;
- persuasive;
- counselling.

- *Media*: Mass, interpersonal and hybrid media are basic devices that help to combine different communication channels for the 'transportation' and exchange of textual, visual, auditive, tactile and/or olfactory signals. Hence, different media can be used in the context of methods and methodologies. We have discussed basic media and their potential in Chapter 12.

The above is summarised in Box 13.2.

Box 13.2 The relationship between methodologies, methods, tools and media.

Levels 1 to 3 (see Table 13.1)	communication for innovation strategies, functions and approaches
Level 4	methodologies
Level 5	incorporate several methods and activities
Level 6	methods can be operated using various tools and techniques
Level 7	methods and tools can involve several media

13.2   Reasons to focus on methods, functions and process management

In the main part of this chapter (sections 13.4 to 13.7) we present a selection of specific communicative intervention *methods* that can be relevant to each of the communication for innovation *functions* mentioned above (see also Tables 2.1, 2.2).

Focusing on methods rather than methodologies

We have various reasons for focusing on methods (as associated with confined activities) rather than on methodologies. The most important one is our experience that methodologies are often followed like a recipe. Frequently, change agents simply follow the steps and procedures described in some kind of manual, applying the methods and tools without reflecting on whether they fit the situation or not. In our view, change can never be achieved so mechanically. The idea of developing or using a pre-defined series of steps to foster change reflects a certain amount of 'blueprint planning' at the process level that we feel is fundamentally flawed (see Chapter 4). As we have argued, change processes are inherently context-specific, messy and conflictual (see section 4.2.1 and Chapter 10), and hence cannot be guided by a fixed series of steps and procedures. Apart from this, most of the existing methodologies turn a blind eye to conflict as an inherent aspect of change, and fail to offer appropriate methods for dealing with it (see Leeuwis, 2000a). In conclusion, the activities of interventionists will have to be developed and designed as the process unfolds. That is, interventionists will have to construct their own contextual and once-only methodologies. For each activity, however, communication workers will have to choose and/or adapt certain methods and tools. Of course, methods too may be

applied 'mechanically', but since their time-span is much shorter than that of methodologies, focusing on activities and methods increases the chances that emergent
dynamics are taken into account. Some basic principles for selecting and adapting
methods are discussed in Chapter 15 on communication planning.

Focusing on functions

Most methods can be associated with general communication for innovation functions such as raising awareness and exploration, and can be useful in several strategies. Hence, instead of discussing methods per communication for innovation strategy,
we mainly group them around functions. We make an exception for some methods
that are particularly useful in 'individual' farm management communication (see
Table 2.1); these are discussed in section 13.3. It makes little sense to have a separate
discussion of methods in relation to the communication for innovation services that
work towards co-ordinated action (see Table 2.1) because in all three cases we are
dealing with a variety of stakeholders who may have different views, values and
interests. Thus, from a methodological point of view, the situations have similarities.
For example, although conflict is the starting point in situations where change agents
aspire to play a role in conflict management, tensions and conflicts are likely to
emerge during the generation of innovations and organisation development as well.
And although organisation development usually involves actors that belong to the
same category of stakeholders (e.g. farmers), many different views and interests
may exist among them. Thus, the difference with the two other services, which are
typically multiple stakeholder situations, is only gradual.

Focusing on process management

We have already suggested in Chapter 10 that collective change and innovation is
best supported through the facilitation of joint learning and negotiation processes in
and around a more or less permanent 'platform' (see section 2.2.1). In connection
with this we have identified several facilitation tasks (see section 10.2.1). Guidelines
and principles for each task will be discussed in the context of *process management*
(see Chapter 14). At that point we will also discuss issues relating to stakeholder
analysis and selection.

   The methods discussed in the following sections are not exhaustive, and methods
are discussed only in essence. However, references are made to other methods and
more detailed descriptions.

13.3   Specific methods and issues related to farm management communication

In addition to the methods discussed in relation to general communication for
innovation functions (see sections 13.4 to 13.7), we discuss, in this section, some
specific methods that may be relevant to farm management communication. As we
have discussed in section 2.2, farm management communication involves *advisory*

communication and *supporting horizontal knowledge exchange* (see Table 2.1). Both strategies have in common that they are suitable in 'individual' farmer's problem situations which do not require collective action beyond the household.

13.3.1 Advisory communication

As described in section 2.2, we refer to advisory communication when a farmer takes the initiative to seek the assistance of a communication worker in solving management problems. Below, we will discuss some specific concerns and methods about this type of communication for innovation service.

Agreeing on an appropriate medium and spatial context

Advisory communication can take various forms, depending on the media used and the spatial context, for example a meeting in a communication worker's office during consultation hours, a telephone consultation, an exchange via the internet (i.e. in 'cyberspace'), or a discussion in a public space or a farm visit. As we have already discussed in Chapter 12, each medium has its potential and limitations. Different clients, and problems with different characteristics (e.g. clarity, scope, history, uniqueness, complexity, sensitivity, etc.), pose different demands on the type of discussion needed, so it is important during a first contact to explore what kind of meeting makes sense. During a telephone conversation, for example, it may become clear that it would be better for a communication worker to visit the farm because it is impossible to give proper advice without additional contextual information. Visiting farmers and their fields often contributes to a proper diagnosis, and has an additional advantage that it may enhance farmers' own observation and problem-solving skills (Benor & Baxter, 1984). At the same time it is important to consider and respect that a farmer may have specific reasons (e.g. to maintain a certain amount of anonymity, cultural preferences, saving money) for establishing contact through a particular medium (e.g. telephone) and may prefer to continue in that way, even if a communication worker might feel another medium was more appropriate. In many cases, however, agreeing on an appropriate medium and spatial setting is non-problematic or even automatic, but it is nevertheless important to be aware of possible tensions. It is important that advisory organisations have a variety of clearly announced modes through which they can be contacted for advisory communication.

Tuning an advisory meeting

We have already discussed some basic ways of structuring and administrating meetings, as well as the importance of several skills for this (see sections 12.2.2 and 12.2.3 respectively). Here we will assume a situation that does not require and/or allow for some kind of directive advice and discussion mode (see section 12.2.2), in other words a situation where communication workers would be wise to adjust their advice and mode of advising to the client and his or her situation in order to prevent mutual frustration. In such situations it is first important to find out during the early stages of a meeting what exactly the farmer expects from a communication worker.

What mode of administrating the advisory meeting does the farmer prefer (see section 12.2.2)? And how far has the farmer already progressed in identifying and solving the problems? Does he or she want a discussion partner to help structure his or her thoughts and arrive at a jointly agreed solution? Or just a second opinion on a solution that has already been identified? Or does he or she want a fully fledged analysis, and advice on how to solve the problem from the advisor's goals and point of view? Furthermore, employing the skills mentioned in section 12.2.3, a communication worker will have to find out the nature of the problems that a farmer experiences, i.e. which tensions exist vis-à-vis which aspirations and priorities (see section 5.2.1). In doing so, one may at times have to probe beyond the original question or problem posed by the client, and look for 'the question behind the question'. At first glance, for example, it may seem that a farmer wants to discuss the pros and cons of different investment options, while the underlying issue is whether to continue farming or not. Moreover, it is crucially important to realise that there is considerably diversity in different farmers' goals, aspirations, priorities and circumstances (see section 5.2.5), so that one cannot simply assume or impose 'standard' goals and aspirations (such as, for example, profit maximisation). A farmer's logic of farming is something to be explored and discussed, which requires considerable empathy and other communication skills (see section 12.2.3).

In principle, the problem-solving model discussed in section 4.1.1 can serve as a checklist for a discussion, provided that sufficient account is taken of its limitations (see sections 4.2 and 5.3.7). In thinking about technical solutions for a 'problem', it is important to consider their social implications and dimensions (see sections 5.3.5 and 8.2.1). More generally, it is essential to consider that, due to the co-ordinated nature of farming, a solution in one farming domain may well create problems elsewhere on the farm. Thus, a holistic outlook is required. Although a problem-solving process usually starts at a certain level and within a certain domain (see section 5.1), it is imperative to regularly '*switch*' the attention from one domain or level to others during a problem-solving process. This means, in turn, that one needs to '*translate*' the problems, practices and solutions in one domain or level to others (Stolzenbach & Leeuwis, 1996). This is by no means a simple task as many uncertainties may exist. Moreover, it requires a flexible mind as well as considerable insight, foresight, experience and intuition.

Written and computer-based memory and feedback applications

In problem-solving processes, it is vital to have access to information about the farm. Although quite a bit of information can be derived from the human memory, the analysis of problems on a farm can benefit immensely from detailed and precise information. Not suprisingly, many farmers collect and record information for future learning and problem-solving (see Leeuwis, 1993). Such information may include information about:

- weather and climatic circumstances (temperature, rainfall, etc.);
- technical practices in various domains (sowing dates, weeding dates, introduction of biological pest controls, use of pesticides, fertiliser use, etc.);

- production (crop growth, animal weight gains, crop yields, quality of products, etc.);
- specific production units (milk yield per cow, veterinary treatments per cow, production per field);
- financial and economic aspects (prices obtained, spending, labour costs, etc.).

It has been recorded that, in addition to financial bookkeeping and administration, some glasshouse horticulturists in the Netherlands registered and stored more than 80 different parameters about their enterprises *per week* (Leeuwis, 1993). With the help of such information, farmers can make calculations and analyse trends over time by comparing information about current performance with that of previous periods. In essence, such information can be used as *feedback* for learning and problem identification, and can play a role in problem analysis. Thus, registering information about one's own farm can be a useful activity, which can be supported and/or encouraged by communication workers in various ways. Communication workers may, for example, play a role in identifying relevant parameters for registration, or in developing appropriate methods for measurement. Moreover, they can engage in discussion with farmers about the interpretation of such information, relating figures to the practices from which they were derived. In general, it is very important for farmers to engage in dialogue and have a 'sparring partner' when trying to make sense of registration materials.

It is important to note here that there are different carriers on which registration information can be stored. These range from a simple notebook or registration form to sophisticated computer equipment. In the latter case, management literature refers to management information systems (MIS), but we prefer to speak of electronic memory and feedback applications (see section 12.3.3). An important difference with search and access applications (see sections 12.3.3 and 13.6.1) is that here the end-users themselves are 'filling' the system and providing the information. Many computer-based memory and feedback applications can do more than just store information. Usually, they incorporate facilities for selecting, manipulating (e.g. calculating) or representing (e.g. graphs) information. These allow farmers to 'play around' with large amounts of information more easily than on paper, which in turn can greatly enhance analytical activity and learning (Leeuwis, 1993). Also, they can sometimes be connected, for registration purposes, with automated devices in production (e.g. an automated milking or feeding machine).

For written and computer-based memory and feedback systems to be of use, it is crucial that relevant information is registered and recorded. This can be a rather time-consuming activity, and it is complicated by the fact that it is difficult to assess in advance what may be relevant information in the future. Here we touch on a major weakness of memory and feedback applications. As we discuss in more detail in section 13.6.2, the information and feedback that people want and need tends to be a 'moving target' since learning progresses and situations change continually. At the same time, memory and feedback applications often make use of an – at least temporary – fixed list of registration variables and parameters, which is almost inherently based on yesterday's expectations and problem perceptions. It is essential that the contents of memory and feedback applications can be flexibly adapted over time.

Computer-modelling and advisory applications

We increasingly see that computer programs (on the internet or CD-ROM) go beyond the provision of feedback, and are used to generate advice. In generating advice, some of these advisory applications make use of information in a memory and feedback application, and in other cases user-specific information needs to be entered through an electronic questionnaire.

A crucial element in such applications (which in management literature are often referred to as decision support systems or DSS), is a model attributed with analytical predictive qualities. There are different forms of such models. In its most simple form an advisory application contains a rather trivial *calculation model* with the help of which one can, for example, calculate how many bricks are needed to build a wall of 24 m^2. Many advisory applications, however, comprise much more complex calculation models. With the help of *simulation models* (e.g. crop-growth model), it is possible to make projections on the basis of hypothetical interventions. A farmer may, for example, ask an integrated economic and crop-growth simulation model how production and income will be affected by adding more fertiliser. Often, such a simulation exercise results in an evaluation or advice concerning the desirability of such an intervention. Other advisory applications encompass *optimisation models* (for example a model for solving logistic problems), which – by means of linear programming or other operations research techniques – generate advice on the 'optimal' allocation of means, given some previously defined set of objectives. A third type of model which frequently underlies advisory applications is *diagnostic models*, for example a model for diagnosing and remedying diseases. In contrast to simulation models and optimisation models (which merely carry out complex calculations), diagnostic models often make use of 'reasoning' (or 'artificial intelligence') techniques. In such 'reasoning' applications, knowledge is not represented in the form of algorithms but in the form of heuristic statements (or rules of thumb) of the type:

IF-<criterion x>-THEN-<condition y>*<certainty factor>

When such systems are aimed at capturing the knowledge of one particular expert, they are usually called 'expert systems'. When the aim is to model all available knowledge in a particular domain they are often referred to as 'knowledge systems'.

It is important to recognise that computer-based advisory applications can have certain strengths, but also many weaknesses. Possible strengths include that the underlying computer models can incorporate and link formalised knowledge from different disciplines, and allow for the making of complex calculations that otherwise would never be realistically carried out. Moreover, with simulation models especially, it is sometimes possible to engage in a series of 'virtual' experiments that could never be carried out in practice due to, for example, time constraints or risks involved (see section 13.5.8). Thus, computer models may in principle provide additional modes of exploration during advisory communication, and help to accelerate learning in certain domains. An important weakness of computer models is that, despite their seeming complexity, they always deal with problems and issues in a simplified way, and often with certain *biases*. Computer models are always select-ive in terms of the knowledge, variables, goals and possibilities for manipulation that they include, and hence they tend to overlook implications for aspirations and domains that are not considered. A simulation model may, for example, advise that certain farming practices are technically and economically feasible, but fail to take into account that their 'social–organisational' feasibility is very limited due to a lack of available labour or inputs. Moreover, computer models inherently include various assumptions (e.g. about biophysical processes and interrelations) that may be valid in the locality of those that developed the model, but which may not be accurate in other contexts (see Chapter 6). Thus, the validity of models in specific contexts is often dubious. Finally, from the perspective of end-users, complex models usually lack *transparency* and remain a 'black-box'. This limits the scope for learning, and makes it difficult for them to verify, understand and/or correct the outcomes of the advisory application.

In summary, advisory applications are likely to reflect implicitly the modes of reasoning, concerns and context of those that developed them (usually scientists), and may fail to anticipate the diverse logic and local contexts of farmers. Of course, human advisors too may suffer from such biases, but due to the interaction in a face-to-face advisory meeting, these may be corrected: communication workers can relatively easily deal with feedback (verbal and non-verbal) from farmers and their farming situation, and can change their repertoire and mindset accordingly. This is much more difficult in interactions between human beings and computer software (see also section 12.3.2). In conclusion, specific advisory applications may provide useful input for discussion in face-to-face advisory meetings, but usually cannot replace communication workers or function in a *stand alone* mode.

13.3.2 Supporting horizontal knowledge exchange

As indicated in Chapter 2, horizontal knowledge exchange among farmers is something that happens in various ways and on various occasions. Communication workers can organise a variety of activities to further stimulate and enhance this. Here we will focus on three basic strategies.

Farm comparisons in groups

An important mechanism by which farmers can become aware of problems and solutions is through comparison with other farms. Differences can be noted between one's own farm and those of others, which tends to pose mental and interpretative challenges. One starts to wonder why other farmers employ different practices and/or have different results, and whether or not there is reason to rethink one's own practices. Farm comparisons can take place more or less accidentally as a con-sequence of passing by other farmers' fields, through exchange of labour and/or by chatting with farmers in a market place, but it is also an activity that can be deliberately supported and stimulated by communication workers. An important possibility here is to create opportunities for comparison in existing farmer groups (e.g. commodity groups, co-operatives), or specially formed 'study groups'. Activities here may include:

● *Regular visits or excursions to the farms of group members (or to selected farms outside the group) to discuss each other's practices and* performance. Usually, the observation of another person's farm in itself can raise a few questions and issues for debate. Of course, one can also visit farms purposely to discuss a specific matter (e.g. combating a disease that has occurred). In any case, it is wise to keep the agenda for debate sufficiently open.

● *Systematic collection, exchange and comparison of information from different farms.* In section 13.3.1 we discussed the importance of *individual* memory and feedback applications (written or computer-based). The learning potential of such applications can be greatly enhanced when different farmers collect and exchange similar information so that it becomes part of a *collective* memory and feedback facility. When farmers have access to each other's information, they can compare their current results not only with those of previous years but also with those of specific group members, or with the average of the group. Through such comparisons many additional differences between farms can be identified. For the interpretation of such differences, it appears crucial not just to focus on quantitative figures and parameters, but to use these as entry points for further qualitative debate and contextual analysis. Thus, systematic comparison can play an agenda-setting role during farm excursions.

Both types of activity mentioned can contribute greatly to the exchange of relevant experiences and also to innovation in a wider sense. This is shown, for example, in the highly dynamic and sophisticated glasshouse horticulture sector in the Netherlands, where study clubs of growers are a major force in formulating research agendas and in setting the (rapid) pace of change (see Leeuwis, 1993). It must be noted, however, that horizontal knowledge exchange is not necessarily a tension-free process. As we have already hinted in our discussion of diffusion processes (see section 8.1), social differentiation may complicate the formation of groups (see also our discussion below related to farmer-to-farmer communication for innovation). Moreover, exchanging detailed information can be a sensitive process, especially when economic competition is involved. In the Netherlands, for example, openness among growers of potted plants tended to be less common than among vegetable growers, not least because the former tended to sell their product individually and under their own name, while the latter sold their product anonymously in 'blocks' of equal quality produce at a co-operative auction. Hence, the prices obtained by vegetable growers depended partly on the quality of other growers' products.

The exchange of information may also lead to tensions between those who, in the eyes of farmers, 'do well' and others who have become stigmatised as 'bad farmers'. Moreover, tensions may revolve around the quality and validity of information provided by some farmers, and around the uncontrolled spreading of information outside the group (e.g. to the tax inspector) (see Leeuwis, 1993). Issues like group composition and group rules are important points for communication workers if they wish to facilitate farm comparisons in groups. Defining the substantive boundaries of study groups is another issue. One can use various organising principles for delimiting groups; one may have special groups per farming system (mixed farming), per crop (maize), or per theme (crop protection). Each mode of separation between groups has advantages and disadvantages. Decisions will have to be made according to context and taking into account the preferences of farmers.

Games and competitions

As part of more persuasive strategies, the exchange of feedback about individual (or collective) achievements can also be integrated into games and competitions. Interventions aimed at reducing excessive use of pesticides, for example, may include an award or prize for the farmer or group of farmers with the least amount of pesticide used per unit of product delivered. Such competitions may serve to enhance and sustain motivation in various ways. They may show that improvement is indeed possible and feasible, and may help to create 'heroes' that others can identify with. Moreover, games can introduce elements of sport, fun and, especially in group achievements, social control. It must be noted, however, that competitions can also easily 'turn sour', cause tensions and even work counter-productively, for example when people feel they are forced to compete from a disadvantaged position or are not treated fairly by juries or other competitors. Hence such methods need to be handled with great care. In such methods people need not necessarily compete with *others*, but can also be encouraged to compete with *themselves*. Collecting and receiving feedback (see section 9.3) can in itself be a source of motivation to 'do better', even without competition with others. When, for example, the author installed a speedometer on his bicycle, his average cycling speed rapidly increased (at least for some time).

Farmer-to-farmer communication for innovation

Intervention organisations can enhance horizontal exchange among farmers by enrolling farmers to perform communication for innovation tasks in their community. Farmers may, for example, be asked to consult another named farmer when faced with a particular problem at a time when the communication worker is not available (or perhaps even when he or she is available). Similarly, specially 'appointed' farmers may act as facilitators of (study) group meetings in the absence of a communication worker. Through such arrangements, communication for innovation services can be offered or sustained in situations where this might not be possible otherwise. Moreover, farmer-to-farmer communication for innovation can be seen as a way to optimally use the available knowledge, experience and skills of farmers in a community. As Scarborough et al. (1997) point out, the enrolment of farmers in this way has several additional advantages. 'Farmer-communication workers' tend to speak the same language, literally and culturally, as their colleagues, and are faced with similar constraints and problems as fellow farmers, which may enhance the relevance and credibility of their advice and views. Moreover, in the case of temporary projects, using farmer-to-farmer communication for innovation can improve the long-term impact and sustainability of project efforts. In other intervention contexts (e.g. in health promotion) similar advantages exist;

here the strategy of enrolling members of a group to perform communicative intervention tasks is called 'peer education' (Turner & Shepherd, 1999; Backett-Milburn & Wilson, 2000).

Issues of social differentiation discussed can impinge on farmer-to-farmer communication for innovation. In practice, selecting a particular farmer-communication worker will often effectively mean the inclusion of some audiences and the exclusion of others, for example because it may not be culturally or politically acceptable for some farmers to consult a farmer from a different social status, religious denomination, gender or political affiliation. Thus, it may be wise to 'appoint' several farmer-communication workers. In any case, the selection and training of farmer-communication workers is something that requires careful attention. Several selection criteria can be used, and selection can be done by community members, farmers organisations, communication workers, etc. Another sticky issue is whether or not farmer-communication workers need to be rewarded, and if so how and by whom. Selection procedures and reward arrangements can cause tensions within communities, and can significantly affect the types of people who are willing and/or 'elected' to act as farmer-communication workers. Moreover, one can debate whether farmer-communication workers need to be 'mandated' for specific issues only, or more generally. No general answers can be given to such questions; they need to be resolved according to the context. In relation to these and other issues, several concrete experiences and insights are collected by Scarborough et al. (1997).

13.4   Methods related to raising awareness and consciousness of pre-defined issues

As we have seen in sections 9.4 and 9.5, active learning pre-assumes awareness of, and interest in, a problematic situation. Frequently, therefore, communication for innovation efforts are geared towards encouraging farmers and/or other stakeholders to problematise their current situation and become conscious of individual or collective problems and solutions that they were not aware of or interested in before, but which change agents perceive as being relevant to them. In one way or another, such a process of activating learning requires that people receive feedback that is sufficiently 'strong' or confrontational to stimulate mental activity (see sections 9.3 and 9.4). It is relevant to note that such feedback does not necessarily come from change agents. Studies in western countries, for example, have pointed out that *farmers' magazines* can play a significant role (Engel, 1995). For many farmers their first encounter with new ideas and insights takes place through such magazines, and this causes them to problematise their current situation and/or become interested in alternative solutions. Having said that, several methods and strategies have particular potential in this respect as well; these are described in the following sections.

13.4.1   Mass media campaigns

We have already seen that several mass media have particular potential in arousing attention (see section 12.1 and Table 12.3). A series of radio advertisements combined with widely distributed posters, for example, may make clear to a large audience that there is something going on. When audiences are not yet aware of a problem, such mass media campaigns will often have to do more than present 'dry' arguments and insights for deliberation, as people are not yet actively seeking these. Thus, it is important to build in sufficient 'peripheral' triggers (see section 9.2; Petty & Cacioppo, 1986; O'Keefe, 1990), such as attractive and creative presentation and/or the use of well-known 'messengers' with a good reputation. Through such triggers interest may be increased, which may eventually lead to active learning and deliberation (Van Woerkum, 1991). To the same end, one can use strategies for fostering emotional arousal in much the same way as commercial advertising campaigns do. By using music, images, story-lines or non-verbal cues, one can, for example, try to foster the idea that 'dealing with problem A' (compare: 'buying product A') will make you feel happy and cheerful, and conversely that not dealing with it will cause pain and sadness. However, the use of such strategies raises ethical questions (Van Woerkum, 1991; Singhal & Rogers, 1999). Van Woerkum (1991) argues basically that the use of emotional appeals is legitimate as long as they are not used to mislead people; the emotions aroused must indeed be valid, and backed up by sufficient empirical evidence made available to receivers for cross-checking. A difficulty here (which by the way also exists when no emotional appeals are made) is that what counts as 'valid' and 'empirical evidence' may differ across communities of actors (see Chapter 6 and the discussion of ethical aspects in Chapter 3).

13.4.2   Entertainment–education

Linked to the idea of using peripheral and emotional triggers is the idea of entertainment–education (Bouman, 1999; Singhal & Rogers, 1999). The basic idea here is to 'wrap' educational messages in forms of amusement. This idea is certainly not new. Storytelling, poems and songs, for example, are forms of entertainment that have existed for thousands of years, and have had educational functions. Nowadays, there many examples both in northern and southern contexts where educational messages are built into pop songs, radio or television soaps, cartoons, theatre performances and the like (for examples see Singhal & Rogers, 1999). Through audiences' identification with characters in, for example, a soap or story, they can become aware of and interested in the problems that these fictitious characters (or real role models such as movie and rock stars) experience. Typically, entertainment–education is used in connection with issues that large audiences can easily identify with (e.g. health issues, family planning, intercultural conflict, farm succession, struggle for resources), and not in relation to problems that cannot be easily 'dramatised' (e.g. pest control). Entertainment–education is especially useful in relation to problems and issues that are not easily discussed openly and/or in a personal way, as it allows people to debate such sensitive issues with reference to others (e.g. the characters in a play). Potentially, entertainment–education can have a great influence in setting people's agendas and mobilising interest, at least temporarily. If it is part of a well-elaborated wider communication plan, it can be a trigger for active learning and behaviour change (see Singhal & Rogers, 1999 and Chapter 14 on process management).

13.4.3   Visualising what is difficult to see

In many instances awareness of problems can be restricted because the processes involved cannot be observed. Sometimes problems are difficult to detect (e.g. pollution of ground water), while in other cases it may be clear that there is a problem (e.g. disease, erosion, plagues) which is difficult to further define or analyse since the underlying processes are largely invisible. In such cases awareness and interest in a problem can often be stimulated with a variety of techniques for visualisation. A well-known example relates to biological crop protection. Using transparent nettings, small areas in Indonesian farmers' rice fields were demarcated as 'insect zoos' in which farmers could observe the behaviour and development stages of various insects. Through regular observation of the interactions between plants and insects, they became more aware of which insects were to be considered plagues, and which could be regarded as natural enemies of such plagues. Moreover, they became conscious of the fact that chemical pesticides tended to not only kill plagues but also many beneficial insects, and hence that curative and preventive spraying tended to destroy natural resilience, necessitating ever more subsequent sprayings (Van de Fliert, 1993).

The above example represents a 'tangible', and also labour intensive, mode of visualisation, which in principle could also be provided (perhaps more cheaply) by video films, cartoons and/or forms of computer animation. Due to the directness of the experience, however, the original is likely to have a higher impact in terms of experiential learning. Where direct experience is very hard to organise, these less direct tools can still be useful.

13.4.4   Result demonstrations/demonstration experiments

Demonstrating the results of particular practices is another strategy for raising awareness which makes use of visualisation. The basic idea behind such *result* demonstrations (as distinct from *method* demonstrations; see section 13.7.1) is that farmers can see things for themselves by comparing different treatments, and become interested in new practices without having to fully understand the processes involved. Often, such demonstrations take the form of experiments or trials, but their purpose is not so much to generate new insights (for such experiments see section 13.5.8) but rather to communicate existing knowledge and experience. A classic example here is fertiliser trials, which allow farmers to observe and compare the consequences of different fertilisation practices during the growing season. In a similar vein, one can demonstrate the different potentials of various rice or maize varieties and cultivation practices, or compare the qualities of different pieces of equipment and farm machinery under similar circumstances. A less conventional example is the mechanical 'rain simulator' described by Hamilton (1995) which essentially tried to demonstrate visually how different forms of soil tillage influenced the impact of rain on soil erosion. With the aid of this machine a shower of rain could be mimicked on farm plots, and measurements taken of how much water ran off (causing erosion) and how much was absorbed by the soil. By repeating this demonstration experiment on different plots, farmers became more aware of the negative consequences of widely used tillage methods (Hamilton, 1995). Until then farmers had not been fully aware of this, since they tended to stay inside during heavy rains.

The demonstration experiments described above are largely restricted to specific practices, which implies that they can easily be conducted at the field level. In some cases, communication workers might like to show the results of a totally different farming system, including alternative crop rotations, special farm infrastructure, etc. Clearly, such demonstrations cannot be administered at the field level, but require a demonstration farm. Similarly, one may want to show the results of new forms of community organisation (e.g. in relation to marketing or pest management) and make use of 'demonstration communities'.

In principle, demonstration experiments can be a very powerful strategy in fostering awareness due largely to their direct experiential nature which is beneficial to discovery learning. However, when designing demonstration experiments a number of issues need to be carefully considered, including questions like:

- *What is to be demonstrated and why?* It regularly happens that intervention organisations conduct demonstration trials on issues where little lack of awareness exists. In many situations, for example, farmers are aware of the potential of fertilisers and/or new varieties, but have all sorts of other reasons for not using such technologies (see Chapter 5). In such cases, fertiliser trials and/or demonstrations of new varieties do not make sense. Before designing demonstrations, therefore, communication workers need to analyse carefully whether or not there is a lack of awareness about certain problems and solutions, and identify the precise issues on which lack of awareness exists (see sections 5.2.1 and 13.5.7). In addition, it is important to interact with farmers during the design of demonstrations to ensure that result demonstrations are within the limits of what is, or may become, realistic in the context. It is important to not just involve a single variable (e.g. production level) in demonstrations, but to make visible other possibly relevant parameters (taste, labour requirements, impact on cash flow, etc.).

- *Do we need to organise special demonstrations?* Demonstrations do not always need to be organised, as it may be possible to make use of the existing diversity in farming practices (see also section 13.3.2 on farm comparison). In community level issues this will be a necessity, but also when one is dealing with issues at farm level it may be more practical to use existing farms as demonstration farms (provided this is culturally and politically acceptable; see below).

- *Where and under what conditions are demonstrations to take place?* In principle, demonstrations can be administered at research facilities, on individual farms, or at collectively managed communal fields. Generally, demonstration experiments outside research facilities – that is, under realistic farming and management conditions – tend to be much more convincing for farmers, as research facilities constitute different 'localities' (see Chapter 6), characterised by different conditions and modes of operation. In situations where farmers work on individual farms, the same may be true, albeit to a lesser extent, for demonstrations in communal fields. Here too it may be difficult to demonstrate under realistic management conditions, while tensions can easily emerge on issues like the distribution of work and/or the division of revenues. Moreover, as we discussed in connection with farmer-to-farmer communication for innovation (section 13.3.2), selecting individual farms for demonstrations may involve cultural and political sensitivities and the exclusion of certain audiences. A solution may be to have several farms for demonstration in a community, which also takes account of diversity in farming (in terms of farming style, size, etc.). In some cases it may be possible for communities to 'elect' farms for demonstration purposes, while in other cases the choice may be better left to communication workers or community leaders. Specific solutions are required for each situation.

Finally, it is important to recall that some results and aspects can be notoriously difficult to demonstrate directly – income effects, consequences on soil biology, future consequences, etc. Here visualisation techniques as described in section 13.4.3 may be useful. Also, computer simulations and the like can have some potential (see sections 13.3.1 and 13.5.10).

13.5   Methods related to the exploration of views and issues

For every communication for innovation strategy (see Table 2.1) it is important that communication workers and/or other stakeholders gain access to the ideas and perspectives of relevant actors, in order to help clarify, identify and/or redefine problems and 'needs'[1] and/or work towards solutions[2]. Although the exploratory methods discussed in this section also contribute to awareness raising, they differ from the methods discussed in section 13.4 in several ways: (a) there is more emphasis on improving the awareness of interventionists and/or on developing a shared awareness among participants and interventionists; (b) the issues and problems considered are much less pre-defined; and (c) they mostly pre-assume that participants already have an active interest in participating (except for the first method mentioned). In section 12.2.3 we discussed some general skills (i.e. empathy, active listening, posing questions) that may be helpful to communication workers. However, making optimal use of such skills requires an appropriate methical context in which people are invited and stimulated to speak out. Several methods can be useful in this respect, as described in the following sections.

13.5.1   Analysis of everyday talk

Much can be learned about people's views by noting and analysing the way they talk about something outside an interview situation (Darré, 1985; Te Molder, 1995; Van Woerkum, 2002). Agricultural matters, for example, are discussed at markets, in bars, and during funerals, work parties, community meetings, etc. Communication

---

[1]   For reasons explained elsewhere (see section 14.2.3) we prefer the term 'problems' rather than 'needs' when talking about exploration.
[2]   See Table 14.1 for a more detailed overview of areas that may require exploration in an interactive process.

workers and/or agricultural researchers may have the opportunity to be present at such occasions and carefully note or record what people say. By carefully looking at everyday conversations, and by wondering why people make certain statements (including humour and jokes[3]), communication workers may be able to identify the things people find important, values and interests that are at stake, worries people have, existing knowledge gaps, etc. (Note that this type of analysis may be aided by subsequent interviews as described below.) For example, by analysing how farmers talk about genetically modified seeds, one may be able to gain insight into, for example:

- the ethical concerns that farmers' have in relation to these technologies;
- farmers' technical understanding and beliefs regarding genetic engineering;
- whether or not such technologies are compatible with farmers' religious beliefs;
- the short and long-term consequences and risks that farmers associate with these technologies;
- the extent to which they trust agro-industrial producers of such seeds;
- the criteria that farmers consider when evaluating such a technology;
- the questions farmers have in connection with the technology.

It is important to remember that in everyday conversation people are not necessarily consistent in their views, and that the things they say may vary according to, for example, the identity they take on (e.g. farmer, parent, consumer, community member), self-presentational purposes (see section 6.3) and relational interests during the communication process (see section 5.2.1). In his role as a farmer, for example, a person may tend to evaluate genetically modified seeds quite positively, whereas in his role as consumer or member of a religious community he may have strong reservations. Thus, such variations and contradictions can help to gain more insight into the dilemmas that people face in dealing with certain issues; they are informative and enriching and should not be regarded as negative. For exploratory purposes, the analysis of everyday talk can be done intuitively and pragmatically. However, everyday talk can also be analysed in a much more systematic and academic fashion; here we would refer to the academic field of discursive psychology (e.g. Potter & Wetherell, 1987; Edwards & Potter, 1992; Te Molder, 1995).

13.5.2   In-depth interviewing

One way of gaining access to views and perspectives of people is through in-depth interviews with individuals or groups. Although it can be useful here for the interviewer to raise some specific questions and themes as entry points for the discussion, it is important to allow plenty of time for the interviewees to elaborate their views and raise new issues, which can then be followed up by the interviewer. Highly pre-structured interviews are unsuitable for exploring people's views, as they confine the discussion to what the interviewer finds relevant. In in-depth interviews questions

---

[3]   Humour and jokes can be especially interesting to look at in many situations (see section 16.4.1 for elaboration).

and probing about the reasons behind certain practices and views are very important. In other words, one does not only want to know *what* people think and do, or *how*, *when* and *where* they do things, but also *why*. Conducting good quality in-depth interviews is an art in itself, for which we refer to other specialised literature (Denzin & Lincoln, 1994; for practical guidelines see Pretty et al., 1995). In practice, regular use is made of *focus group interviews*. Here a small group of key informants is brought together for an in-depth discussion of a specific issue (for details see Mikkelsen, 1995). In general, a risk with interview methods is that people may tend to give socially desirable responses; that is, in order to please the interviewer (or others), they answer what they think the interviewer (or others) would like to hear. Thus, it is important that interviewers reflect continuously on the presence of such biases (which by the way can in themselves be informative), and do not encourage these by asking leading questions (like 'Do you use hybrid maize? instead of 'What type of maize seeds do you use?').

13.5.3   Metaplan cards

A 'metaplan kit' consists of a large number of empty cards in different shapes, colours and sizes, as well as markers, pins and one or more pinboards. These simple ingredients can be a great help in managing exploratory group discussions (with 10 to about 60 people) in literate societies. In response to any question, individuals or small groups can be asked to write their responses (in telegram style) on separate cards, and stick them on a board. In some cases, cards of different colours or shapes can be solicited for different kinds of responses (e.g. agreements or disagreements with a statement). In this way, many inputs can be collected, made visible and stored in a short time, and verbally less strong individuals or groups have a more or less equal opportunity to make their points. After collecting the various inputs, they can be grouped in clusters of similar statements. In some cases, clusters can be partly pre-defined (e.g. in an earlier round) and cards can be placed under the appropriate header by the author. In order to prevent cards being misinterpreted, it is important to carry out (or at least discuss) the clustering with the group, so that the original writers have a chance to explain what they meant. A basic rule when grouping cards is that the author eventually decides in which cluster a card belongs. The outcomes of a clustering effort can be used in various ways, for example as an agenda for further debate, for further prioritisation in a ranking exercise (see section 13.5.6), for inclusion in a socio-technical problem tree (see section 13.5.7), etc.

13.5.4   Open space technology

Open space technology (Bunker & Alban, 1997; Owen, 1997) is a specific way of organising exploratory discussions in large (e.g. 100 people) or very large (e.g. 300 people) groups, while providing maximum space to participants to discuss whatever they want. Although there must be a complex problematic situation or issue that all participants are interested in (albeit in different ways), there is no agenda for discussion beforehand. The idea is that participants, after a first introduction of the method, identify topics that they feel passionately about and that they would like to discuss with others. Each topic is announced centrally (literally, as participants are usually seated in concentric circles) by those who propose it, written down, and then put in a schedule of x number (e.g. 8) parallel sessions and y discussion rounds (e.g. 4) that is posted on the wall (in this case allowing for 32 topics).

Afterwards participants can provisionally indicate which topics they would like to discuss in each round (of flexible duration, but usually 1 to 1¹/₂ hours). Then they split up to different locations and have a debate on the chosen topic, with one of the participants preparing a short report immediately afterwards. During the discussion every participant is allowed to leave if he or she feels the discussion is no longer interesting, or is negatively affected by dominant people, and if he or she feels unable to do something about it (this is referred to as 'The Law of Two Feet'). After leaving, participants can join another discussion. This ensures that people talk about topics they want to talk about, for as long as they wish, and in a way that they find acceptable. In the end, participants and organisers end up with a report of all discussions and brainstorms, which may provide all sorts of creative ideas. When open space sessions last longer than one day, one usually tries to identify some priority issues and make follow-up arrangements (for more details see Owen, 1997). Open space technology can work well in situations where people are used to speaking out, and are not afraid that what they say will be used against them. It provides a quick way of eliciting many ideas and encourages networking, while requiring relatively little facilitation effort.

13.5.5   Visual diagramming and mapping

In addition to written and verbal forms of exploring views and issues, it can often be useful to ask participants to represent their views and experiences in a visual mode, for example in a map, diagram, drawing/cartoon, or table/matrix. Sometimes one picture, map or graph can say more than a thousand words. Other potential advantages (adapted from Pretty et al., 1995; Chambers, 1992) are that:

- the process of making visual representations may *invoke interesting and creative discussion* among participants, and when the representations are complete they can be helpful in generating debate;
- making visual representations helps to *focus attention* among members of a group;
- visual representations can help to *capture and store* the discussion in a group, and thus make it easier to cross-check, summarise and communicate it;
- making visual representations can *stimulate people's memory* about the past and present;
- visualisation provides a *shared language* that can help to facilitate communication between the literate and non-literate, while some visualisation techniques (but certainly not all) can help to resolve communication problems (based on language and culture) between relative outsiders and insiders;
- comparing visual representations of distinct groups (men, women, young, old, etc.) can yield interesting *insight into relevant diversity among participants*.

In view of such advantages, visual representation has become an important tool for situation analysis in widely used participatory methodologies such as Participatory

Table 13.1 Examples of frequently used forms of mapping and diagramming in exploratory processes. The references in superscript refer to where operational descriptions can be found: Pretty et al. (1995)[1], Van Veldhuizen et al. (1997)[2], Engel & Salomon (1997)[3], Jiggins & De Zeeuw (1992)[4], PID (1989)[5], Rambaldi & Callosa-Tarr (2002)[6].

Type of map/diagram	Gives insight in what (different) groups perceive to be
Village/town sketch maps[1]	relevant physical and/or social entities and boundaries in their spatial environment
Village transects[2]	relevant differences in agro-ecological characteristics and human activy of different sub-zones of the community (e.g. according to height or distance from the village)
Farm maps[4]	relevant geographical and land-use characteristics of their farms
Three dimensional (3D) spatial models[6]	relevant spatial boundaries, entities and resource use patterns in a region with different elevation levels (e.g. a mountainous water catchment area, fish grounds)
Time lines[5]	relevant historical episodes and occasions
Seasonal calenders[1]	relevant variations over a year with regard to variables such as rainfall, activities, labour requirements, plant diseases, expenditure, market prices, etc.
Trend lines[5]	relevant variations over a range of years with regard to variables such as soil fertility, population size, income, land-use, resource availability, etc.
Village institution (Venn) diagram[2,5]	relevant village institutions and their characteristics and interrelations (in terms of, for example, size, hierarchy, overlap, roles, contact, etc.)
Internal organisational (Venn) diagram[1]	relevant organisational units and their characteristics and interrelations (in terms of, for example, size, hierarchy, overlap, roles, responsibilities, contact, etc.)
Dynamic flow diagrams[1,2]	relevant dynamic (causal) interrelations and sequences in connection with, for example, nutrient cycles, water distribution
(Actor) systems drawings[3]	relevant actors in what is considered a system and the patterns of linkage between them
Problem identification 'map' or drawing	relevant aspects of a problematic situation

Rural Appraisal (PRA; Chambers, 1994a), Participatory Learning and Action (PLA; Pretty et al., 1995), Participatory Technology Development (PTD; Van Veldhuizen et al., 1997) and Rapid Appraisal of Agricultural Knowledge Systems (RAAKS; Engel & Salomon, 1997). Various forms of mapping and diagramming are frequently used for exploratory purposes (see Table 13.1). It is important to recall that each map/diagram is a *mental* construct, which can tell something about how different groups of people see reality. Large differences can be expected between the maps/ diagrams made by different groups. When women are asked to draw a map of important elements in their village, for example, they usually draw a rather different map from men because men and women are often engaged in different activities and networks and have different aspirations in life. At the same time, both maps would probably differ significantly from one made by a soil scientist or a professional cartographer. Most maps/diagrams can be used for purposes of historical reconstruction; communities can, for example, be asked to make maps of their village geography or institutions for different historical periods. Such an exercise can be helpful for the identification of trends and for encouraging historical consciousness (see Sadomba, 1996). The making of each map/diagram itself requires different methods and techniques. The making of a village transect may, for example, require a transect walk accompanied by discussion with farmers, while the making of a historical time line may require discussion with elderly people and archival analysis. For more background and detailed operational suggestions on the use of mapping and diagramming, see Chambers (1994a), Pretty et al. (1995), Van Veldhuizen et al. (1997); Defoer & Budelman (2000) and Engel & Salomon (1997)[4].

The types of maps and diagrams described above require only simple materials such as markers and paper, or even just a stick and soil surface suitable for drawing. However, in specific contexts graphic computer facilities can also be useful. In relation to mapping, computerised Geographical Information Systems (GIS) may offer special opportunities. A GIS is essentially an electronic geographical map, where different information can be stored in relation to each geographical location. By selecting variables to be visualised on a map (e.g. infrastructure, soil type, height, vegetation, crops, human activity, rainfall, etc.), a large number of different maps can be created and represented graphically. Due to their flexibility and processing abilities (see section 12.3), such maps can have an added value compared with conventional maps, especially in relation to issues that transcend local boundaries. Moreover, the precision and high-tech status of GIS maps may be beneficial when interacting with formal bureacracies and/or legal institutions. An example of this is given by Gonzalez (2000), who describes how satellite images, and manipulations of these by means of a GIS, have helped to create mutual awareness among interventionists and local stakeholders about problems relating to water catchment management in a mountainous region of the Philippines. In this case, the various maps helped to identify and visualise effectively existing incompatibilities between the 'natural' boundaries of sub-catchments and political–administrative boundaries and arrangements. This in turn led to an awareness among administrators that the latter needed to be adapted to facilitate catchment management (Gonzalez, 2000). In the case of hilly or mountainous regions, the use of GIS may be combined with (or preceded by) the making of three-dimensional spatial models (see Rambaldi & Callosa-Tarr, 2002). Such models are much more tactile and tangible than GIS computer images, and the building of relief models is an activity in which a large number of community members can be involved, while working with computer images and prints is a more exclusive exercise.

It is important to bear in mind that the making of diagrams, pictures and three-dimensional models should not be regarded as an end in itself. They can help to put issues on the agenda for further discussion and debate, but without such further discussion and debate visual diagrams are not likely to lead anywhere.

---

[4] More visualisation tools, specifically geared to the exploration of 'agricultural knowledge systems' or other 'human activity systems', are presented in Table 17.1.

13.5.6   Ranking and scoring techniques

Ranking is a tool particularly suited to exploring criteria that different groups of people find relevant when evaluating different options or items (e.g. crop varieties, conservation measures, soils, communication workers) (see Jiggins & De Zeeuw, 1992; Chambers, 1994a; Pretty et al., 1995). In the case of *preference ranking*, for example, different varieties of a crop (say six) are compared in pairs. Thus, an informant is given 15 combinations of two seeds and asked which one he or she prefers most and why. The results are listed, along with those of other informants, in a matrix which has the six seeds listed on both the x-axis and the y-axis. By counting the number of times a seed was preferred above another, one can get an indication of what, on average, is the most preferred seed variety within a group, which is the second best, etc. Of course, one can also make several matrixes for different groups (e.g. men and women) and compare their preferences. From the discussion that ensues, much can be learned about the criteria that people employ. *Matrix ranking or scoring* is a more refined method, whereby farmers are asked beforehand to identify a number of relevant (positive) criteria in judging, for example, a seed variety (e.g. productivity, amount of crop residue, taste, resistance to disease, etc.). Then the seeds are put on the x-axis and the criteria on the y-axis. Participants can then indicate which variety is to be ranked first, second, third, etc. on each criterion, or alternatively can distribute a fixed number of counters (e.g. stones) per row to indicate the different value of the seeds per criterion (scoring). For more background information, special applications (e.g. wealth ranking) and detailed operational suggestions on ranking, the reader is referred to Pretty et al. (1995).

Like visual diagramming, ranking is a visual exploration tool whereby the discussions in the process of arriving at rank orderings usually provide more insight than the eventual matrix produced (see also Pretty et al., 1995). While we (following Chambers, 1994a and Jiggins & De Zeeuw, 1992) look at ranking mainly as an instrument for getting a better understanding of a situation, others emphasise its significance as a tool (or voting procedure) for taking 'democratic' community decisions about priority problems, solutions or action strategies in the context of project planning (PID, 1989). In participatory projects we regularly see the latter mode of ranking the most used (e.g. Khamis, 1998; Zuñiga Valerin, 1998). However, we emphasise that there is no guarantee of sufficient political support and backing within communities to work on specific problems and solutions, even if they emerge at the top of a list in a ranking exercise. Typically, ranking exercises give an impression of the *average* priorities that groups of people may have, and as such they run the risk of making invisible significant differences of opinion and interests. It may well be that a significant minority in a community – or even one or two influential people – have difficulties with priority problems and solutions decided upon through ranking, so that there eventually appears insufficient community support and agreement (e.g. Zuñiga Valerin, 1998). In our view, therefore, community decision-making can never take place on the basis of ranking only, but must be accompanied by wider negotiation efforts (see Chapter 10).

13.5.7   Socio-technical problem tree analysis

Problem tree analysis is a technique frequently used by interventionists to get a better understanding of a situation (GTZ, 1987; Van Veldhuizen et al., 1997). The original idea was to jointly identify, and graphically connect, a central problem (the stem of the tree), its 'causes' (the roots) and its 'consequences' (the branches and leaves). Thus, a hierarchy of problems was identified in a more or less interactive mode, which could be used for the identification of project goals (GTZ, 1987). In this section we present an adapted version of this technique which is especially suited to exploring the relationships between the technical and social dimensions of complex problem situations. In terms of the original technique, it focuses especially on the roots of the tree. The idea is to take three basic steps:

(1) Identify a *central problem* that participants are willing to take as a starting point for the discussion. The choice in itself is rather arbitrary, and different stakeholders may have different initial ideas about what the central problem is. Thus, one may choose to draw several trees, starting from different problem perceptions.

(2) Unravel what *specific technical and social practices*, by different stakeholders, contribute to the problem. This can be formulated in terms of what people do, or do not do. In problem analysis these are two sides of the same coin. In an intervention context, it is usually easier to make a tree when focusing on alternative practices that are apparently not applied.

(3) Identify the *reasons* different stakeholders have for not using alternative practices (or for reproducing existing practices). Here the variables of the model presented in Chapter 5 (relating to what people 'believe to be true', 'aspire to achieve', 'think they are able to' and 'think they are allowed/expected to do') can be used as a checklist.

An example of possible results is given in Figure 13.1.

Making such a socio-technical problem tree can be a useful method of visually integrating and connecting insights from different academic disciplines and/or stakeholders. In addition, a socio-technical problem tree can be helpful for making intervention decisions. Depending on the analysis of one or more trees, one may decide the selection of problems on which a project may need to focus. Moreover, the identification of reasons behind practices (using the model from Chapter 5) may be very useful for choosing appropriate communication for innovation strategies and policy instruments (see also sections 5.2.5 and 5.3.2) in relation to problems. In view of its specific set-up and analytical purpose (i.e. dissecting relevant social and technical issues), this method is not meant to be used during exploratory sessions with societal stakeholders. Rather, it may be used by multi-disciplinary teams and project staff to integrate and debate findings arrived at with the help of other exploratory methods.

Some points have to be kept in mind when working on such a problem tree. It is important to realise that a socio-technical problem tree is likely to be actor and theory-dependent. When men and women are asked to make a separate tree, for

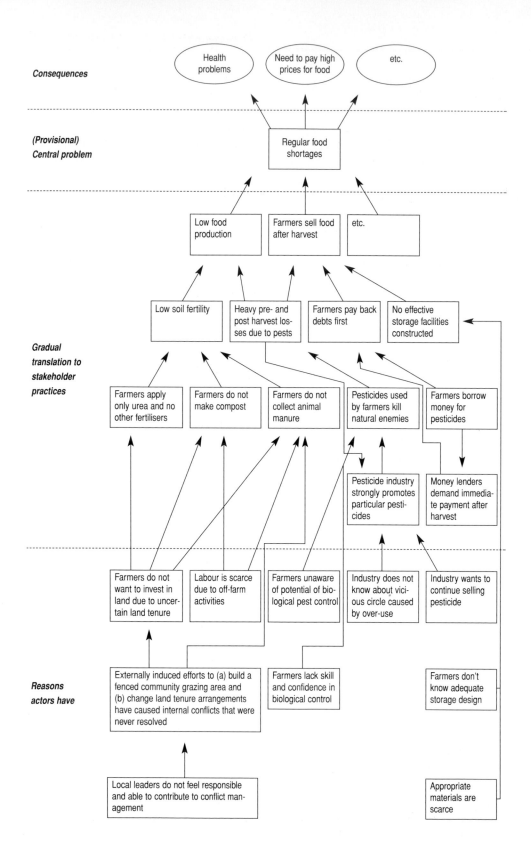

Consequences

Health problems

Need to pay high prices for food

etc.

**(Provisional)
Central problem**

Regular food shortages

Low food production

Farmers sell food after harvest

etc.

Gradual translation to stakeholder practices

Low soil fertility

Heavy pre- and post harvest losses due to pests

Farmers pay back debts first

No effective storage facilities constructed

Farmers apply only urea and no other fertilisers

Farmers do not make compost

Farmers do not collect animal manure

Pesticides used by farmers kill natural enemies

Farmers borrow money for pesticides

Pesticide industry strongly promotes particular pesticides

Money lenders demand immediate payment after harvest

Reasons actors have

Farmers do not want to invest in land due to uncertain land tenure

Labour is scarce due to off-farm activities

Farmers unaware of potential of biological pest control

Industry does not know about vicious circle caused by over-use

Industry wants to continue selling pesticide

Externally induced efforts to (a) build a fenced community grazing area and (b) change land tenure arrangements have caused internal conflicts that were never resolved

Farmers lack skill and confidence in biological control

Farmers don't know adequate storage design

Local leaders do not feel responsible and able to contribute to conflict management

Appropriate materials are scarce

example, one can expect rather different trees, as aspirations, problems and life-worlds are likely to differ. Similarly, scientists from different disciplines (e.g. sociology, gender studies, soil science or entomology) and/or scholars with a different theoretical orientation (e.g. a Marxist versus an actor-oriented sociologist), are likely to come up with different interpretations of reality. Hence, when making a socio-technical problem tree it is important to select those stakeholders and scientists that are likely to have a relevant background in a given context. Furthermore, although the making of a tree requires valid insights, it would be wrong to approach it as an exercise in establishing scientifically underpinned 'cause' and 'effect' relationships. Apart from practical limitations and the fact that these terms are problematic in the social sciences especially (see section 19.1), the purpose of making a problem tree is different. It is useful primarily as a discussion tool, and for organising, visualising and storing different thoughts. At the same time it may help to identify knowledge gaps and hypotheses that require further exploration and testing. Thus, making a socio-technical problem tree can take place over a period of time.

13.5.8   Joint research and on-farm experimentation

Although innovation cannot be equated with 'doing research' (see section 11.2), forms of joint investigation can play a significant role in exploratory processes. Joint research can also play a significant role in conflict management, as it may help in the development of shared understandings (see section 14.2.4). It can take various forms, and sometimes may even be limited to literature research only. In many cases, however, active experimentation can be very useful. In this section, we focus on on-farm experimentation as an activity that can be supported by communication workers.

The nature of farmer experimental activity

When aspiring to engage in experimental activity with farmers, it is important to realise that farmers are likely to engage already in 'experimental' activities, even if this may not be immediately clear and visible to outsiders. This is because farmers may not relate to their activities as 'experiments' or 'trials', but in other terms such as *shifleli* (in parts of Mali, see Stolzenbach, 1994) or *kuturaya* (in parts of Zimbabwe, see Agritex, 1998). Perhaps even more importantly, farmers' experimentation can take many forms, which usually deviate to a large extent from the ways in which scientists think about experiments. This relates to the issue of different epistemological cultures discussed in sections 6.5 and 6.7. In connection with this, scientists may well fail to recognise farmer experimental activity. Let us discuss some important characteristics that farmer experiments may have:

● *Different horizons in comparing treatments*: Farmers do not always 'run' different experimental 'treatments' (including a control treatment) simultaneously. Instead of comparing simultaneous treatments (as scientists usually do), they may

Figure 13.1   (Left) Example of a fictitious, partial socio-technical problem tree. Note that for reasons of presentational ease farmers are treated as a homogeneous category, while in reality there is often a lot of diversity.

well compare different 'treatments' over the years. And instead of having their own 'control treatment' they may well use other farmers' farms and practices as a point of reference. Thus, farm comparison (see section 13.3.2) is in many ways a form of farmer experimentation.

- *Ex-post reconstruction*: In connection with the above, farmers' experiments – unlike scientists' experiments – are not necessarily designed deliberately and planned prospectively. Experiences may well become constructed *as* experiments in retrospect (Richards, 1994). By comparing one's own practices and results with those of others or from previous periods, for example, one can come to think about observed differences as the outcome of an 'experiment'. Similarly, experiments may happen accidentally, for example when two household members carry out the same task in a slightly different way, or when two fields are handled in the same way but at a different time.

- *The role of improvisation*: Although farmer experiments may often be carried out from interest, farmers may sometimes also be 'compelled' to engage in 'experiments' in the face of external conditions, such as the non-availability of inputs used normally. Here, experimentation takes the form of improvisation.

- *Multiple 'independent' variables*: Farmer trials do not usually take place under controlled conditions but rather in the context of wider farming activity. Due to the carefully co-ordinated nature of farming practices (see section 5.1), uncontrollable conditions, and the different horizons of comparison that farmers may apply, there are usually several 'independent' variables at the same time (whereas scientists often prefer to isolate *one* independent variable). This is especially so when the horizon of comparison is a previous year. When, for example, a farmer tries out a new maize variety there will usually be other relevant differences with previous years besides the variety (e.g. weather, sowing dates, etc.). This may also be the case in simultaneous comparisons.

- *Holistic evaluation and measurement*: Even if scientists usually consider several dependent variables when evaluating an experiment, farmers are likely to take into account an even wider range of variables. In a fertiliser experiment, they may not only evaluate yield, cost effectiveness and pest-infestation, but also taste, marketability, crop residue, labour demand, etc. Moreover, while scientists usually prefer precise operationalisations and measurement of variables, farmers may use less tangible (i.e. tacit) modes of evaluation, such as impressions, intuitions and feelings (see also sections 6.2 and 6.7).

In view of the above it is perhaps better to speak of farmers' *experimental experiences* rather than of farmers' *experiments*, as the latter suggests a degree of deliberateness and demarcation that is often absent.

Modes of supporting farmer experimental activity

In our view, engaging in experimentation with farmers should not be equated with 'turning farmers into scientists' or 'imposing scientists' epistemological culture'. Research activity may have a rather different meaning and purpose for farmers than for scientists. As already pointed out in section 6.7, it is often impossible and/or inefficient for farmers to delay new practices until scientists are fully convinced. They may want and need to go ahead when they have sufficient evidence that something works, even if such evidence does not live up to scientific standards. Rather than replacing current modes of investigation and farmer research, the support of farmer experimentation could build on existing practices in various ways:

- *Explicating and exchanging existing experimental experiences*: Many of the existing experimental experiences may not yet have been explicated and shared among farmers. Hence, identifying, collecting and exchanging existing experiences may already contribute much to problem solving and innovation. Several of the other exploratory methods mentioned in this section (e.g. in-depth interviews, historical mapping, etc.) can be useful here.

- *Supporting interpretative debate in groups*: Due to the nature of farmers' experimental experiences, it is often not easy to draw clear conclusions as there tend to be a number of possible explanations for certain phenomena. One way of improving the capacity to draw valid conclusions is through talking with people who have similar experiences. Organising group discussions around such experiences can therefore be useful.

- *Improving measurement, memory and feedback*: Often the capacity to draw inferences from experimental experiences can be enhanced by adapting modes of measurement, and by the collection and storage of information about regular and experimental activities (see our discussion of farm comparison and memory and feedback applications in section 13.3).

- *Identifying issues and adding options for deliberate experimentation*: Outsiders can organise group debates and analytical activities geared towards identifying areas that require experimentation. Forms of joint socio-technical problem analysis and priority ranking can be useful here. In addition, outsiders can be useful in suggesting new options and opportunities for experimentation and/or can provide farmers with insights that lead them to adapt their research agenda (see also Van Veldhuizen et al., 1997). As agricultural innovations frequently emerge from accidental experiences or from experiments that neither farmers nor scientists considered very promising initially, it may be useful not only to think about 'the obvious' but also to solicit and consider seriously 'crazy' and/or unconventional ideas and solutions (Chambers, pers. comm.). This avoids falling into the trap of 'blindness' that accompanies existing knowledge (see Chapter 6).

- *Including social-organisational 'experiments'*: Often the focus in on-farm experimentation (and in strategies to support it, e.g. Van Veldhuizen et al., 1997) is solely on technical experiments and issues. Given the experience that innovation usually requires new social–organisational arrangements as well (see sections 2.1 and 8.2.1), this is a rather one-sided approach, which may lead to technically sound solutions that can never be applied (e.g. case 4 in section 10.1.1). Thus, in many instances it can also be relevant to experiment deliberately with, or work towards, alternative social–organisational arrangements (see Van Schoubroeck, 1999). Such alignment activities (see section 8.2.1) may involve social tensions, more than with technical experiments only, and hence require efforts to manage conflict.

- *Debating the design and management of deliberate experiments*: When making plans for new on-farm experiments, the design of experiments is obviously an area of debate with farmers. Design issues include the precise area of interest, different 'treatments', evaluation criteria, measurement procedures, etc. Without necessarily imposing scientific modes of experimental design, scientists' concerns and insights on systematic experimentation may still be valid in such a debate. Sometimes small changes in the design of farmers' experiments can lead to a considerable increase in the potential to draw accurate conclusions. In line with what we have discussed in relation to *demonstration experiments* (see section 13.4.4), it is pertinent also to discuss where to conduct and how to administer experiments. One need not necessarily arrive at a single design and location; it can be enriching to make use of the existing diversity in farmers' preferences and views, and run several on-farm experiments at the same time.

- *Reducing risks*: Sometimes potentially interesting experiments involve pro-hibitive risks and uncertainties (see sections 5.2.1 and 8.2.3). Farmers may, for example, be wary of experimenting with reduced use of pesticides, due to fear of crop losses. In such cases, outsiders may provide insurances and resources that allow farmers to experiment anyway.

- *Co-ordination and interaction with formal research*: It has been documented that on-farm experimentation, and research in facilities for formal scientific research, can enrich, inspire and complement each other (e.g. Van Schoubroeck & Leeuwis, 1999). In general, carrying out similar experiments in several locations tends to lead to different experiences and discoveries 'by chance'. Formal on-station research can also provide a back-up to on-farm experimentation in several ways. Farmer experiments may 'fail' for a variety of reasons (related to natural con-ditions, technical practices or social–organisational affairs) and comparison with on-station research may at times provide clues about such reasons. In addition, formal research facilities often allow for more in-depth exploration of underlying mechanisms, provide some 'free creative space' for scientists to follow their gut-feelings and intuitions, and allow more rigorous and frequent data collection. As Van Schoubroeck indicates, complementary experimentation is more easily achieved when the same persons are involved in both on-farm and on-station research.

- *Enhancing scope through 'virtual' experimentation*: Progress in exploration is sometimes restricted by the fact that only a limited amount of experimentation can be conducted at any one time, and because it can take several production cycles before final conclusions can be drawn (Rossing et al., 1997). Furthermore, some experiments at the farming system level or beyond (see section 5.1) can be impractical. It sometimes happens that farmers abandon certain experiments not only because of high risks, but also because of legal or other social constraints. In some cases such problems can be overcome partly by conducting 'virtual' instead of 'real' experiments, with the aid of computer simulation techniques

(Rossing et al., 1999). Due to the inherent shortcomings of computer-modelling (see our discussion in relation to advisory applications in section 13.3.1), the outcomes of such experiments can never be taken at face value. However, especially when the models are made transparent, it is possible to critically discuss the validity of findings and arrive at interesting and revealing insights.

More detailed discussions of issues related to on-farm experimentation can be found in Okali et al. (1994) and Van Veldhuizen et al. (1997).

13.5.9 Public debates

In relation to controversial policy issues (e.g. concerning new technologies) policy-making institutions may organise public debates among a variety of societal stakeholders and specialists. Such debates can be held face-to-face, via the internet, or using a combination of both. The purposes of such public debates, from the perspective of the organisers, can vary according to the different ideas they have about the role of government in policy making (see below). Basically we can distinguish between three types of debate:

- *Exploratory debates*: These debates are primarily geared towards generating a rich picture of ideas, opinions, concerns and solutions in relation to a specific issue. Usually, participants are asked to react to initial propositions and questions, and later to each other. The views expressed in the debate are summarised in a document. The idea is that policy institutions can use them in their policy formation process; the institutions interact with stakeholders but remain in charge of policy making (interactive government).

- *Converging debates*: Here the purpose of the debate goes further than just exploring perspectives. The idea is that the stakeholders reach some sort of consensus and conclusion on the direction that policy-making should take. Thus, the debate must not only be summarised, but must evolve into a productive joint learning and negotiation process (see Chapters 9 and 10). The associated governance model here is 'network-steering', in which policy-making institutions delegate to some degree the responsibility for policy-making to a network of societal stakeholders.

- *Opinion polling debates*: Sometimes a debate is merely used as a source of inspiration for arriving at relevant questions in a voting or opinion poll procedure. The idea is that, for participants, the debate serves as a process of opinion forming, and helps to identify the most important problem definitions, solutions and/or political choices. To conclude the debate, opinion polls and voting are used to assess the support for particular views and opinions. This type of debate is inspired by ideas about 'direct democracy' and referenda, where the argument is that competent citizens should take direct political decisions, instead of being represented indirectly by political parties and/or parliaments.

When public debates are conducted face-to-face, typically a series of intensive informative and discussion meetings is held over a period (e.g. 6 months), in which a facilitator uses various discussion formats (discussion in small groups, forum discussions, plenary debates, etc.). In the recent past, public debates on the internet tended to make use of *mailing lists*, with participants receiving the contributions to the debate in a chronological order by e-mail. Nowadays, discussions are organised mainly with the help of more advanced discussion software on the world wide web. Participants can connect to a website where contributions to the debate are usually arranged in *discussion lines* with a hierarchical structure. Each discussion line starts with an opening statement or theme (first level), to which participants can respond (second level); subsequently participants can react to these responses and create a third level, then a fourth level (responses to third level responses), and so on. When they are online (or after they have downloaded the site), participants can open contributions to the debate, and respond to them if they wish.

In western countries especially, the expectations of internet applications for public debate (when compared to interpersonal forms) were quite high, because it was expected that more people could participate in such debates, with less time pressure, more time for reflection, a lower barrier to response, better information provision and – as a result of a degree of anonymity – a greater chance for openness in debates on sensitive issues (see Leeuwis, 1999c). However, practical experience led to several sobering insights. It appears far from easy to encourage participants who only share an 'electronic space' to really respond to each other in an in-depth and interactive (see section 7.1) manner, let alone to reach agreement on a sensitive issue. People tend to present their own views, rather than to discuss the merits of other participants' contributions (see Leeuwis et al., 1997). Also, any debate on a significant scale tends to be very time consuming (both for participants and facilitators), and participants are often far from representative (Leeuwis et al., 1997; Jankowski et al., 1999). For an exploratory debate, however, these limitations are somewhat less significant; even if there is no in-depth debate they can help to identify diverse, creative and unconventional views and ideas from which policy makers can benefit. In public debates, communication workers can be particularly useful in compiling relevant information preceding and/or during the debate. They can also play an active role in developing an overall set-up for the debate, and in facilitating/moderating the debate as it unfolds.

13.5.10   Future explorations

To facilitate different actors and stakeholders developing some common ground, it can be useful to abstract, at least temporarily, from current problems, concerns and issues, and instead focus on a point in the relatively distant future. There are several participatory, large group methodologies and approaches which make use of this principle; well known examples are Future Search (Weisbord & Janoff, 1995) and Search Conferences (Emery & Purser, 1996). For detailed descriptions of these we refer readers to those authors; here we indicate some basic ideas and similarities related to these approaches.

Typically, a large number of stakeholders (e.g. between 50 and 100) and/or stakeholder representatives are brought together for about 3 days to discuss a complex multi-stakeholder issue, with the aim of working towards a shared vision on the future as a basis for further action. As shown in Box 13.3, both Future Search and

Box 13.3   Basic steps/tasks in Future Search and Search Conferences (based on Weisbord & Janoff, 1995; Emery & Purser, 1996; Bunker & Alban, 1997)

Tasks in Future Search	Steps in Search Conference
• review the past • explore the present and current trends • discuss 'prouds and sorries' (+ and –) of stakeholders • create ideal future scenarios • identify common ground • make action plans	• discussion of our turbulent environment • our system's history • analysis of our current system • the most desirable system in 5 to 20 years • action planning • implementation

Search Conferences try to encourage people to think about the relationships between past, present and future. Looking at the past helps stakeholders to analyse how the present has been shaped, which phenomena are persistent, and which larger trends can be identified. When combined with an assessment of what different stakeholders find positive and negative about the present, looking at trends may help them speculate on more and less desirable characteristics of what is likely to happen in the future. Often, this helps to foster a general sense that 'something must be done', even if stakeholders still disagree about what that might entail. From there, the focus shifts towards generating creative ideas on how the future could ideally look in 5 to 20 years, in contrast to the scenario that is deemed likely. When different stakeholders can identify sufficient commonly attractive elements in such ideal scenarios, they can start to reason back to the present ('backtracking') by asking: 'What is it we can/must do now to improve the chances of arriving at a more desirable future?' This is where co-ordinated action planning becomes significant. In all, the basic idea is that an intensive process of joint analysis improves the chances of arriving at a shared understanding about a desirable and possibly realistic future, as well as agreement on what action the various stakeholders must take 'today' in order to arrive there.

For each of the steps indicated in Box 13.3 various discussion techniques and tools (e.g. metaplan) can be used (see Weisbord & Janoff, 1995; Emery & Purser, 1996). During the process, the large group is supposed to split up regularly into smaller, relatively homogeneous or heterogeneous groups, which report back to the plenary meeting. Also, several modes of visual diagramming and mapping, including time lines and trend lines, are suggested.

According to Weisbord and Janoff (1995), the following principles must be followed in order to have productive meetings of this kind (see also Bunker & Alban, 1997):

• get representatives of the 'whole system' in the room, including all relevant stakeholders and management levels;
• analyse the global context, and act locally;
• focus on common ground and the future, rather than on immediate conflicts and problem-solving;
• let small groups direct their own internal processes;

- participants must attend the whole workshop;
- provide sufficient time to include enough 'break points' (i.e. include two over-night stays);
- let participants publicly declare the actions they will take next;
- meet under healthy conditions in terms of food, space, daylight, etc.

Computer-based explorations[5]

Although the computer is not mentioned as an explicit element in the above methodologies, it must be noted that, in agriculture and resource management, computer simulations are frequently used as a method for exploring future scenarios at the field, farm or regional level (e.g. Rabbinge et al., 1994; Rossing et al., 1997, 1999; Ten Berge & Riethoven, 1997). Computer models are used, for example, to calculate likely and/or technically feasible options for future land use, using different sets of objectives. For example, such models were used to calculate what agriculture could look like in different parts of Europe (in terms of area under cultivation, employment, use of pesticides, cropping systems, etc.) given a required level of overall production, and under the various constraints posed by different policy orientations (e.g. striving at 'environmental protection', 'nature and landscape conservation' or 'free market and free trade') (see WRR, 1992; Rabbinge et al., 1994; and on a wider scale IFPRI, 1995). Scenarios and options generated by modelling may provide interesting input to multi-stakeholder debates about the future.

In principle, such computer-based explorations have similar potential strengths (knowledge integration, additional opportunities for experimentation) and weaknesses (ignoring social–organisational feasibility, questionable validity, lack of transparency) as explained in our discussion of advisory applications (see section 13.3.1). When applied in multi-stakeholder situations additional strengths and weaknesses are relevant. A potential strength is that modelling exercises may help to quantify not only trade-offs between different goals and interests (e.g. increased production versus environmental pollution), but also so-called win-win situations (see section 10.2). Such information can be very useful in a multi-stakeholder learning and negotiation process about the future. In addition, working as a group with a model, or building it, may lead to the identification and explication of knowledge gaps (see Van der Werf et al., 1999) and areas of uncertainty. This may stimulate stakeholders into joint fact-finding, which can be a positive force in learning and negotiation (see sections 13.5.8 and 14.2.4).

A weakness is that – in association with earlier discussed substantive biases (see section 13.3.1) – computer-based explorations typically generate only a limited number of articulate options or scenarios, and inevitably exclude a range of other possible options. There is a risk that the debate between stakeholders is drawn almost by default towards those options and scenarios that are articulate, while alternative options are much more difficult to discuss, propose and/or underpin. Thus, *the scope of debate* may be limited undesirably.

---

[5]   Parts of this section derive from an earlier publication by Rossing et al. (1999).

Another possible bias is that exploratory land use studies may help to foster a climate in which talking about pros and cons becomes the default option. This may be attractive to academics, but it is contrary to key recommendations for the facilitation of negotiations among different stakeholders. Fisher and Ury (1981), for example, stress the importance of talking about interests rather than positions and arguments, as arguments are always secondary to interests; that is, stakeholders tend to select and construct those arguments that fit their interest, so that talking about arguments becomes unproductive and inefficient (see also Aarts, 1998 and Chapter 14). An *over-emphasis on arguments* may well occur in debates where exploratory models are used as a resource, as stakeholders may assume – for example, on the basis of their associations with computers as powerful and 'magic' devices – that complex problems can be solved through rational argument and deliberation inside the computer model. Such partly inherent biases may easily jeopardise the position of particular stakeholders and societal interests, as it is likely that some significant issues and options can be less easily articulated, discussed and/or made credible than others. Also, during integrative negotiations (see Chapter 10), such an unintended narrowing of the area of discussion may be at the cost of finding creative win-win solutions and trade-offs which fall *outside* the scope covered by the computer model. In some cases, stakeholders may, wittingly or unwittingly, use the inherent selectivity of computer models to manipulate and influence a policy debate. This is especially so when the skills and resources required to use or understand computer-based explorations are not equally distributed among stakeholders, i.e. when the transparency of the underlying models varies across participants.

Finally, due to inherent validity problems – especially at higher levels of integration (see Rossing et al., 1999) – there is a risk that the validity of models itself becomes a major issue of debate among stakeholders. Such a debate may be productive and insightful (even if it can never lead to a validated model), but may also lead to endless argument about whether or not an outcome is meaningful and significant.

In summary, there is a risk that debates about the future become *biased* in various ways when complex computer models are allowed to play a dominant role.

More generally, explorations of the future (with or without computer modelling) may foster and/or express an unwarranted and unrealistic belief that future developments and scenarios can be predicted and controlled (see section 4.2). At the same time, it is inevitable that assumptions will be made about the future when developing policy and/or striving for co-ordinated action. Thus, it is important to be aware of the limitations of such exercises, and to make sure that there is sufficient opportunity for regular critical reflection on, and adaptation of, action plans developed on the basis of explorations of the future.

13.5.11 A caveat: be aware of ritualistic use of exploratory methods

The developers and advocates of several exploratory methods argue that these should be used in a way that contributes to human capacity building and empowerment, and not just to extract information on behalf of outsiders (e.g. Chambers, 1994b; Pretty et al., 1995 in relation to visual diagramming in the context or PRA and PLA). New knowledge and improved insight (that may result from exploration) can indeed 'empower' people in that it may enhance agency (see section 6.6). However, in practice exploratory methods are often used rather ritualistically. In the context of PRA, for example, we often witness project staff enter an area or community, make sure the whole battery of visual diagrams and ranking exercises are completed as quickly as possible, and then leave again without developing a better, shared and/or self-critical understanding of a situation. At the same time, communities are well able to make sure that a prepared 'shopping-list' finds its way into the outcomes (see Brown et al., 2002; Pijnenburg, 2003). Similarly, the outcomes of exploratory public debates (see section 13.5.9) are frequently ignored in the policy arena (Pröpper & Ter Braak, 1996; Te Molder & Leeuwis, 1998). Participants and/or organisers can have many reasons for going about exploratory processes in this way. Frequently the purpose of using exploratory methods is not genuinely 'to explore', but rather to gain experience with methods (Leeuwis et al., 1997; Van Arkel & Versteeg, 1997), to meet donor demands or satisfy bureaucratic needs (Pijnenburg, 2003), to help create a positive organisational or community image (Pacheco, 2000), and/or to give the impression that public concerns are being considered (Van Woerkum, 1997). In other cases, a lack of skill and understanding of exploratory methods plays a role; this may include the misconception that methods and methodologies can be treated as mechanical procedures for inducing change (see section 13.2). It is important to reflect on the the underlying rationale whenever one considers using exploratory methods. If there is no genuine interest in exploration it would probably be better not to go ahead with it. When this is impossible (e.g. the donor strongly insists on a PRA, while project staff and the community already basically agree on what must be done), it would be fair to be open with participants so that everybody knows they are going through the motions (see also Chapter 14 on process management).

In addition, one needs to be critical about the exploratory method chosen. As we have seen in this section there are various methods available, each with their own potential advantages, labour requirements and costs. And even within methods various options exist; in visual diagramming, for example, one can choose between a large range of maps. It is important to select or combine the methods and tools that best fit the context in which they are to be used. As every situation is different, no recipes exist.

13.6   Methods related to information provision

When people are actively involved in learning, they are quite likely to look for information from outside sources. Communication workers can play a considerable role in making such information available. This involves not only adapting and translating insights from various sources into a language and terminology that audiences can relate to, but also ordering and grouping together information in such a way that they can find it. In agriculture, it is not only advisory organisations that can provide such services; agricultural publishers, for example, sometimes prepare agricultural handbooks and the like. From a communicative intervention point of view, the most relevant methods are written and/or electronic search and access facilities. Although the technologies used for these two are very different, the basic ideas and problems are similar, which is why we discuss them simultaneously here.

13.6.1   Written and computer-based search and access facilities

In agriculture, examples of written search and access facilities include agricultural handbooks and series of leaflets on a variety of agricultural topics. Farmers can search for information by looking at the table of contents and/or the index of a book or leaflet, or by scanning pages or a rack in which leaflets are displayed. Computer-based search and access facilities involve internet websites or CD-ROM databases (e.g. electronic encyclopedias). Here search facilities often include menu-structures (comparable with a table of contents) and choosing or entering search words to identify a selection of electronic pages or websites that meet specific search criteria; such selections can often be further refined through adding new search criteria and/or looking at overlap with other selections. When compared to a book index, the user has more possibilities to determine search words, make combinations, and search many 'books' at the same time.

An important pitfall with search and access facilities (written and computer-based) is that the ordering of the information by means of tables of contents, menu structures or pre-defined search words may reflect the logic of the information supplier, rather than that of the user. A book, series of leaflets or website on agriculture may, for example, be structured along the lines of specialised academic disciplines or organisational divisions, and not according to the problem areas that farmers and/or communication workers experience. Farmers may think, for example, about pasture management, whereas a website or book may not have a specific section on this, but provide a variety of largely unconnected insights into the subject under the headers of 'soil fertility', 'plant ecology', 'animal feeding' and 'weed science'. Similarly, a book or website may use the scientific terminology (e.g. on fertilisers, insects, soil types, etc.) as search words, rather than locally used names, making it difficult for users to find what they are looking for and/or to sensibly connect different information found.

As mentioned earlier, an important role for communication workers and specialists here is to prevent such mismatches from happening. Also, they can play a role in maintaining and updating leaflets, handbooks and websites, as these may become outdated rapidly. In relation to search and access internet applications in particular, communication workers can help to combat biases and information overload. While searching the internet one often finds large numbers of websites where the quality and relevance are rather dubious. At the same time, highly relevant information may be missed because specific organisations are not on the internet, or do not wish to make information freely available. On the websites of communicative intervention organisations, communication workers can act as information brokers by providing web-links to high quality websites elsewhere, and/or by pointing out omissions on the internet, thus saving information searchers a lot of time and trouble.

13.6.2   Information-needs assessment

In order to develop useful search and access facilities, it is important that communication workers gain insight into their clients' 'information-needs'. Methodologically speaking, assessing such needs is far from easy. One cannot simply ask someone what his or her information-needs are, or what they would like, for example, a search and access internet application to do for them. It is difficult for anyone to analyse their own needs, basically because we tend to think more in terms of the problems we face than of well-defined information requirements. It is even more difficult to evaluate the potential added value of a relatively unknown technology like the internet in meeting such needs. People are often not aware of the needs that are already fulfilled (precisely because these are already catered for), and will express no 'need' for information which they do not know exists. In a way it can be argued that an information-need only emerges when it is about to be fulfilled, i.e. when particular information is within reach; and only if one has access to particular information does it become possible to evaluate whether or not there was a real need for it.

In view of this complex relationship between information supply and demand, it is far too simplistic to argue that communication workers must cater for the information-needs of clients. These difficulties are compounded by the fact that information-needs tend to be a 'moving target'. Farmers, for example, continuously learn, solve problems and identify new ones in an ever-changing environment, which means that information requirements can change rapidly. If one identifies an information-need there is a chance that the need has altered or been resolved even before a leaflet or website is ready. Obviously, this problem is somewhat less significant when one deals with large numbers of clients in a relatively stable environment, because one can expect that different clients will require similar information at different times.

Given that it is often not productive to simply ask people about their information-needs, several alternative strategies remain (e.g. Davis & Olson, 1985), including:

- analysing the way in which already existing facilities for information provision (leaflets, articles, websites) are used;
- observing and analysing current information-seeking behaviour of clients;
- registration and analysis of questions and requests that communication workers are asked in interpersonal and group contacts;
- discovering information needs during the interactive and iterative development of search and access facilities;
- deducing information-needs from what others (e.g. scientists, subject matter specialists, agri-business) think the information-needs of clients *should* be.

The fourth strategy above, and also the first, build on the experience that information users tend to become aware of their 'real' or more precise needs only when they are confronted with a particular information supply (Vonk, 1990; Leeuwis, 1993). Making use of this, one can for example make a rough version of a leaflet or website, and solicit comments from prospective audiences. On the basis of such feedback and subsequent discussions, a new 'prototype' can be created, and this procedure repeated until a final and satisfactory version is obtained. This procedure is called prototyping (Vonk, 1990). The fifth strategy above is rather risky because no attempt is made to connect and integrate the concerns of outsiders with the rationalities, circumstances and perspectives of clients. In combination with prototyping, however, this may still be a useful strategy given that information from external sources can be useful in a context of innovation (see section 8.2.2).

13.7   Methods related to training

When there is agreement in a change process that audiences lack certain technical, social and/or organisational skills, communication workers may organise training activities or courses directed towards transferring specific knowledge and skills. For specialised insights on educational approaches, course didactics, curriculum development, distance education and the like, we refer readers to agricultural education science (Blum, 1996). Here we will discuss briefly some important methods that can be useful when interacting with farmers outside a classroom or distance education setting.

13.7.1   Method demonstrations

While *result* demonstrations (see section 13.4.4) are useful for raising awareness, *method* demonstrations can play a role when people are already interested in a particular solution. Method demonstrations essentially show people *how* to perform certain tasks and practices, while result demonstrations are more geared to visualising, by means of trials, *that* certain practices may be worth considering. Once farmers have developed an interest, one can, for example, demonstrate to them how to perform a particular pruning technique, how best to count 'good' and 'bad' bugs, or how to operate a specific piece of equipment. Thus, method demonstrations involve showing people how to do something, with the expectation or hope that they may copy it. This may involve a demonstrator being physically present, but it may also include video footage or animations on tape or CD-ROM. In modern garages, for example, car mechanics are shown how to adjust the ignition of different car models on a computer screen. Whenever they need to, they can consult a database of video extracts demonstrating a variety of model specific maintenance and repair practices. As is the case with all training materials, it is important to make sure that the demonstration content is adapted to the context of the audience. Relating to the example of cars, it makes little sense to demonstrate how an ignition can be adjusted with the help of sophisticated electronic measurement tools, if very few garages own such equipment.

13.7.2   Experiential practicals

In many cases, just *showing* something will not be sufficient to ensure that people become comfortable with new practices and/or adjust them to their situation. From an experiential learning point of view (see section 9.1) it makes sense to create situations in which people can gain experience with new practices, with the possibility of getting feedback from others on their performance. Such situations can be called *experiential practicals*. One may, for example, arrange for a group of farmers to 'play around' with a newly designed plough after its use has been demonstrated.

Apart from technical practices, experiential practicals may also involve experiential learning of social and analytical skills. One may, for example, use role plays to enhance the participants' negotiation skills. Similarly, one can develop 'protocols' through which farmers can gain experience of analysing their farms and/or the agro-ecosystem in which they operate. It may require considerable imagination and creativity to develop experiential practicals for less tangible areas of learning.

Questions for discussion

(1) Select a few methods discussed in this chapter that are not being used currently by your organisation (or another organisation that you are familiar with). What is the relevance and potential of these methods in view of the purposes of the organisation?

(2) Discuss the advantages and disadvantages of the authors' choice to focus in this chapter on methods rather than methodologies, and group them according to communication for innovation functions instead of strategies.

(3) An organisation uses visualisation techniques (e.g. tools as used in Participatory Rural Appraisal) to explore problems in a community. The team decides to rank problems using a paired matrix ranking on the ground, in which the community selects the most important problem out of a series of paired problems. Because there are many problems (so the matrix is big), the ranking exercise becomes quite lengthy and complex. All members of the team become involved in making sure that the exercise is completed, and that the ranking results are transferred to paper. The procedure becomes rushed and the community resorts to simply raising their hands in deciding which problem is the most important out of the two presented.

  What do you think of this procedure? What important principles of visualisation and/or ranking are being violated here?

(4) An organisation wants to organise a campaign to stimulate healthy behaviour. They are considering building messages on health related issues into a popular television soap (entertainment–education). What are the possible strengths and weaknesses associated with this method that you would like the organisation to take into account?

(5) An organisation decides to use on-farm result and method demonstrations and on-farm experiments as their main strategy to support innovation in a rural community. What are the possible strengths and weaknesses associated with this approach?

14 The management of interactive innovation processes

So far this book is based on two important premises: (a) that change and innovation must be regarded as multi-actor processes and that they evolve over time in a relatively unpredictable and uncontrollable manner (see Chapter 4); and (b) that communicative interventions aimed at stimulating change and innovation are best inspired and organised on the basis of theories of social learning and negotation (see Chapters 9 and 10). These premises reflect a rather historical and evolutionary outlook on innovation. We are aware that this may at first sight not always correspond with the situation that communication workers face.

In line with the above, sometimes communication workers are asked to engage in a process with a longer time-scale, for example when asked to facilitate a shift from government-managed to farmer-managed irrigation schemes. On other occasions, however, they are confronted with a government request to organise a campaign of limited duration to promote the use of, say, a new maize variety. Still, we feel that even in the latter case a longer term process perspective is essential, because the viability of such a limited campaign can only be assessed by analysing the history of the proposed technology and the government request. Relevant questions, for example, are: Who designed the technology, where, how and why? What are the experiences of different categories of farmers so far? What have proved to be conditions for successful use?

In addition, a communicative campaign can only be designed by taking into account a closer look at earlier communications regarding the same or similar topics. Has this technology been advocated before? What do farmers already know and think about it? How do they perceive the government's interest in promoting it? Do farmers regard us as a trustworthy source in connection with this topic? How does the technology relate to future aspirations of farmers? These are just some of the many questions that can be asked in relation to seemingly 'limited' communicative intervention assignments, and which essentially put the issue into a wider historical perspective by asking: What processes have been going on? Where are people heading? And what does this mean for our role at this time? (Where are we now in which process?). In other words, we feel it is essential to look at prospective communicative interventions, even if limited in scope, as elements or micro-episodes in wider 'historical' processes.

As we have already indicated at various points, we do not believe that change processes can be planned and/or controlled by prefixed goals, procedures and methodologies (see sections 4.2 and 13.2). Thus, facilitating a change and innovation process requires the weaving together of different strategies and activities (incorporating various methods) flexibly and contextually. However, this does not imply that individual communication for innovation activities (or micro-episodes) need not be carefully and systematically prepared and designed. On the contrary. While it makes little sense to prepare and plan detailed processes over a long period, it is important to think very carefully and systematically about individual activities, and place them in the context of a broader process. Following this line of thinking, we discuss general guidelines and insights on process management in section 14.2. Subsequently, Chapter 15 deals in some detail with the preparation of individual activities. Before this, we address the issue of how our approach to the management and facilitation of interactive processes relates to more widely advocated notions of participation and participatory development. When discussing the interactive model of communicative intervention (in section 4.2) we used the terms 'interactive' and 'participatory' interchangeably. Moreover, throughout this book we have drawn on literature about 'participation'. Nevertheless, although it is evident that we are inspired by the idea of 'participatory development', we prefer to use the phrase 'interactive innovation'. This is mainly because the term 'participation' is laden with certain connotations that we are uncomfortable with, as they are incompatible with our proposal to look at innovation processes as social learning *and* negotiation processes.

14.1   Some limitations of conventional thinking on participation

Let us first re-emphasise that there are solid arguments for interacting with societal stakeholders and/or prospective beneficiaries when the purpose is to stimulate change and innovation. On the basis of participation literature, we have identified in Chapter 4 four categories of argument in relation to this (see section 4.2.2):

- pragmatic arguments;
- ideological and normative arguments;
- political arguments;
- accountability arguments.

These arguments are relevant to most situations. However, we feel that the way in which participation has been put into practice is not without problems (see section 4.3). Earlier in this book we touched on some important issues in this respect. To the extent that participatory intervention follows predefined steps, procedures and methodologies, we feel that it falls into the same trap as 'top–down' approaches by assuming that change is something that can be planned (see sections 4.1 and 13.2). Many participatory projects are in fact about participatory *planning* (i.e. going through planning stages with participants), which in our view is unfortunate. As we have emphasised in Chapters 8, 9 and 10, interactive trajectories are better thought of in terms of learning, network building and conflict management. Clearly, different 'stages', 'steps' or 'tasks' are more relevant to such processes than the classic steps in planning models (compare, for example, sections 9.1, 9.5 and 10.2.1 with 4.1.1). Moreover, we have argued in section 10.1.2 that conventional participatory literature and methodologies are particularly weak in conceptualising and dealing with conflict as an inherent component of change; hence, our attention to negotiation in Chapter 10. In this section we will discuss a third worry (related partly to the previous ones), which is that there is often a large gap between participatory *rhetoric* (including theoretical discourse) and participatory *practice*.

14.1.1   Defining 'participation'

When defining 'participation', we are faced with the same problem encountered when defining extension (see section 2.1.2): that it can be defined either normatively, descriptively or literally. Literally, for someone to 'participate' means to 'take part in' or 'to be involved in'. In this sense, everything people do is 'participation'. Hence, such a literal definition does not help much to inform interventionists on how to 'involve' stakeholders in innovation processes. The same is true for more descriptive approaches to defining participation, which could result in a definition like:

'Participation is everything that interventionists label as participation.'

Studying participation in this way can yield an interesting insight into, for example, the different meanings and practices that people attach to the term, the various reasons they have for doing so, and the consequences that ensue from interventions that are labelled 'participatory'. For intervention practitioners, however, such descriptive findings are useful especially if they are translated into guidelines on 'how to improve', which requires in essence that 'participatory' practices need to be mirrored against a normative framework that defines 'good' and 'bad' participatory practice. Not surprisingly, therefore, participation is often defined in normative and prescriptive terms, indicating that certain criteria must be met in order for something to count as 'participatory'. Along these lines, the World Bank definition on its website in 2001 was:

'Participation is a process through which stakeholders influence and share control over development initiatives and the decisions and resources which affect them.'

From this definition it can be derived that a process cannot be labelled 'participatory' if 'influencing' and 'sharing' of 'initiatives, decisions and resources' do not occur[1]. Similarly, participation literature suggests numerous normative 'principles' that must be adhered to during participatory processes (e.g. Chambers, 1994a; Pretty et al., 1995; Fals Borda, 1998a). From these it can be derived that:

- all relevant stakeholders should be involved in the participatory process;
- participants must have equal opportunities to speak out;
- participants need to be able to speak freely;
- the multiple perspectives (including values, interests, local knowledge and 'needs') of stakeholders must be explored and taken into account;
- 'ownership' needs to rest with participants as much as possible;
- participation must lead to the 'empowerment' of the participants;
- power imbalances among stakeholders need to be rectified as far as possible;

---

[1]   Note that the wording in this definition (in particular the terms 'share' and 'affect') suggests that the initiative to develop comes from outside, i.e. it is stakeholders who participate in a relative outsider's project. Starting from the idea that development initiatives exist anyway (Hounkonnou, 2001) and that external agencies (like the World Bank) are the ones that might want or need to participate in them, a rather different normative definition would result.

- it is illegitimate to intervene in a 'top–down' mode during participatory processes;
- the role of interventionists is mainly to facilitate critical learning and dialogue;
- participatory processes must be flexible and context specific;
- participatory processes must proceed on the basis of joint agreement and mutual respect.

The 'gap' between participatory rhetoric and participatory practice that we referred to in the introduction lies in the phenomenon that participatory practice (as shown, for example, by descriptive studies) hardly ever matches the criteria formulated by normative theories and definitions of participation (e.g. Guijt & Cornwall, 1995; Pijnenburg, 2003). In our view this discrepancy points not only to poor application of participatory principles (which indeed can be frequently observed), but also to fundamental flaws in the theoretical rhetoric (including principles) itself. In particular, we have worries about the phenomenon that the rhetoric helps to foster the belief among practitioners that during participatory processes 'decision-making authority' and 'control' must be 'handed over' to 'all stakeholders' as much as possible and at all times (see also section 14.1.2). We will refer to this overriding idea as 'maximum participation'.

14.1.2   Types and levels of participation

The idea of 'maximum participation' connects to a notion that there are different levels of participation. Widely used typologies and classifications of forms and levels of participation (e.g. Pretty, 1994[2]) seem to be based on three dimensions: the distribution of (a) information input and (b) decision-making authority between participants and interventionists in relation to (c) different key functions in development planning, such as situation analysis, problem identification, goal setting and implementation (see section 4.1.1). Other authors (e.g. Paul, 1986; Biggs, 1989; referred to in Okali et al., 1994:48–9) also use the level of involvement in decision-making as a basis for classifying different types and degrees of participation. With regard to information input and decision-making authority, the levels typically include (in ascending order, and adapted from Pretty, 1994 and Biggs, 1989):

(1) *Receiving information*: Participants are informed/told what a project will do after it has been decided by others.
(2) *Passive information giving*: Participants can respond to questions and issues that interventionists deem relevant for making decisions about projects.

---

[2] Pretty (1994:41), for example, distinguishes between 'passive participation' (people are informed about what is going to happen), 'participation in information giving' (people can respond to pre-defined questions), 'participation by consultation' (people can give their own views), 'participation for material incentives' (people participate because it gives them access to resources), 'functional participation' (people participate by creating conditions that are favourable for an external project), 'interactive participation' (people participate in joint analysis and decide on follow-up) and 'self-mobilisation' (people take their own initiatives).

(3) *Consultation*: Participants are asked about their views and opinions openly and without restrictions, but the interventionists unilaterally decide what they will do with the information.

(4) *Collaboration*: Participants are partners in a project and jointly decide about issues with project staff.

(5) *Self-mobilisation*: Participants initiate, work on and decide on projects independently, with interventionists in a supportive role only.

While some authors indicate that there is no 'best' level of participation (e.g Okali et al., 1994:21), others emphasise that only higher levels of participation can lead to sustainable results (Pretty, 1994:40) and that it is pertinent to 'hand over the stick' as much as possible (Chambers, 1994a, b). Among practitioners, the latter view seems to be most popular, as most of those I speak with argue, and write in project documents, that they strive to do things in a 'fully participatory' manner. On closer questioning, they tend to mean that they want to organise a project in such a way that all the major decisions are taken by, or jointly with, stakeholders. Although such comments are sympathetic and stated with the best of intentions, such a mode of operating is often not workable and/or productive. Moreover, the very suggestion that it *is* possible and desirable tends to mislead both interventionists and participants, and shifts the attention away from important areas of consideration. We elaborate on this below.

14.1.3 Is 'maximum participation' possible and desirable?

Questions that needs to be addressed are whether popular descriptions of levels of participation are useful, and whether it makes sense to strive for 'maximum participation'. In relation to the former question, our main concern is *conceptual*, as explained here.

The limited conceptual relevance of the notion of 'decision-making'

By defining levels of participation largely in terms of decision-making authority, it is suggested that 'decision-making' is indeed the central process in a participatory innovation trajectory. This is, for example, explicit in the earlier quoted World Bank definition of participation. This emphasis on 'decision-making' is closely related to the central concern with planning in many participatory projects, as formal planning and decision-making models strongly resemble each other (see section 4.1.1). Moreover, decision-making is often thought of as being *about* the different planning stages. Although of course eventual 'decisions' are important, we have argued that they do not normally result from a rational decision-making process, but rather they 'grow' from a gradual learning process (section 9.2). Also, 'decisions' in projects are shaped by a variety of social processes, including negotiation (see Chapters 5 and 10). Thus, one could argue that it would make sense to measure the level of participation in terms of involvement in learning or negotiation, rather than in decision-making. This would considerably change the outlook; after all, it is feasible that specific stakeholders have had a big influence on decision-making but have not learned anything and/or have not contributed in any way to the removal of tensions and obstacles to innovation. In short, the focus on decision-making is rather arbitrary, and trying to enhance stakeholders' decision-making authority is by no means a guarantee of successful innovation.

Apart from this conceptual concern, there are several reasons why it may not be very *productive* to always strive for 'maximum participation' as described in section 14.1.1. These are explained below.

Conflict management may require 'top–down' intervention and stakeholder exclusion

We have argued that meaningful changes hardly ever occur without social tensions and conflicts (see section 10.1). As we have elaborated in section 10.3, conflict management may well require affirmative action by interventionists. In order to create conditions for negotiation, for example, certain stakeholders may at a given time have to be excluded from the process. And during interactive processes, people may have to be put under pressure in the interest of the process (see section 10.3.5). It is very difficult to imagine how such crucial process interventions can be decided on in a 'fully participatory' mode, and in agreement with all stakeholders; such interventions are inherently 'top–down', at least for some.

The significance of agreed upon rules

In connection with the previous point, one could perhaps say that facilitators need to acquire a mandate in a participatory fashion, but that such a mandate will often need to include the possibility of taking unilateral process decisions if deemed necessary. In a course on participatory methods a student once came up with the analogy of a jazz orchestra and a symphony orchestra (Christian Gouet, pers. comm.) to illustrate this point. In his view, the participants of both musical ensembles are faced with the challenge (or problem) of creating a piece of music attractive to a particular audience. In a jazz orchestra, improvisation is often an important principle, which means that the musicians may not know in advance exactly what the piece of music will sound like. A basic musical scheme (e.g. a blues scheme) may have been chosen by the band leader or agreed upon jointly, and starting from there a piece of music is improvised and created on the spot, with the various musicians taking turns to lead and/or have their solo moments. A symphony orchestra tends to work in a very different way. There is usually a detailed 'script' for every musician, which needs to be followed closely, and there is someone who directs the musicians in a fairly authoritarian way, and who is in command of the way in which the piece is interpreted and performed. A question discussed during the course was whether or not a jazz orchestra was inherently more 'participatory' than a symphony orchestra. Under the assumption that the musicians of both orchestras joined on a voluntary basis and had a say in the musical repertoire, the conclusion was that this was not the case. By joining either orchestra, the musicians had subscribed to certain basic rules on how to produce a particular kind of music, and in both cases the rules could be seen as enabling and constraining. The analogy shows that it is too simplistic to equal all forms of leadership and authority as 'non-participatory'.

Innovation may require strong leadership within communities

Meaningful change usually does not occur without leadership; that is, without people pushing, pulling and sticking their necks out. Thus, projects that aspire to stimulate change cannot avoid working with existing community leaders (who are sometimes more interested in maintaining the status quo), or with other people who have leadership aspirations and qualities. There may be tension between, on the one hand, supporting and/or fostering leadership and, on the other, striving for 'full' and/or 'equal' participation. This is because leadership is indeed about 'taking the lead' even if others would like to move more slowly or differently.

The need to alternate between instrumental and participatory intervention

We have discussed in Chapter 11 that instrumental and more 'interactive' forms of communicative intervention can alternate successfully. Instrumental forms of intervention can, for example, contribute to creating appropriate conditions for a productive interactive process (see section 10.3). Similarly, an interactive process may give legitimacy to, and/or add to the quality of, more instrumental interventions. Instrumental campaigns may, for example, play a role in the further promotion of interactively developed innovations and policy solutions (see Chapter 11, also for limitations). It is important to note that, from the point of view of cost-effectiveness, it is often not feasible to fully 'repeat' interactive processes with every new audience or group of potential beneficiaries.

Alternating different forms of intervention is often also needed in connection with crisis situations, such as the outbreak of contagious animal diseases. Once an outbreak has happened, a government may have to impose all sorts of stringent measures in a fairly non-interactive manner. However, such measures are more likely to be effective when they are based on earlier interactions with stakeholders (e.g. farmers, slaughtering houses, veterinarians, animal transporters) about what to do in the case of an emergency (also in the light of previous experiences).

Participation is a scarce resource for participants

It is important to realise that, like learning, participation is a 'scarce resource' from the participants' point of view; it requires time, energy and other resources. Thus, similar factors to those affecting people's motivation to learn may also affect their motivation to participate (see section 9.4). There are instances where many people are only willing to invest limited time and effort, or where only a few participants are in a position to spend a lot of time without being remunerated or compensated for 'opportunity costs'. In practice, this scarcity implies that participatory activities must be well prepared so that optimal use can be made of the available time. 'Being well prepared' is not the same as 'having a rigid schedule'. Rather it means that flexibility is built in and that different scenarios are anticipated. Nevertheless, 'being well prepared' inherently implies that a number of choices need to be made about, for example, the methods and facilitation strategies that will be used, the agenda of debate and the order in which things will be addressed, the budget, etc. Again, this is at odds with the idea that 'everybody should participate fully in everything'.

In addition to the above reservations about being 'fully participatory' at every point, there are often constraining circumstances surrounding participatory traject- ories, which cannot be avoided. Several *practical* limitations may exist, as follows.

Boundaries posed by the politics of intervention and development

In social life, 'participation' is a far from neutral term. It has become a buzz-word and the central term in a language of 'participation speak' in which many related concepts play a role. Being able to speak the language well is likely to increase the chances for NGOs and development organisations to attract resources and funds from large donors, which can be used to further certain aspirations. Thus, 'participa- tion speak' is used as a strategic asset for mobilising scarce resources in the higher levels of the development arena. The process of distributing and/or gaining access to such resources often involves making project proposals in fixed formats, and the use of log-frames (GTZ, 1987; Shields, 1993), tendering procedures, etc. In view of such 'rules of the game', it is inevitable that NGOs, farmers' organisations, government bodies, etc. form coalitions and strike deals with each other and with funding agen- cies. During these preparatory stages of projects, this usually happens largely behind the back of prospective beneficiaries. In the process all sorts of boundaries and conditions are posed to later activities, e.g. in the form of goals, definition of sup- posed beneficiaries, budgets, methods, time planning, regions involved, evaluation criteria and procedures, etc. Even if a number of open ends and flexibilities are built in, it is undeniable that such maneouvres limit the scope of a project and restrict possibilities for participants to later move into different directions. Moreover, once a participatory trajectory takes off at local level, it implies again the creation of a certain 'space' for influencing further development activities and for further alloca- tion of resources.

Not surprisingly, therefore, different actors within and around communities (e.g. project staff, representatives of specific groups in the community, local government, local leaders, etc.) often have a keen interest in influencing the outcomes of parti- cipatory deliberations, and are likely to put pressure on others to move in certain directions (see section 5.2.4). In doing so, resourceful people may, for example, use their resources to stimulate some people to show up and participate and others not, and to get certain issues raised while others remain hidden. Such local struggles and politics can hardly be avoided[3], and again pose limitations to openness, critical learning and 'maximum participation' as prescribed by normative principles.

In summary participatory projects usually take place in a complex institutional environment. As Craig and Porter (1997) indicate, a participatory project may mean different things to the various parties involved; that is, they may 'frame' a project differently and put forward different ideas as to what it should be about. Clearly, the 'frames' that become dominant are bound to constrain local participants.

---

[3]   Other authors have pointed out that it is misleading to think of communities as homogeneous and harmonious entities (see Guijt & Shah, 1998).

Constraints may derive from pre-existing local agendas, responsibilities and arrangements

As outlined above, participatory interventions never take place in a social vacuum. In almost all communities there are forms of organisation and leadership that, formally or informally, have responsibilities and take action on issues of change and development. Often, participatory intervention projects have a tendency to overlook or deliberately ignore such existing institutions, and create their own forms of organisation (e.g. separate committees, meetings, etc.), because the latter are considered more 'democratic' and conducive to 'maximum participation' (e.g. Pijnenburg, 2003). However, it may well be that members in a community prefer to work through their existing institutions (even if these are less 'participatory'), or prefer to avoid getting into conflict with these, and hence 'refuse' to participate. It is significant to note here that for many 'to participate seriously' also implies 'to accept responsibility for'. In some cases people may find this inappropriate in view of, for example, cultural norms, or undesirable because it shifts the responsibility away from where people feel it belongs. When our university was reorganised in the face of budget shortages, for example, few senior staff felt an urge to participate actively in developing an alternative proposal in which disciplinary groups should be abolished and/or staff laid-off. Most people felt that this was an issue for the board of the university, as they were the ones hired to do that difficult job. Thus, apart from the time and resource-related reasons referred to earlier, people may have 'political' reasons for not *wanting* to participate fully.

Conclusion: It is not useful to deny the intervention and political dimensions of participation

It is important to realise that the above mentioned limitations to 'maximum participation' (as defined conventionally, see sections 14.1.1 and 14.1.2) are likely to be relevant to situations where those involved in initiating or facilitating participation are *genuine* in their intentions, i.e. where there is a real concern for 'bottom–up' policy-making, innovation and development. In addition, other motives frequently lead people to use the language and methods of 'participation'. As we have already hinted in section 13.5.11, the rhetoric of participation may be used solely to gain access to resources, or for organisational image management (window-dressing), or to provide legitimacy to already pre-conceived policies. Similarly, lack of resources, skill and understanding of participatory principles may lead facilitators to engage in ritualistic participation practices (see section 13.5.11). Unfortunately, it is not always easy to distinguish between more and less legitimate ways in which the space for participation is narrowed during interactive processes. This is not only because they may look similar in practice, and because it is notoriously difficult to look into people's minds, but also because different opinions may exist among stakeholders and intervention staff involved.

In any case, the more genuine forms of narrowing space stem from the fact that (a) development and change are inherently political phenomena, and (b) the participatory trajectories that we are interested in always remain at the same time '*interventions*'.

There is a tension between, on the one hand, the assumptions which underlie the idea of 'participation', and, on the other hand, the rationale of 'intervention'. Many participatory approaches assume that human beings are active and knowledgeable agents who can make a crucial contribution to their own development. In contrast, the basic idea behind intervention is that actors *lack* crucial 'ingredients', so certain means and facilities must be brought in from outside (see also Long & Van der Ploeg, 1989). Hence, 'participatory intervention' falls between two opinions when it comes to the evaluation of actors' endogenous development. In our view, it is unhelpful to deny this 'participation paradox' (see Leeuwis, 1995), and to act as if participation has (or should have) nothing to do with intervention, and as if particip-ants are the only ones in command. Similarly, pretending that one can ban politics from participatory trajectories is not very productive either. One can only deal with the risks and problems that arise from this if one fully recognises that they exist and why. Moreover, one cannot instill realistic expectations in participants if one does not communicate honestly about – and reflect critically on – externally imposed boundaries and conditions, political realities, leadership needs, etc.

14.1.4   Reservations at the theoretical level: strategic versus communicative action

In the conventional literature on participation and interactive processes, the exist-ence of politics and conflicts of interests is not denied (e.g. Pretty et al., 1995:70). However, it is suggested, implicitly or explicitly, that frictions between stakeholders can be resolved through the development of a shared understanding of a situation, as a result of joint learning and improved communication (see also section 10.1). The trainer's guide by Pretty et al. (1995), for example, provides numerous tools that essentially aim at enhancing communication within groups; these include exercises for improving group dynamics, listening, observation exercises, visualisation, etc. In philosophical terms, these ideas can be traced back to Habermas' (1981) notion of communicative action, as distinguished from instrumental and strategic action (e.g. Fals Borda, 1998b; Bawden, 1994; Röling, 1996; Maarleveld & Dangbégnon, 1999).

Instrumental action, according to Habermas, is behaviour which involves follow-ing technical prescriptions – based on nomological knowledge[4] – in order to achieve certain previously defined goals (Habermas, 1981:385). Strategic action is still ori-ented towards the realisation of specific goals, but the actor recognises other actors as equally strategic opponents, rather than as 'objects' that obey certain nomological rules. In this strategic process, actors bring in various power resources to further their aims. Finally, Habermas speaks of communicative action (and communicative rationality), when actors aim to reach agreement or consensus on a shared definition of the situation as a basis for co-ordinating their activities. This type of action is distinguished from instrumental and strategic action in that the co-ordination of action does not arise from egocentric goal-oriented 'calculation' by self-interested

---

[4]   Nomological knowledge is knowledge based on empirical laws and regularities (see Koningsveld & Mertens, 1986:12).

actors, but from an open process of argument in which any claim (including normative ones) is subject to critical debate.

According to Habermas, communicative action requires an 'ideal speech situation' in which undistorted communication can take place (Habermas, 1970a, b, 1981). In such a situation conflicting situation-definitions (based on diverging interests) can be solved by the 'peculiarly unforced force of the better argument' (Habermas, 1973:240) rather than by bringing in power resources. In essence, it is this kind of 'power free' situation that advocates of participatory approaches aspire to achieve. Hence, our reservations about mainstream participation thinking go deeper than the somewhat uncritical belief in 'maximum participation'. Many of the points made in section 14.1.2 already reflect that we have practical and theoretical reservations about the notion of communicative action as a 'model' for organising participatory trajectories. At this point we will raise some additional concerns that relate specifically to this notion.

The wishful thinking dimensions of communicative action

Although the idea of communicative action is sympathetic, we feel that the reasoning applied for advocating it as a conflict-resolution strategy is flawed. In a typical situation of over-exploitation of natural resources, for example, the argument for communicative action and social learning boils down to the following: (1) There is a problem because people do not act in a communicatively rational manner (in the collective interest), but act in a strategic mode (following self-interest). (2) As a solution, they must engage in a process of communicative action (facilitated by a third party), so that they can act in a communicatively rational manner in the future.

The difficulty here is that the solution proposed is in fact a negation of the problem. What is lacking in such a 'wishful thinking' solution is an analysis as to why, in a given situation, people act in a mode which is deemed undesirable by policy makers. In the case of resource use, the implicit assumption is apparently that people cannot behave in a communicatively rational manner because there is no one who is able and/or willing to organise and facilitate such a process. Even if in many cases this factor may play a role, it is unlikely to be the main one. The use of natural resources like land and water is often a critical element in different actors' livelihood strategies and opportunities (Manzungu & Van der Zaag, 1996). Given the increased scarcity of natural resources, conflicts of interest and different opportunities to defend such interests are the rule rather than the exception (e.g. case-studies presented in Gronov, 1995; Mosse, 1995; Manzungu & Van der Zaag, 1996). In such cases, it is quite possible that the problem is not so much that no external facilitator is available, but rather that some actors, for a variety of reasons, are simply *not willing or able* to seriously take part in communicative platform processes (see also section 10.1).

The untenable separation between strategic and communicative action

The celebration of communicative action as a key model for organising participatory trajectories becomes questionable in view of some conceptual difficulties related to Habermas' separation of communicative and strategic action. Elsewhere (Leeuwis,

1993:347–86; Leeuwis, 1995), I have shown that the occurrence and outcomes of interactions that in themselves might be termed 'communicative action', can only be adequately understood if one recognises that they are at the same time strategic actions in relation to *other* communities of actors. Whether or not actions are strategic or communicative, therefore, seems to depend on where one draws the boundaries between communities of actors in time and space, which considerably blurs the distinction. This also calls into question a related assumption which underlies the social learning model, namely that the motor for future societal progress is consensus and shared understanding. Evidence shows that this is only partly true since effective consensus among some (i.e. consensus that leads a specific set of actors to generate tangible 'progress') is frequently based on conflict and competition with others (e.g. case 1 in section 10.1.1).

Incompatibilities with insights from constructivism

By emphasising social learning and communicative action, advocates of participation tend to make cognitive change the prime prerequisite for behavioural change and conflict resolution. By focusing on cognition and argumentation as *the* entry point for change, authors on participation – even if inspired by constructivism (Pretty, 1994; Röling, 1996; Fals Borda, 1998b) – seem to disregard one of its major insights, namely that differences in (i.e. conflicts of) perception are not just cultural or accidental, but are intricately intertwined with social interests and practices (see section 6.3). In other words, social practices and interests shape perceptions as much as the other way round (e.g. Aarts, 1998). In this context, Habermas' idea of making 'the better argument' count seems rather impractical, as in practice people are likely to differ on the issue of what is 'the better argument'. In theoretical terms too, this notion is quite problematic[5] (see also Chapter 6). Moreover, by focusing on cognitive processes as an entry point for inducing change, advocates of participatory methodologies tend to disregard a range of other strategies or 'policy instruments' (Van Woerkum, 1990a; see also Chapter 4) that may help to change practices, interests and, eventually, perceptions in the context of conflict management.

Practical difficulties in realising an 'ideal speech situation'

Apart from theoretical problems, there are considerable practical problems in creating the conditions for communicative action (i.e. 'an ideal speech situation'), if one succeeds in bringing all relevant stakeholders together in, for example, a platform for social learning (which already is something that often proves difficult to achieve, e.g. cases 4 and 6 in section 10.1.1). It will not be easy to make actors set aside their conflicting personal and/or institutional interests during the process. Even if one

---

[5] Although Habermas does not believe in the existence of objective (in the sense of politically neutral) knowledge, he eventually still seems to rely considerably on science in distinguishing between 'good' and 'bad' arguments (for a more detailed discussion see Leeuwis, 1993:95–9). In our view, therefore, there is a conceptual tension between, on the one hand, accepting and advocating communicative action, and, on the other, embracing constructivism (e.g. Röling, 1996).

assumes that the actors abstract temporarily from their strategic interests and power positions, they will usually have different resources at their disposal (e.g. in the form of knowledge, time, access to certain sources) with which to make and criticise certain validity claims concerning truth, normative rightness and authenticity (see Habermas, 1981). Thus, even if the opportunities to speak out are equal, the possibilities of making claims and criticising them are not.

In the face of these and other difficulties (see section 14.1.2) we feel that it is neither possible nor productive to ban strategic action (in the Habermas sense) from participatory trajectories. In line with what we have concluded in section 14.1.2, we feel it is necessary to develop an approach towards participation that does not negate, conceptually and methodologically, the significance of strategic action and conflicts of interest by rendering them 'normatively undesirable'. Rather, we need an approach that starts from the assumption that actors are likely to act strategically in relation to existing and emerging conflicts of interests, and attempts to make this productive to solving societal problems (see also Edmunds & Wollenberg, 2001). In Chapter 10, we have already suggested that the notion of 'integrative negotations' – i.e. negotiations that incorporate social learning – may serve as an alternative model for shaping interactive processes.

14.2 Guidelines for the facilitation of interactive processes

With the term 'facilitation' we refer to the more or less deliberate use of communicative strategies and methods in order to enhance social learning and negotiation towards innovation in a multi-stakeholder setting. This implies that facilitation may involve a broad range of activities, geared towards, among others, creating 'platforms' (see section 2.2.1), improving insight, explicating tacit knowledge, managing conflict, creating productive group dynamics and bringing about co-ordinated action. As we have elaborated in Chapter 10, facilitation is complex, far from easy and needs to be understood as having 'political' dimensions (see sections 10.2.2 and 10.3.5). Based on the idea of 'integrative negotiations' we have already identified several facilitation tasks in section 10.2.1. For each of these a number of recommendations can be made. Below we present some guidelines and factors that may affect the progress made. This overview is based partly on a literature study that was conducted mainly by Van Meegeren, and published in Van Meegeren and Leeuwis (1999:210–15). The guidelines should be treated as a selection of tips relevant to tasks that are important within integrative negotiation processes. They should not be looked upon as 'steps' that need to be followed in a linear fashion, as all tasks remain relevant continuously (see sections 10.2.1 and 10.2.2). A creative, selective and purposeful application of such tips increases the chances of a balanced and productive learning and negotiation process, but is still not a guarantee of success. The course of human interactions is too capricious and unpredictable for that.

14.2.1 Task 1: Preparing the process

The involvement of communication workers in interactive processes can begin in different ways. Sometimes, interventionists are asked by societal actors to play a role in furthering local initiatives for innovation and problem-solving. In other cases, new 'projects' evolve more or less 'organically' from previous involvement with particular networks or communities of actors. Also, interventionists themselves frequently undertake project identification initiatives in communities or networks that they have not worked with before. Clearly, the need for engaging in deliberate preparatory activities may differ accordingly. Such activities may require considerable time and attention since one cannot start an interactive process 'out of the blue'. Several preparatory issues may require attention, many of them relating to whether or not conducive conditions exist or can be created for an interactive process with outsider involvement.

Reconnaissance of existing or past initiatives and local innovation capacity

When dealing with problematic situations, one can safely assume that actors have already taken and/or thought about initiatives that could help to improve the situation. It is important to identify such, often 'hidden' (see section 8.2.6), endogenous initiatives and experiences, and see how they have developed and what obstacles have transpired during such problem-solving efforts. It is equally important to explore existing social arrangements and forms of organisation, and their capacity to contribute to innovation and change. As Hounkonnou (2001) has shown, development initiatives may well originate from organisational forms (e.g. sports clubs) that interventionists do not usually associate with innovation and development. Hence, endogenous forms of organisation (i.e. potential carriers of, and/or stakeholders in, interactive innovation processes) may be as invisible to outsiders as existing or past initiatives.

A range of exploratory methods and tools (see section 13.5) may be helpful for this kind of reconnaissance, which is particularly useful for developing a broad understanding of where (if at all) external intervention might fit in, and how it might be linked with existing initiatives and forms of organisation. An analysis of earlier intervention experiences may also help in this respect. It is often advisable here to use 'low profile' media and methods (e.g. in-depth interviews and small focus group discussions; see section 13.5.2), since there may be very little to tell yet about whether external intervention is desirable and feasible, and what this might entail. Despite their widespread use, large community meetings are usually not very conducive to getting an in-depth insight into existing experiences and initiatives, and may well raise expectations that cannot be met later. Similarly, in-depth exchanges and bilateral or small group meetings make it easier for interventionists to show genuine interest in people, as a first step in a process of *building trust*.

Preliminary stakeholder analysis and conflict assessment

Problematic situations usually have various layers and dimensions (e.g. sections 5.1 and 10.2.3), which means that endogenous and/or external initiatives and innovative solutions have implications for various actors at different societal and/or governance levels. When preparing an interactive process, it is important to identify those whose interests are 'at stake' (i.e. the stakeholders) in maintaining and/or changing a multi-dimensional problem. This is not always as straightforward as it seems, as people can always be categorised in different ways. In a given context, for example, one may wonder whether small and large farmers, male and female farmers, or young and old farmers constitute separate 'stakeholders' (see section 15.2.1 for elaboration on assessing relevant diversity). Once stakeholders have been identified, it is significant to acquire historically sensitive insight into, for example, the different aspirations and interests of various stakeholders, the nature and quality of relationships between them, the various forms of co-operation and conflict (see section 10.2.3) that exist among them, the resources and capacities that stakeholders may mobilise to influence outcomes, etc. In order to gain insight into such matters, specific stakeholders can, for example, be asked to review and analyse current and past experiences during in-depth interviews or small group discussions. As sensitive issues may be involved, it makes little sense to bring all stakeholders together for this purpose. At a preparatory stage, one cannot expect actors to speak openly about their experiences with others when these others are present.

Identifying broad areas and boundaries for interactive intervention

When engaging in explorations as mentioned above, it is important that interventionists do not immediately limit the scope for debate as a result of pre-conceived problem definitions or narrowly defined organisational mandates. However, at a certain point outsiders will have to reflect on how emerging insights relate to their own capacities and mandates. In this context, interventionists need to reach a common understanding with at least some of the stakeholders on where, roughly speaking, they can play a meaningful role. Questions that may need to be answered in this process are whether or not it is desirable and possible to stretch the organisational mandate and capacity, and/or to which other institutions issues can be referred. Similarly, before eventual participants can be selected (see following paragraphs) it is not only necessary to agree on a substantive area of concern (i.e. what the project is roughly about), but also on sensible preliminary boundaries in terms of time (e.g. duration of the external involvement) and space (e.g. geographical limitations). As we have discussed in section 14.1.3, it is often unavoidable that intervention poses certain restrictions and limitations on the choice of boundaries (in terms of area of concern, actors, means, time and space). In so far as such restrictions are known in advance, they need to be discussed openly with prospective participants. This is not only to give people a fair chance to determine whether their participation may be worthwhile, but also to find creative ways to circumvent such restrictions.

Assessing and creating institutional space for using results from interactive processes

Interactive processes often only make sense if there is a fair chance that the wider institutional policy environment will react positively to innovative results from integrative negotiations (see section 10.3.4). This implies that it makes little sense to organise such processes in isolation from relevant policy institutions; they too are stakeholders. Ensuring a good link (e.g. in terms of timing) between interactive processes and formal policy procedures can be essential (Vermeulen et al., 1997). Thus, regular communication with bureaucrats, policy-makers and/or politicians throughout an interactive process deserves careful attention. In many situations, however, one may not want such people to be present as participants during the whole process, as this might limit the space for a creative and open process. In many ways, forging linkages between interactive processes and the policy environment can be regarded as a negotiation process of its own. In particular, attention must be paid to reaching agreement on the status of interactively developed solutions. In other words, it must become clear to what extent there is a commitment by institutions in the policy environment to follow up the outcomes of interactive processes.

Interdependence as a criterion for selecting stakeholder participants

As we have discussed in section 10.3, a crucial condition for overcoming pre-existing conflicts and emerging tensions in an interactive innovation process is that participants feel mutually interdependent in solving a problematic situation. Hence, we feel the eventual stakeholder participants can best be selected on the basis of this criterion. This may effectively result, at least temporarily, in the exclusion of stakeholders who are in a relatively powerful and comfortable position (see section 10.3.2). If necessary, feelings of mutual interdependence and urgency can be stimulated, for example, by the introduction of certain external pressures (i.e. carrots, sticks and threats), deadlines and/or the timing of initiatives (see section 10.3.2). Provided that stakeholders feel interdependent, we agree with the warning by Susskind and Cruikshank (1987) not to exclude certain parties in order to smooth the process. In a similar vein, De Bruin and Ten Heuvelhof (1998) point to the positive side of involving diverse stakeholders, as it widens the scope for 'package deals'. Moreover, an inclusive approach offers stakeholders an opportunity to widen their network, which may make participation more attractive. Finally, to allow for balanced representation of interdependent interests in an interactive process, it can be necessary to support structurally and financially less well organised interests (see section 10.3.3 and Susskind & Cruikshank, 1987).

Selecting stakeholder representatives

As there are usually practical limitations to the number of people who can participate actively in an interactive process, one often needs to work with people who represent others. Whether or not interactive processes succeed depends partly on the presence of a certain amount of positive 'chemistry' between those who sit round the table (or under the tree). Hence, the selection of representatives deserves attention. Clearly, interventionists are not always in a position to influence who represents who. If possible, however, it makes sense to stimulate the selection of people who have good communication skills, are trusted by their constituents, and have a certain amount of 'credit'. Furthermore, the productiveness of negotiations can be enhanced when people are involved who are likely to be change minded, and are known to have qualities like imagination, freedom of spirit and empathy. Such people are often better able to act as brokers between their own constituents and others than those who are good at defending certain interests from a narrow perspective. Similarly, in situations where there is a history of conflict, it may be wise to select representatives who have experienced earlier conflicts in a less personal manner.

14.2.2   Task 2: Reaching and maintaining process agreements

To get the various stakeholders into the process, it is often necessary to pay attention to procedural issues. Agreements on procedures may be required to enhance trust among participants that the interactive process will be organised fairly and/or yield positive results. Such agreements can relate to the terms of reference (ambitions and overall objective of the process), provisional agenda (what will and will not be discussed), rules of conduct (presence of external observers, format of discussion sessions, dealing with the media, protection of participants, the mandate of representatives, etc.), choice of methods (see Chapter 13) and/or the role of external facilitators and other outsiders. Particularly in the case of pre-existing or emerging conflicts, procedural issues tend to be important, not only for the participants but also for a facilitator. For the latter, previously agreed ground rules can provide some of the power, authority and legitimacy that may be needed to put a deadlocked learning and negotiation process back into motion (see also section 10.3.5 on the political dimensions of facilitation). During an interactive trajectory, process agreements need not only to be maintained, but also adapted and developed, as one cannot predict in advance how the process will unfold. Experience and literature suggest various recommendations, as follows.

Ensuring sufficiently flexible terms of reference, agenda and procedure

An important condition for people to participate in a process is often that there is a sense of purpose and direction. Moreover, without a clearly stated ambition and overall process objective, it is impossible to facilitate an interactive trajectory. Depending on the problematic context and agreement among stakeholders, overall process objectives may vary in nature and also with regard to the ambition to move beyond learning, and work towards new forms of joint action and commitment. Thus, examples of process objectives are: diagnosing a problematic situation, generation of creative ideas and solutions, improving relationships between stakeholders, overcoming conflict, designing and/or realising socio-technical solutions, enhancing co-ordinated action among stakeholders, etc. Although preliminary terms of reference are important at the start of a process, it may be necessary and/or productive to change them during the process in view of unforeseen developments and changing stakeholder ambitions. Particularly in situations where relationships are disturbed and/or trust among stakeholders is low, it may be a useful strategy to set fairly modest process objectives initially (e.g. diagnosing a problematic situation) in the hope and expectation that more ambitious objectives can be set later. Given broad terms of reference, the setting of a more detailed provisional agenda can start with an inventory of all the issues and themes that participants would like to raise at the outset (Susskind & Cruikshank, 1987:111). De Bruin and Ten Heuvelhof (1998) insist that the agenda for debate must remain sufficiently open, and warn against too many restrictions at the process level, so as to ensure that all stakeholders can bring forward their views and interests, and influence the outcomes. Widening the agenda for debate also increases the opportunities for making comprehensive 'package deals' (see section 14.2.5).

Clear and sufficiently wide mandates

In many interactive processes the participants represent others. When this is the case, it is important to clarify the mandate of different representatives, i.e. to investigate to what extent the representatives are free to operate without consultation with their constituencies. It can also be necessary to set some minimum criteria in this respect, as too many restrictions placed on representatives can lead to a complicated discussion structure, which could cause an increase in the bureaucracy involved in the process. When the overall objective of a trajectory is restricted to exploration only, it is sometimes more productive to invite participants on a personal basis only. This usually allows actors to speak more freely as they have to worry less about whether their organisation or constituents agree with what they say (e.g. Leeuwis et al., 1997).

A right to withdraw from earlier proposals

As a way of protecting participants it may be sensible to reach agreement on the status of intermediate proposals and negotiation outcomes. Participants may be offered the right to withdraw from early negotiation results (e.g. after consultation with their constituents), in order to prevent people feeling increasingly trapped as the process unfolds. When such arrangements are not in place, stakeholders are likely to be more cautious and hesitant than is desirable for the process.

Creating 'protected space' and conditions that prevent 'disempowerment'

On the issue of 'protection', we have seen in section 8.2.3 that innovation processes require a certain amount of 'protected space' to allow socio-technical solutions and new arrangements to mature, and mistakes to be learnt from. Thus, it is important in interactive innovation processes to develop appropriate arrangements in the form of funding, insurances, etc. (see section 8.2.3). Another issue that requires careful attention is the use of information and knowledge provided by stakeholders in an interactive process. In the context of natural resource management, for example, participants may be asked to give very detailed information about the location of specific resources (e.g. water, fish, biodiversity, etc.), occupation patterns, resource use, agro-ecological conditions, social networks, local conflicts, etc. Moreover, such information may be 'stored' in a GIS or three-dimensional model (see section 13.5.5).

Although it has been documented that the collection and visualisation of such information may strengthen the position of local communities in their communication and negotiation with outside institutions (e.g. Gonzalez, 2000; Rambaldi & Callosa-Tarr, 2002), one can easily imagine that outside institutions may use such information *against* particular communities or stakeholders as well. Once government institutions *know* where people fish, cut wood, live and graze cattle, for example, they may use that information to effectively impose restrictions. Thus, due to the inherent connections between knowledge and power (see section 6.6), the insight and knowledge generated through explorations can be 'empowering' or 'disempowering'. Of course it is very difficult to foresee in advance whether the sharing of information with others will prove to be 'empowering' or 'disempowering'. Nevertheless, it is important to create conditions in which people feel safe to share information. In this context, agreement may be needed on rules (and enforcement mechanisms) regarding how and where information will be stored, who is allowed to have access to it, and under what conditions.

Division of facilitation tasks, and defining the role of 'outsiders'

Facilitation should not be looked upon as needing to be done by one person only, or to be performed necessarily by outsiders (for elaboration see sections 10.2.1 and 11.2). Thus, it is important to reach a common understanding on the roles and responsibilities of various insiders and outsiders (including facilitators and scientists) in furthering the process. Again, it is important to be flexible, and monitor how tasks are being performed and whether or not changes are desirable (see section 11.2.3). In particular, when interactive processes are not very conflictive, build on local initiatives, and involve well-established forms of organisation, the role of external facilitators may be rather limited. In opposite circumstances, a much more prominent role may be required.

For any innovation process to be sustainable, however, it is clear that process management responsibilities will eventually have to rest primarily with those who can be expected to make the innovation work in the long run (i.e. the stakeholders themselves). Thus, it may sometimes make sense to have professional facilitators play a fairly significant role initially, and then withdraw gradually once positive outcomes start to take shape. It should always be kept in mind that 'playing a significant role' is not the same as 'being the centre of attention' and or 'running the show'. Rather, it means creating conducive conditions for *others* to make progress, such as through supporting the use of creative working methods and maintaining process agreements. Interactive processes are not likely to be productive when participants do not have, or develop, the feeling at an early stage that it is somehow *their* process, for the success of which they carry considerable responsibility. In other words, facilitators should be careful not to take over the *ownership* of a process (on the issue of ownership see also Hounkonnou, 2001).

14.2.3   Task 3: Joint exploration and situation analysis

Exploration is a crucial element in interactive processes as it forms the starting point for the changes in perception and cognition (i.e. learning, unlearning, reframing) that are needed to bring about innovation (see Chapter 9). Depending on the situation, the specific issues being explored may vary. In more general terms, explorations may centre usefully around the issues in Table 14.1.

As can be noted from Table 14.1, we do not associate exploration primarily with '*needs assessment*', a term that is used rather frequently in participation literature.

Table 14.1   Areas that may be explored usefully in interactive processes.

Keyword	Elaboration
Context (task 1*)	analysis of the agro-ecological system, physical infrastructure and socio-economic institutions in historical perspective (e.g. trends and stabilities)
Initiatives (task 1*)	identification of existing and past initiatives, the way they have evolved and why
Innovation capacity (task 1*)	analysis of existing social arrangements and forms of organisation and their capacity to contribute to innovation processes
Stakeholders (task 1*)	identification of relevant stakeholders and their history of interaction, e.g. in terms of conflicts, co-operation, competition, feelings of interdependence, etc.
Problem perceptions	eliciting stakeholders' different social and technical problem perceptions with respect to the current situation, trends and/or likely scenarios
Current practices	identification of the social (including economic and organisational) and technical practices through which stakeholders contribute to and/or reproduce the problematic situation
Current reasons	analysis of the various reasons that different actors may have for engaging in existing practices (i.e. in terms of what they 'believe', 'want', are 'able' and 'allowed/expected' to do; see Chapter 5)
Conflicts/tensions	identification of tensions and conflicts in terms of underlying values, norms, aspirations, interests, knowledge, etc. (see section 10.2.3)
Commonalities/ overlap	identification of commonalities and overlap in terms of underlying values, norms, aspirations, interests, knowledge, etc. (see section 10.2.3)
Solutions/ opportunities	identification of possible social and technical solutions and alternative scenarios to the problematic situation
Alternative practices	identification of the alternative social and technical practices that stakeholders may need to engage in to help realise solutions and alternative scenarios
Obstacles/reasons	analysis of the various reasons that different actors may have for engaging or not engaging in alternative practices (i.e. in terms of what they 'believe', 'want', are 'able' and 'allowed/expected' to do; see Chapter 5)
Policy instruments	identification of 'policy instruments' that may help to overcome obstacles to the realisation of solutions (see sections 4.1 and 5.2.2)
Knowledge gaps	identification of areas of limited understanding that need further investigation
Process	identification of stakeholders' views on how facilitation of the process can be improved
Consequences (task 7†)	monitoring of the way in which agreements are translated into practice, and the intended and unintended consequences of that

* See section 14.2.1
† See section 14.2.7

This is not only because there are many things besides 'needs' that may require exploration (see Table 14.1), but also because the idea of 'needs' itself is somewhat problematic. Parallel to what we have argued in connection with the identification of 'information needs' (see section 13.6.2), we believe that 'needs' cannot be simply 'assessed' at a particular time, but need to be discovered (and in many cases agreed upon) in a creative learning and negotiation trajectory that includes a confrontation between different stakeholders, insiders and outsiders. Like the idea of 'problems', the concept of 'needs' refers to a tension between actors' aspirations (i.e. wants) and perceptions of the current situation (see section 8.2.5). Despite this overlap, we feel that 'problems' form more useful entry points for exploratory activity, as they can be more easily linked to further analysis of underlying perceptions, practices and reasons. Such problem analysis, in turn, is more likely to lead to a redefinition of what exactly are the problems that innovations must address. In contrast, the idea of 'needs' already carries with it the idea of a specific solution and thus might hinder additional reflection. At the same time, we feel that 'needs assessment' tends to reinforce the idea that something from outside must be provided, rather than that existing innovation and problem-solving capacity is to be mobilised and supported. We can conclude that the focus can only shift usefully to concrete needs *after* sufficient problem analysis and exploration.

In section 13.5 we have already discussed a range of methods and strategies that can be used for exploratory purposes. Depending on the context, facilitators and participants will have to decide on appropriate methods and tools. While the emphasis in section 13.5 is on methods and strategies for learning, we will at this point add some insights from negotiation literature in particular.

Make exploration a continuous process

It is important to recall that exploration, like other tasks, is important throughout interactive learning and negotiation processes. In other words, when working on other tasks (e.g. preparation, process agreements, joint fact-finding, communication with constituencies, etc.) exploration is a vital condition for arriving at productive and creative outcomes. We emphasise this because we have often witnessed that exploration is 'boxed' in a narrow time frame (e.g. two weeks at the identification or start of a project), and largely 'forgotten' afterwards. During innovation processes, however, one is likely to regularly encounter new challenges, problems and obstacles. Overcoming these requires equally regular exploration.

Focusing on disturbed relations first

According to Fisher and Ury (1981), conflictive situations are usually characterised by two types of conflict: relational conflicts (i.e. damaged human relationships) and substantive conflicts (e.g. water distribution). In relation to the former, considerable estrangement, distrust, protectiveness and stigmatisation may exist at the outset between individuals and/or groups. Stakeholders may see things from an 'established' versus 'outsiders' (or 'us' against 'them') perspective (Elias & Scotson, 1965; Aarts, 1998). In such cases, the image of the entire group of outsiders ('them') is modelled on the 'worst' characteristics of the 'worst' part of that group. In contrast, the self-image of the established group ('us') is modelled on the most exemplary part, on the minority of its 'best' members (Aarts, 1998:33). Fisher and Ury suggest that it is advisable to work on the improvement or restoration of human relations first, before tackling the substantive problems. In working towards a more constructive atmosphere, it may be necessary to first create some space for the expression of feelings of mutual reproach regarding events in the past. It is only after stakeholders have expressed what is really bothering them that they may find mental space to pay attention to different perspectives. Such 'letting off steam' may have to be encouraged by a facilitator, but often also happens by default when one brings parties in a conflict together. After some time it becomes important to make the point that the conflict is not going to be resolved by 'shouting at each other', and to introduce alternative methods for exploratory communication (see section 13.5). To improve human relationships and the atmosphere in an interactive process, it can also be helpful to create opportunities for informal communication and positive joint experiences. Here we can think of choosing a nice working environment, organising good food, sufficient time for breaks, drinks, sightseeing, etc.

Joint reflection on group/process dynamics

Whenever a group of people work together in an intensive process, personal irritations and frictions are likely to emerge between some of the participants. These may relate, for example, to the styles in which individuals operate, the way they react to critical remarks and suggestions, etc. Such frictions can be a threat to interactive processes, since they may spoil the atmosphere and undermine the formation of collective spirit. Hence, it is important that such tensions are recognised, and dealt with before they grow to unmanageable proportions. Creating regular opportunities for people to give feedback on the process as a whole, or to individuals within the group, may help to identify problems and/or 'clear the air' at an early stage. However, as giving and taking feedback are not always easy (see section 9.3), it may be important to pay some attention to enhancing participants' skills in this respect, so as to prevent the exchange of such feedback itself becoming a source of conflict. An implication of our earlier discussion on feedback (see section 9.3) is that it is important to encourage people to provide both negative and positive feedback on the process as a whole, or on individuals' mode of operating in it. Another basic ground rule is that people would be wise to 'stick to themselves' when formulating confrontational feedback. This can be achieved by using a format like: When *you* do X (e.g. shout at me), it makes *me* respond Y (e.g. I can no longer say what I really think), *because* Z (e.g. I feel threatened and not taken seriously). Someone presenting feedback in this way clearly makes his or her point, but the feedback may be seen less as an attack, as the sender effectively speaks more about himself/herself than about the other person. In balancing positive and negative feedback, it can also be useful to realise that weaknesses are frequently 'strengths carried to the extreme' (e.g. Ofman, 1992). A person may be very energetic and a sharp thinker, and fall into the trap of being dominant and silencing everybody else in a group meeting. When criticising the dominance, one can still welcome the energy.

In situations where people may not feel safe to provide personal feedback, it may be a useful strategy to ensure that feedback is given more indirectly and/or anonymously; for example, by asking for more general remarks regarding the process (e.g. strengths and weaknesses) and/or by using metaplan cards (see section 13.5.3). The chances are that those who should take specific notice of remarks will draw their own conclusions, even without being addressed directly and personally. A last guideline in this respect is that it often makes sense for facilitators to limit the space for personal responses to feedback, because people may react in a defensive mode and/or try to justify why they could not do otherwise. Discussions on 'right' or 'wrong' are not very fruitful. It is more important to stimulate people to reflect on feedback for a while, and see what can be done to improve the participants' well-being in the group.

Looking underneath the arguments

During negotiation processes stakeholders often have a tendency to present arguments and counter-arguments in favour of or against certain problem perceptions, solutions, etc. When discussing the social construction of knowledge (in section 6.3), however, we have already noted that arguments are never neutral, but tend to be 'invented' as 'weapons in the struggle' to realise or defend certain interests and aspirations. Thus, it makes little sense in negotiation processes to endlessly discuss the (non-)validity of certain arguments, without considering the underlying norms, values and interests that are at stake. During explorations, therefore, it is important first to identify such aspirations (Aarts, 1998). Talking about stakeholders' fears and 'doom-scenarios' can be a useful entry point for eliciting these. By listening to other parties' concerns, underlying interests and goals (rather than arguments), it becomes easier both to understand each other's rationale and to examine the possibilities that exist of satisfying one another's interests (Fisher & Ury, 1981). Moreover, participants may not only gain sharper insight into what separates them from others, but may also discover unexpected commonalities at the level of norms, values and aspirations. In the process, they may gradually develop an understanding for others, come to realise that their own perspective is not the only possible one (Vermeulen et al., 1997:47–8), and even adapt their own views.

Ensuring sufficient content input from outside

Our view that arguments are secondary to interests and aspirations, does not mean that arguments are irrelevant during interactive processes. The idea is that, by focusing the discussion on underlying interests and aspirations first, discussions on arguments are more easily resolved. Once stakeholders have the feeling that their underlying interests are taken into account and respected, it often becomes less threatening for them to accept the validity of certain arguments made by others. Nevertheless, there will always remain knowledge disputes that cannot be resolved in this manner. It is relevant to note that even if arguments presented by stakeholders are never fully objective or neutral, one can still distinguish between arguments that are well or less well underpinned by experiences and evidence, given the standards of particular epistemological cultures (see section 6.5). Thus, it can be relevant to organise forms of exploration in which arguments are reflected upon, tested, debated and scrutinised on their socially constructed dimensions (i.e. 'deconstructed') (e.g. section 6.3). In doing this, several exploratory methods and learning tools (see sections 9.3 and 13.5) can help generate better insight into substantive processes, and/or provide feedback on the consequences of possible measures, agreements and solutions. It can also be useful to look for information inputs from sources that can be considered relatively independent and/or trustworthy. Such independent outsiders may also help to expand the bandwidth within which alternative solutions are sought. Joint fact-finding (see section 14.2.4) is another powerful strategy in resolving knowledge disputes and conflicts about arguments.

Focusing on commonalities to identify win-win solutions

As we have seen, exploratory activities may lead gradually to the identification of commonalities at the level of underlying norms, values and aspirations. This process may be supported by looking first for common views regarding the more distant future (see section 13.5.10) and/or relatively abstract expressions (see section 8.2.7), and then working towards identifying more immediate and concrete areas of agreement. Moreover, it is a useful facilitation strategy for recognising existing differences, but also for emphasising regularly what the stakeholders agree on (Aarts, 1998). Furthermore, one can stimulate stakeholders to identify problems in connection with such shared aspirations. Such new problem definitions can form the basis for future co-operation, and the identification of win-win solutions. An example here can be derived from Aarts (1998) who describes how Dutch farmers and nature conservationists were initially involved in a struggle over the use of agricultural land. Gradually, both parties realised that farmers' unsustainable practices were related to inadequate government policy and an unwillingness by consumers to pay a higher price for food produced in an ecologically sound manner. Thus, the two parties had discovered shared problems and a basis for joint activities and lobbying from which both parties would benefit. The identification of win-win solutions requires an environment in which new creative ideas get a chance. Susskind and Cruikshank (1987:118) expressed the opinion that a process agreement that allows the generation of ideas without initial commitment ('inventing without commitment') can be stimulating.

14.2.4   Task 4: Joint fact-finding and uncertainty reduction

From the confrontation between different perspectives and aspirations it can become evident that a lack (or conflict) of knowledge and insight remains regarding certain aspects of the problematic situation. This may result in an impasse in the development of agreed solutions and co-ordinated action. In such a case, it may make sense to see if agreement can be reached on a process of generating more knowledge and insight instead. This may involve joint literature study, forms of collaborative experimentation (see section 13.5), and/or consultation of external experts. Making such research and fact-finding efforts a joint process is not only important as a stimulant for arriving at shared understandings and a common knowledge base, but also for improving relationships between the parties. Doing something together may help to create a bond. In conflictive situations especially, the setting up of such joint fact-finding activities forms a negotiation process in itself. For example, in a conflict situation regarding water distribution in an irrigation scheme, there may be disagreements, knowledge disputes or uncertainties about how the flow of water may be affected by different water control structures and management practices. Experimentation with these may be a useful form of joint fact-finding, but in order to stimulate outcomes of experiments that are credible to, and acceptable by, the various stakeholders, several issues may need to be agreed in advance. These include: What control structures and management practices will be included in the experiments? What criteria will be used to evaluate experiments? How and where will measurements take place? Who will do the measurements? What standards will be used to decide whether something 'works' or not? How will the costs and risks involved be shared and/or distributed? How many seasons will the experiments last before final conclusions must be drawn? These are not trivial issues, and it may take some preparation time to arrive at a viable design for experimentation.

14.2.5   Task 5: Forging agreement

As we have argued in Chapter 6, coherent innovations require new forms of co-ordinated action within a network of interrelated actors. Thus, it is important in an interactive process to reach practical agreement on what different stakeholders will do or not do with regard to, for example, the further development and/or use of novel technical devices and the establishment of new social–organisational arrangements (including the employment of supportive policy instruments). In some cases, exploration and joint fact-finding may result in agreements on co-ordinated action in a fairly smooth way. In other instances, a process of 'give and take' may transpire in which stakeholders typically take up certain negotiating positions, state claims regarding solutions, bring pressure to bear in order to move others towards concession, create impasses and break them. Such maneouvres are notoriously difficult to manage, not least since informal 'behind the scene' interactions between stakeholders are likely to play a significant role. Moreover, whether or not the participants reach a constructive agreement depends to a considerable extent on the quality of the preceding process. Nevertheless, during 'give and take' maneouvres, facilitators can do much to stimulate a conducive atmosphere, an exploratory attitude, clear communication and adherence to process agreements. In addition to these more general guidelines, some specific tactics may be employed.

Making people talk in terms of proposals and counter-proposals

To create a constructive atmosphere, and to prevent stakeholders merely criticising the arrangements and solutions forwarded by others, a facilitator may encourage stakeholders to speak in terms of proposals and counter-proposals (Mastenbroek, 1997), or even set this as a rule. Instead of stakeholders being allowed to elaborate on why they feel a proposal is 'good' or 'bad', they can be invited only to indicate under which conditions a proposal would be acceptable, and/or which adjustments and additions are desired. In this way one or several original proposals may be discussed and filled out until an acceptable set of arrangements is arrived at.

Working towards 'package deals'

In interactive processes, the stakeholders usually have several interests at the same time, so integrative negotiations often take place on multiple issues (see also section 10.2.3). Instead of dealing with each issue separately, facilitators may usefully stimulate several themes to be dealt with at the same time, even if they are only indirectly related. This is because it may be easier for stakeholders to give in on one issue (e.g. access to land) when this is linked to gains in another area (e.g. water rights). Similarly, participants may be more keen on reaching agreement if they have a number of parallel interests besides the conflicting ones: in other words, if mutual interdependence will be high. Thus, facilitators can work towards a so-called 'package deal', which is essentially a cohesive package of problem-solving measures and arrangements that can be either accepted or rejected as a whole (see Van der Veen & Glasbergen, 1992:231). In many cases, it makes sense to officially record the eventual agreements arrived at, for example by putting them in writing. This helps ease communication with constituents and gives a shared point of future reference.

Do not work out details for the distant future but create variation

Since many unanticipated developments may take place as the future unfolds (see also section 4.2), Aarts (1998) suggests that it does not make much sense to try and reach detailed agreements on specific goals and co-ordinated actions for the relatively distant future. The likelihood that such agreements will need to be revisited is high. Similarly, there is a chance that such agreements create rigidity, and prevent a flexible and creative mode of dealing with coincidences and unforeseen developments. In this context, Aarts (1998) argues that 'coincidence must be given a chance' in integrative negotiation processes. This can be done by defining detailed goals and actions for the immediate future only, while formulating more abstract goal orientations as a shared vision for the longer term (Aarts, 1998). Moreover, when negotiating towards concrete actions and means to achieve short-term goals, it is often wise to leave space for several routes to reach similar ends. This is important not only for anticipating relevant diversity among and within stakeholders (see section 15.2), but also to create a degree of variation that can be beneficial in view of the evolutionary nature of innovation processes (see section 8.2.2).

14.2.6   Task 6: Communication of representatives with constituencies

In cases where the participants in an interactive process represent others, regular communication between the representatives and their constituencies is important. If insufficient attention is paid to this, it is quite thinkable that representatives and constituents 'grow apart' during the process, and that the latter reject (i.e. are not prepared to sanction) the eventual agreements reached. What happens essentially in such cases is that, due to their direct interactions with other stakeholders, the representatives are subject to a different and more intensive learning process than their constituents, which causes them to take into account perspectives and realities that others are not yet prepared to accept. Mastenbroek (1997) proposes that, in conflict situations, the interactions between representatives and their constituencies must be regarded as separate negotiation processes. Some people even go so far as to say that the key to successful conflict resolution is to be found predominantly in the willingness and ability of representatives to negotiate with their own constituency (Mastenbroek, 1997:66).

We can conclude that it is worthwhile for change agents to stimulate and/or facilitate interactions between representatives and constituents throughout the process (Susskind & Cruikshank, 1987:105). This is not only a way of 'transferring' the representatives' learning processes to others, but also of keeping the representatives 'connected'. Here too various media and methods can play a role, and in some cases it may even be possible to mobilise the press to report on developments in the interactive process. Clearly, this has risks as well, as the press may not present things in the way representatives and facilitators find productive. In the case of sensitive issues especially, it is often advisable to ensure that significant developments and agreements are first communicated to constituents by their own representatives, to avoid constituents interpreting these as being imposed on them by other stakeholders (e.g. Veldboer, 1996). Much of what has been said on the interaction between representatives and their constituents may also be valid for the interaction between the process participants and the wider institutional policy environment (see section 14.2.1). These parties too need to be given the opportunity to 'grow' towards the eventual results of an interactive process (see also Tatenhove & Leroy, 1995).

14.2.7 Task 7: Co-ordinated action

The whole idea of an interactive process is to reach agreement eventually on co-ordinated action, as an element of coherent socio-technical innovations (see section 8.2). After agreements have been secured, the time has come to put into practice what has been agreed on. Along with this, the need arises to monitor whether stakeholders do what they have promised to do, and whether or not this yields the expected outcomes. For the monitoring of progress, it is important to clearly outline what different stakeholders are expected to do, and put in place clear procedures for monitoring (Susskind & Cruikshank, 1987:131). Who will do the monitoring? Of what? When? And how? In conflict situations especially, confidence in the process may be enhanced by involving mutually trusted outsiders in supervising the fulfilment of agreements. If problems in the implementation of agreements are identified by such monitoring, a quick response can often contribute to continued support for the agreements. It can also be wise to clarify procedures for re-opening integrative negotiations on previously agreed issues (Susskind & Cruikshank, 1987:133). This implies the fostering of ex-ante agreements on the circumstances under which issues can be put on the agenda again (altered insights, unexpected developments or undesired side effects, for example).

It is probably fair to say that the guidelines discussed in this section are culturally biased, as they are derived mainly from western literature and experiences. Hence, they may need to be adapted to different 'learning and negotiation cultures'. Nevertheless, it is relevant to note that paying attention to negotiation may provide a range of tools and guidelines that have so far been largely overlooked in conventional discussions of participation.

Questions for discussion

(1)   In section 14.1.1 we have mentioned several normative principles that are often associated with 'good' participatory practice. Select some of these principles, and discuss their value and appropriateness, taking into account our practical and theoretical reservations regarding 'maximum participation'. Which of these principles are still valid? What adaptations and/or conditionalities would you like to propose?

(2)   We have suggested several facilitation tasks and guidelines in this chapter. In connection with these, what are important requirements, qualifications and characteristics for external facilitators?

(3)   Identify two problematic situations that you are familiar with, and discuss which facilitation guidelines are most significant for each of them, and which are less relevant and/or applicable. What contextual differences have played a role in setting your priorities?

(4)   We have argued that the various facilitation tasks mentioned require continuous attention in the course of change trajectories. What are the implications of this for the planning of intervention programmes and projects?

15 The planning of individual activities

While it makes little sense to create a detailed long or medium-term 'process plan' or 'programme' for change and innovation trajectories, it is important to carefully prepare separate activities in the short term. At the same time, one cannot prepare an activity without a rough idea of what might follow, and how it fits into a wider process. But in order to contribute to change, flexibility through time is essential, so we have decided not to discuss programme planning, as it creates an illusion of control and often fosters inflexibility (see Chapters 4 and 13).

Drawing on literature on communication planning (e.g. Windahl et al., 1992; Van den Ban & Hawkins, 1996; Van Woerkum et al., 1999) we suggest that for each activity one needs to find a careful balance between the following:

- *purpose* of the activity (in the context of wider goals and strategies);
- *stakeholders/target audiences* (the characteristics of the participants involved);
- *content* (what is being discussed/presented);
- *media and methods* (the media, methods and techniques used in the activity);
- *organisation* (location, time management, venue, materials, budget, etc.)

Finding such a balance requires an iterative process of thinking about these elements, as one cannot hope to get everything right at once. That is, one will frequently find it necessary to adapt earlier, provisional choices (e.g. regarding purpose) on the basis of later reflections (e.g. regarding content). Clearly, such thinking should not stop at the point when activities are unfolding. Thus, process facilitation is never simply about 'implementation'. In all, the process of developing an activity can be visualised as in Figure 15.1.

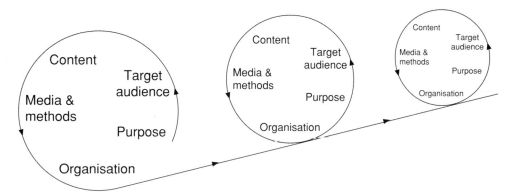

Figure 15.1 The spiral model of developing a communication for innovation activity (adapted from Van den Ban and Hawkins, 1996).

In order to make an adequate plan for an activity, the developers need a variety of knowledge and information (originating from different sources) in relation to the different elements (Bos, 1974). For choosing appropriate media and methods, for example, communication workers may not only need knowledge of media potentials, methods and facilitation (e.g. section 14.2 and Chapters 12 and 13), but also information on the media use and preferences of several stakeholders. Similarly, decisions regarding the content of activities may be informed by academic literature on, say, biological pest control, but may also require insight into local experiences and the problem perceptions of target audiences. Thus, preparing an activity must be accompanied or preceded by, considerable investigation and exploratory activities (see sections 13.5 and 14.2.3).

In the remainder of this chapter we discuss several specific issues and considerations in relation to the 'ingredients' of individual activities.

15.1   The purpose of an activity

One cannot usefully organise and prepare an activity without an idea of where it fits in, and what needs to be achieved. This implies that one needs to assume, at least temporarily, broader purposes that are rooted in past interactions. Thus, the purpose of a single activity needs to be looked at in the context of:

● the wider *intervention approach*;
● *communication for innovation strategies*, as connected with different *intervention goals*;
● *general communication for innovation functions* relevant to communication for innovation strategies (see also sections 2.2 and 13.1).

In the case of an interactive approach, for example, an activity may be geared specifically towards (a) discovering existing initiatives; (b) reaching process agreements; (c) exploring differential social and technical problem perceptions; and (d) improving the atmosphere in a group, or any other process goal that can be derived from section 14.2. As an outcome of (or element in) the process, it may be agreed at a certain point that activities need to be organised to support horizontal knowledge exchange, that it is wise to use a persuasive (i.e. instrumental) strategy to promote the adoption of a particular innovation, or that another of the communication for innovation services mentioned in section 2.2 needs to be provided. As an element of such strategies, it may be decided (again on the basis of a thorough analysis of the problematic situation, see section 5.3.2) that it makes sense to focus on raising awareness of a specific issue (and pay less attention to other general communication for innovation functions, see section 2.2). And when the focus is on raising awareness, one may, depending on the situation, choose to strive for the visualisation of processes that are difficult to see, or try to enhance people's identification with a problem by means of entertainment–education (see section 13.4.3 and 13.4.2 respectively).

15.1.1   Goal hierarchies

From the above it transpires that in every intervention context there is a complex 'hierarchy' of goals, where lower level goals are in fact the proposed means to achieve the higher level ends. Individual activities, then, can be seen as a further operationalisation (in terms of methods, organisation, actors/audiences, etc.) of lower level goals that are relevant at a particular time. We can distinguish between the following levels of goals and means that interventionists may have (and which may or may not be shared by others):

- *ultimate objectives*: wider societal objectives that interventionists aim to contribute to, but cannot realistically achieve with the resources available and/or anticipated;
- *overall intervention (or project) objectives*: objectives that interventionists aspire to and expect to achieve with the resources available and/or anticipated;
- *intermediate intervention objectives*: broad objectives that derive from the overall intervention objective in a dynamic context;
- *specific aims*: detailed aims that derive from intermediate intervention objectives in a dynamic context;
- *actor aims*: specific aims translated into objectives with regard to specific stakeholders or target audiences (e.g. what would we like this actor to learn, tell us, agree to, do, etc.);
- *activities*: specific and actor aims integrated with other aspects (i.e. media, methods, organisation, target audiences and content).

In contrast to what conventional project planning formats (as solicited by donors) suggest, we feel that it is not very useful to define in advance the details of lower level objectives, aims and activities. They need to be tuned continuously to emerging dynamics. Higher level objectives too must be regularly reconsidered and adapted in view of ever-changing circumstances and insights. In Tables 15.1 and 15.2 we give two (incomplete and fictitious) examples of such goal/means hierarchies. The first starts from an interactive mindset (or at a moment that an interactive approach is deemed necessary), while the second is more instrumental in nature. The illustrations are connected to the (equally fictitious) socio-technical problem tree presented in Chapter 13 (Figure 13.1).

15.1.2 Communicative and other means or objectives

Tables 15.1 and 15.2 say little about *how* goals and aims may be realised. That aspect will become more clear as other aspects of activities (e.g. media and methods; see section 15.4) are discussed. However, Table 15.2, especially, illustrates that communicative intervention is not the only intervention strategy that may be used to bring about change in actors' perceptions and practices. This relates to our earlier elaboration of different 'policy instruments' (see section 4.1), and also to the explanatory model presented in Chapter 5. Both discussions indicate that different types of social pressures (e.g. in the form of rewards, sanctions, laws, law enforcement, subsidies, fines, support systems, facilities, coercion, etc.) can be employed to influence human action. Some objectives and aims – often called 'communication objectives' – may be strived at primarily through communicative means. Here communication is used to stimulate awareness and/or active learning (i.e. cognitive change) through central and/or peripheral routes (see also sections 9.2 and 9.5). In connection with other objectives, non-communicative policy instruments may be expected to play a dominant

Table 15.1 Fictitious and incomplete hierarchy of goals and means starting from an interactive mindset. For background see Figure 13.1.

Ultimate objective	increase food production and food security in region Y
Overall intervention objective of project Z	develop appropriate socio-technical solutions that enhance food security in pilot village E
Intermediate intervention objective at t = x • specific aims	identifying existing and past initiatives and their history • establishing contact and trust with relevant actors • getting relevant actors to talk about initiatives
Intermediate intervention objective at t = x + 1 • specific aims	identifying stakeholders' social and technical problem perceptions • make stakeholders think about their problems • identify practices that reproduce the problematic situation • identify reasons behind practices
Intermediate intervention objective at t = x + 2 • specific aims	reach agreement on the way forward • agree on problem areas that need further attention • selecting eventual participants/stakeholder representatives • reach process agreements of various kinds
Intermediate intervention objective at t = x + 3 • specific aims	enhance social conditions for improving soil fertility with the help of already known technical solutions • negotiate more secure land tenure arrangements • help resolve conflicts regarding the aborted fenced community grazing area
Intermediate intervention objective at t = x + 4 • specific aims	explore and select appropriate technical and social arrangements for biological pest control • identify potentially effective biological pest control strategies • identify social-organisational conditions for technical strategies to work • reach agreement on the most feasible socio-technical packages
Intermediate intervention objective at t = x + 5 • specific aims	enhance social conditions for implementing chosen biological pest control strategies • improve individual skills in biological pest control • enhance organisational capacity of the community to co-ordinate biological pest control activities

role in inducing change (see also section 4.1). It is important to note, however, that in such cases communicative aspects also deserve careful attention. The introduction of a new law or regulation can be interpreted as 'a message in itself' and can take on certain meanings for specific audiences. In a particular context, for example, new regulations for surface water extraction may be perceived by some as 'undermining autonomy' or 'unjust', and hence be ignored whenever possible. For such policy instruments to yield desired effects, therefore, it is often crucial to monitor their communicative effects and take additional action through communicative means. More generally, communication is often necessary to ensure that people hear about the existence of laws, subsidies, provisions, etc. in the first place. Finally, in an interactive process especially, communicative intervention strategies may be required to *bring about* new or adapted regulations, provisions and subsidies as elements of coherent innovations. In all, we can conclude that in any intervention trajectory

Table 15.2 Fictitious and incomplete hierarchy of goals and means starting from an instrumental mindset. For background see Figure 13.1.

Ultimate objective	increase food production and food security in region Y
Overall intervention objective of project U	promote the adoption and adaptation of the biological pest control package developed in village E to a range of similar villages in the region
Intermediate intervention objective at t = x • specific aims	raise awareness about the problems associated with current pest control practices • make farmers aware of the relationships between pesticide use, the killing of natural enemies, increased pest infestation and yield losses • make farmers aware of the health risks involved with pesticide use
Intermediate intervention objective at t = x + 1 • specific aims	raise awareness about the existence of alternative biological pest control strategies • communicate experiences (social–organisational and technical) from village E • make farmers aware of the principles of biological pest control
Intermediate intervention objective at t – x + 2 • specific aims	enhance knowledge and skills of farmers to apply biological pest control • improve knowledge and skills for recognition of pests and natural enemies • improve knowledge and skills for agro-ecosystem analysis • improve knowledge and skills for risk assessment
Intermediate intervention objective at t = x + 3 • specific aims	discourage the use of chemical pesticides • create mechanisms to make pesticides more expensive • reduce or prevent the local availability of pesticides
Intermediate intervention objective at t = x + 4 • specific aims	create stimulating conditions for adopting biological pest control • create mechanisms to give added value to pesticide-free products • establish special support to farmers who reduce pesticide use • introduce competitive elements for reducing pesticide use (make it a 'sport')
Intermediate intervention objective at t = x + 5 • specific aims	ensure easy access to reference information and expertise on biological pest control • establish local information points • select and introduce farmer–communication workers

towards change and innovation, a careful balance needs to be worked out between communicative and non-communicative or less communicative strategies.

15.2   Stakeholders, audiences, and targeting

When preparing an activity, one must have a fairly clear idea about *who* one is organising it for. In other words, one cannot effectively anticipate an audience without having some insight into those who make up that audience (see also section 7.3). The other side of the coin is that actors and audiences with different characteristics need to be approached in different ways (i.e. with different kinds of activities) in order to help realise similar[1] outcomes. As Van der Ploeg and others (e.g. Ellis, 2000) have shown, diversity is indeed an important characteristic and asset in farming (see also section 5.3.4). Distinguishing different actors and audiences in order to develop tailor-made activities and strategies is called *targeting*; a term borrowed from marketing approaches (see Kotler, 1985; Röling, 1988). Thus, the purpose of targeting is to define relatively homogeneous categories of people who can be approached in more or less the same way (or through a single representative). Forms of targeting are needed in both interactive and instrumental settings, although the issue may present itself differently in each context. In the case of an interactive process, one may wonder who the relevant stakeholders are, and what characteristics they have that may be relevant for setting goals, choosing methods, etc. (see also section 14.2). At a more instrumentally-oriented moment, typical questions centre around who the different audiences are that one wants to communicate with, how they can be reached, and with what 'message'. If one does not pay deliberate attention to targeting, audiences and stakeholders are likely to 'select themselves'. That is, interventionists are likely to be in contact only with a subset of actors who are comfortable with the activities developed. As a result, the chances are high that one misses out on actors who are crucial to furthering overall intervention and innovation objectives.

Quite frequently communication workers justify missing out on certain audiences by arguing that some categories of people 'cannot be reached'. In our view the idea that some people are 'unreachable' is a myth as well as a gross simplification. All people on this planet can be 'reached' provided that one utilises methods, media, locations, timing, people and 'messages' that are feasible, attractive and/or meaningful to them. Thus, failure to connect with certain audiences has often more to do with sloppy and unimaginative communication planning, lack of priority and/or inadequate institutional resources, than with the inherent characteristics of an audience. One aspect of sloppy communication planning, then, is that interventionists have insufficient insight into the characteristics of the people they aspire to communicate with, including their relevant diversity.

15.2.1   Characterising relevant diversity

A difficulty or challenge here is that there are always numerous ways in which a community or population can be segmented into 'stakeholders' or 'target audiences'. In most situations, for example, one can differentiate between people who have distinct demographic, agricultural and/or socio-economic characteristics. Along such lines, one could distinguish 'stakeholders' or 'target audiences' according to age group, class, gender, family life cycle, religion, farm size, ethnic group, assumed economic viability, agro-ecological zone, education level, farming system, off-farm income, etc. Interventionists frequently resort to such broad segmentations. However, one cannot automatically assume such classifications to be relevant to a given innovation context. 'Relevant', in this context, means that a classification can serve

---

[1] 'Similar' here is meant at an abstract level, e.g. in terms of 'improved sustainability'. Clearly, the latter may mean something different for different actors, and may be realised through different socio-technical practices.

adequately as a basis for developing different intervention activities and strategies for different audiences, or helps to get a sharp insight into which stakeholders need to be involved and/or represented in an interactive process. When working on pest management problems, for example, it is not immediately evident that organising separate activities for 'female farmers' and 'male farmers', or for 'large farmers' and 'small farmers' makes sense. This is because the differences among people *within* the same category (i.e. female or male farmers) may in fact be more important than the differences *between* the two categories. In the case of pest management, it could well make more sense to distinguish between those who use large amounts of pesticide and those who do not, or between those whose religion forbids the killing of insects, and those who follow other denominations. Similarly, in the case of water distribution, it may be fruitful to differentiate between those farmers located upstream and those downstream, or between those with or without formal water rights.

As we have elaborated in Chapter 8, the much used classification into 'adopter categories' (innovators, early adopters, laggards, etc.; see section 8.1.1) tends to suffer from the same problems as the broad segmentations mentioned above. Moreover, using it for purposes of targeting would imply a denial of the idea that diverse audiences may require different socio-technical solutions to problems. This is because the classification basically starts from the assumption that all farmers should move in the same direction (see section 8.1.2). In that sense, classifications into adopter categories are incompatible with the mere idea of targeting[2].

In essence, our point here is that *relevant diversity* is context specific and needs to be explored time and time again (e.g. with the help of exploratory methods, see section 13.5). When searching for relevant diversity in a community of actors (or within a broad category like 'farmers'), two types of difference tend to be of primary importance. The first type relates to differences in what people *do* or *do not do*, and *why*, in creating and/or reproducing a situation that is deemed problematic (see Chapter 5). Thus, such *problematic context related* diversity may be characterised according to:

(1)   diverging (patterns of) *practices* (e.g. actors who do or do not make extensive use of chemical pesticides);
(2)   diverging *reasons* underlying practices, in terms of:
   —   different beliefs and perceptions ('know');
   —   different aspirations ('want');
   —   different opportunities and capacities ('be able to');
   —   different social pressures ('be allowed/expected to').
(3)   diverging *perceptions* regarding *problems* and *solutions* (e.g. pesticide users who recognise or do not recognise that pesticide use causes problems).

The second type of difference that interventionists may need to take into account relates to the possibility of interacting and communicating with particular actors. Such *intervention related* diversity may be identified along the lines of:

---

[2]   In another sense, namely the idea that 'early adopters' may play a significant role in 'diffusion', the classification can at times still be useful. However, to avoid the inclusion of (uni)linear connotations (see section 8.1.2), we would prefer to talk about 'farmers with certain experiences' who may play a role in 'horizontal knowledge exchange'.

(4)   diverse patterns of media use and media preferences;
(5)   differences in awareness, interest and motivation to learn (see sections 9.4. and 9.5);
(6)   different involvement in social networks and/or contact with organisations;
(7)   different quality of relationships with intervening parties (e.g. in terms of trust);
(8)   different capacity and willingness to express views and defend interests in a learning and negotiation setting (see also section 14.2.1).

When looking for relevant diversity through the exploration of differences as outlined above, one may find in some cases that the most significant differences can be explained with reference to crude classifications like farm size, age group or gender. Different pest management practices, for example, may be explained by gender or farm size reasons. But in many instances one will discover that such overlap hardly exists, and that relevant differences cut across simplistic categorisations. Similarly, one may find that relevant diversity cannot be captured while taking into account just one or two dimensions of diversity. One may find, for example, that different levels of pesticides used relate clearly to (a) different degrees of awareness of and insight into agro-ecological consequences; (b) diverging religious convictions; (c) integration into different marketing networks; and (d) different levels of labour availability, etc. Apart from discovering that those who use an excessive amount of pesticide do so for different reasons, one may find that some of them can be reached through local meetings or media, while others reside elsewhere during part of the season, or never participate in community meetings since they have a different political affiliation from the current community leadership. As this example shows, the result might be a five or six-dimensional matrix characterising diversity, which might involve 32 or 64 different relatively homogeneous categories of farmers (assuming only two levels or options per variable). This is, of course, unworkable in the sense that it is not practically feasible and efficient to develop different activities for so many categories. In fact, working with only two or three distinct target audiences already poses quite a challenge in most situations. An important conclusion, then, is that although targeting is important, one can never assume that one is dealing with a truly homogenous audience, not even when one defines it in relation to one specific intervention objective (e.g. reduction of pesticide use). In connection with this, Röling (1988) prefers to speak of defining 'optimally heterogeneous' rather than 'relatively homogeneous' target audiences. In all, an important question remains as to how we can take diversity into account, without resorting to overly simplistic classifications that hide rather than clarify relevant diversity.

A first point to keep in mind is that taking into account relevant diversity cannot be equalled to developing one single, fixed classification of stakeholders or audiences into relatively homogeneous categories. In fact, our argument has been that different problematic contexts may require a different way of describing and anticipating diversity. Thus, it is perfectly legitimate to think in terms of a variety of classifications, and to use a different one according to the issues and problems (e.g. marketing, pest management, animal health, management support, etc.). But even then, our example on pest management shows that diversity may still be characterised along different dimensions and variables (e.g. degree of awareness, religion, labour availability, place of residence, etc.). Several possibilities exist for dealing with this. First, one can *select* those variables and dimensions that one feels are most relevant, in that they explain a large amount of variation, help to reach a large number of people, have practical advantages, or seem to have the greatest implications for the development of separate activities. Each of the selected dimensions can then be used in a different way. For example, one may decide that those who belong to a specific religion and those who are non-resident form only a small minority, and hence that investing in separate activities is not efficient. One might argue, furthermore, that the difference in level of awareness and insight cannot be ignored, as it has inherent implications for content development and choice of methods. Similarly, developing separate activities for those who belong to different political affiliations may be a practical necessity, and at the same time may be cost-effective. Finally, one may, on the basis of exploratory activity, conclude that the diversity in terms of labour availability and participation in different marketing networks has implications with regard to the socio-technical solutions that are required, but that this diversity can relatively easily be tackled *within* certain activities, for example by dealing with different issues and topics in meetings, brochures or demonstration experiments. In essence, separate dimensions of diversity can be taken into account in different ways; that is, they can be selectively:

- ignored for practical (or other) reasons;
- used as a basis for developing separate activities and strategies;
- used to anticipate diversity within one composite activity.

Deciding how to take into account which dimension of diversity, then, is a matter of critical reflection, pragmatic reasoning and/or political choice (see Chapter 3).

A second solution is that one tries to integrate several variables on which people differ into a single classification of diversity. This can, for example, be done through advanced statistical procedures that take into account covariance between different variables or dimensions of diversity. Thus, clusters of variables can be identified that are associated with each other according to some pattern. For the purpose of this book we will not go into the statistical and methodolagal details. Suffice it to say that, depending on whether such clusters can be interpreted sensibly, interesting insights and classifications of diversity may be arrived at. This is shown, for example, by research on farming styles (see also section 5.3.4) in which such statistical methods are combined with procedures of self-classification by farmers (e.g. Leeuwis, 1993 for methodological details). Figure 15.2 shows a farming styles classification of Dutch dairy farmers, while Table 15.3 gives an indication of the kind of strategies involved.

From the targeting point of view, a shortcoming of the farming styles classification in Figure 15.2 is that it is again fairly general and not really problem-specific. In other words, when using it for purposes of targeting, one more or less assumes that *one* classification of diversity provides the sharpest possible insight in relevant diversity, regardless of the issue at hand. Clearly, this is not likely to be the case (see also Leeuwis, 1993). However, when a specific problematic context is taken as the starting point, these kinds of approaches to arriving at a target can be useful.

In many intervention contexts, however, the first solution to dealing with multiple dimensions of diversity is more feasible. In any case, it is relevant that 'relevant

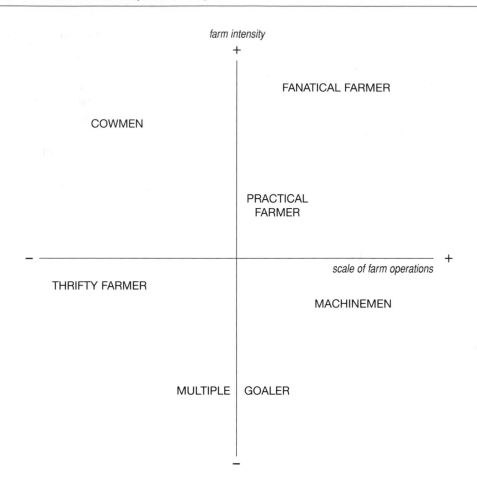

Figure 15.2 Social map of farmers on the basis of labels expressed by them in qualitative interviews (see also Roep et al., 1991:9; Leeuwis, 1993:200).

Table 15.3 Farming styles and the essence of the production strategies that are related to them (adapted from Leeuwis, 1993).

Farming style	Essence of strategy
Multiple goaler	self-sufficiency through low external input
Thrifty farmer	monetary balance
Practical farmer	practical balance, especially in labour organisation
Cowmen	reaching a high milk yield per cow through labour-intensive practices
Machinemen	mass production through labour-extensive practices
Fanatical farmer	gaining a competitive advantage over others

diversity' in connection with a problem field may change over time under the influence of the intervention process itself. Thus, segmentations into stakeholders and target audiences tend to have a temporary value.

15.2.2   Different types of target audiences

A final issue in connection with targeting relates to the different types of audiences one may have. In the context of an interactive process, the inherent idea is to bring together a variety of mutually interdependent stakeholders. However, in more instrumental campaigns too it can be useful to distinguish between target audiences who are expected to play different roles. Thus, one may identify *eventual, intermediate, service* and *alignment* target audiences. When the aim is to combat excessive use of pesticides, for example, one may eventually strive to change the practices of 'individual' farmers (the eventual target audience). However, to communicate with farmers on this issue, one may decide that it is more effective or efficient to work with intermediaries who belong to the eventual target audience, and are expected and/or trained to play a role in contributing to change at the farmer level. Such intermediate target audiences can, for example, be farmers' organisations, farmer-communication workers or 'opinion leaders' (see section 8.1.1). Alternatively, one could try to mobilise others who are in regular contact with farmers, but are not themselves members of the eventual target audience. These are often providers of particular services (e.g. traders, religious leaders, veterinarians), hence the name 'service target audience'. Finally, we have seen earlier that innovation usually depends on co-ordinated change and/or support from several actors. Thus, apart from target audiences who may play an intermediary role, there are usually stakeholders who are important from the perspective of alignment (see section 8.2.2) since they are in a position to either hinder or support processes of change. In the case of pesticide use, this could be the pesticide industry, government bodies, and shopkeepers who sell pesticides. The distinctions between the different types of target audiences are not always clear, and at times they may overlap. Instead of regarding the pesticide industry as an alignment target audience, one could argue that they constitute another eventual audience. Nevertheless, it can be clarifying to distinguish between these categories, and think about different activities (with different objectives) for each type. When identifying intermediate, service or alignment audiences, however, it is important to keep in mind that they may need to differ for different segments of the eventual target audience.

15.3   Content

An activity needs to have content; that is, it must be directed towards eliciting, sharing, generating and/or presenting certain types of ideas, information or messages. What exactly an activity will be about depends almost completely on other aspects of the activity, such as its specific purpose (see section 15.1) and the stakeholders or target audiences involved. Each of the aims mentioned in Tables 15.1 and 15.2, for example, implies communication about different topics and themes.

And in so far as the aims have different implications for different stakeholders and audiences (i.e. go along with diverging actor aims), such topics and issues may need to differ according to whom one interacts with. When thinking about the contents of an activity a number of points may need to be considered (derived largely from Van Woerkum et al., 1999 and Van Meegeren, 1999):

- *Announcement*: Whether or not actors will participate in, or pay attention to, an activity (e.g. in the form of a meeting, leaflet, etc.) can depend to a considerable extent on how the contents are announced and summarised at the outset (e.g. on the title).

- *Introduction to the activity itself*: Once actors have gone through the initial trouble of participating or paying attention, it is often useful to explain at an early stage what the activity is about (in terms of, for example, content, method, duration, etc.), for actors to decide whether or not further attention or participation is justified.

- *Translation*: In order to enhance communication (see Chapter 7) it is usually necessary to ensure that contents that originate from one life-world (or source) are translated to the life-worlds, language and terminology of others (see Chapter 6).

- *Maintaining balances*: In terms of content it can be important to deliberately create and monitor several balances (or at times imbalances) in an activity. These include a deliberate (im)balance:
  — in the input from various actors and sources (e.g. scientists and practitioners);
  — between what actors already know and do not know (see sections 5.2.1 and 6.4);
  — between what actors are already interested in, and what interventionists may want to create interest for;
  — between topics and/or arguments that actors are sympathetic to, and those that they would prefer to ignore (see sections 6.3 and 9.4);
  — between the various steps in a learning cycle (see section 9.1);
  — between positive and negative feedback (see section 9.3);
  — between contents covered and available time and/or space;
  — between central and peripheral stimulants of learning (see section 9.2).

- *Sequence*: It is important to think carefully about the order in which different topics and issues are being discussed and/or presented.

- *Priorities*: One may want to consider which topics and issues deserve priority, and hence need to be allocated more time or space than others.

- *Style*: Related to the issue of central and peripheral stimulants of learning (see section 9.2), it can be relevant to think about the style in which certain contents are dealt with. One may, for example, choose between:
  — a personal versus an impersonal style;
  — a business-like ('cold') versus emotionally rich ('warm') style;
  — a 'discrete' versus a 'continuous' style (e.g. a list of points versus a story);

— an active versus a passive style (e.g. discovery learning versus lecturing, see section 9.1).

Some of the points here already touch on the issue of media and methods that we discuss below.

15.4   Media and methods

Media and methods are crucial ingredients of an activity, and often the most visible ones. In fact, activities are usually identified by the media or methods used; an activity is often summarised as 'a group meeting', 'a leaflet' or 'an on-farm experiment' without mentioning all the other equally relevant aspects. Clearly, the media and methods used must be carefully tuned to the earlier discussed aspects of an activity. In Chapters 12 and 13 we have already elaborated on a range of potential functional qualities associated with diverse media and forms of communication, and on the various communication for innovation strategies and functions that methods may help to perform. In deciding on media and methods these need to be taken into account, and matched with, for example, prospective purposes, aspects of learning involved (see section 9.5), stakeholders, target audiences and contents. As all media and methods have certain potentials and limitations, it can make sense to combine several media and methods, sometimes even within a single activity, so that the weak points of one may be compensated by the other. Over time, of course, it is essential to use different media and methods, as specific aims are bound to change continuously. For guidance and inspiration on the choice of media and methods we refer back to Chapters 12 and 13. Here we will give some fictitious examples that follow on from Tables 15.1 and 15.2 (see Tables 15.4, 15.5).

Table 15.4   Media and methods linked to a fictitious and incomplete hierarchy of goals and means starting from an interactive mindset.

Ultimate objective	increase food production and food security in region Y
Overall intervention objective of project Z	develop appropriate socio-technical solutions that enhance food security in pilot village E
Intermediate intervention objective at t = x	identifying existing and past initiatives and their history
• specific aims	• establishing contact and trust with relevant actors • getting relevant actors to talk about initiatives
• media and methods	• several introduction visits to community leaders and institutions • informal conversations and/or in-depth interviews with those involved with past and present initiatives
Intermediate intervention objective at t = x + 1	identifying stakeholders' social and technical problem perceptions
• specific aims	• make stakeholders think about their problems • identify practices that reproduce the problematic situation • identify reasons behind practices
• media and methods	• socio-technical problem tree analysis during several stakeholder specific group meetings • a joint group meeting with stakeholder representatives to present, discuss and integrate the various trees

Table 15.5 Media and methods linked to a fictitious and incomplete hierarchy of goals and means starting from an instrumental mindset.

Ultimate objective	increase food production and food security in region Y
Overall intervention objective of project U	promote the adoption and adaptation of the biological pest control package developed in village E to a range of similar villages in the region
Intermediate intervention objective at t = x	raise awareness about the problems associated with current pest control practices
• specific aims	• make farmers aware of the relationships between pesticide use, the killing of natural enemies, increased pest infestation and yield losses • make farmers aware of the health risks involved with pesticide use
• media and methods	• a series of radio advertisements on the local radio station • a poster campaign at schools, health centres and other meeting places • visualisation of problems in an exhibition touring around different villages
Intermediate intervention objective at t = x + 1	raise awareness about the existence of alternative biological pest control strategies
• specific aims	• communicate experiences (social–organisational and technical) from village E • make farmers aware of the principles of biological pest control
• media and methods	• a series of radio advertisements on the local radio station • visualisation of solutions in an exhibition touring around different villages • visits by community representatives to village E

15.5   Organisation and logistics

To ensure that prospective activities can materialise, several issues in the sphere of organisation and logistics need to be considered. We mention a few here:

- *Venue and location*: Choosing an appropriate location for an activity is something that requires careful attention, as locations are not neutral. For specific stakeholders or target audiences it may be not be practically, culturally or politically feasible to attend activities that take place in certain locations. Methods and contents too can pose demands on the venue, for example in terms of size and number of spaces, opportunities for informal gatherings and relaxation, etc.

- *Timing*: Owing to other obligations, certain stakeholders and target audiences may not be able to attend activities at certain hours of the day, days of the week, or periods of the year. Hence, the timing of activities is an important issue. In addition to such logistic considerations, timing is also important from another angle. Learning and negotiation processes, for example, take time, and cannot take place in a couple of days, weeks or even months. Thus, activities may need to take place when people 'are ready for them', given the dynamics of a particular learning and negotiation process (e.g. the discussion on mutual inter-dependence in section 10.3.2). Or, as an old Dutch proverb puts it: 'One should work the iron when it is hot'.

- *Staff*: During intervention efforts one needs to find a balance between the nature of activities and the staff involved, because different purposes, contents and methods may pose different demands in terms of the skills, knowledge, experience and social status of a facilitator or outsider contributor (see also section 16.3).

- *Resources*: Activities require the availability not only of a certain amount of manpower, but also of equipment, transport and materials. The need for such resources again differs according to the nature of the activity. Hence, resources must somehow be matched with other elements; that is, adequate resources must be allocated to activities and/or activities need to be adapted to available resources.

- *Language*: In some cases different stakeholders and target audiences are not only separated by different terminologies and cultural meanings, but also by different languages and vernacular. In addition to methodological creativity, several organisational arrangements (e.g. in the form of interpreters) may be needed to overcome such differences.

15.6   Pre-testing elements of activities

As shown in Figure 15.1, developing an activity is an iterative process. As part of such a process, it is important to organise critical feedback on the preliminary choices with regard to target audiences, stakeholders, contents, methods, etc. Such feedback may be solicited directly from colleagues, management staff and/or from (some of) those who are to be involved in the activity. These different points of reference may comment on different issues. Experienced colleagues, for example, may say whether they feel the prospective methods and time planning fit the purpose, while management staff may comment on whether the necessary resources can be realistically provided. Similarly, prospective stakeholders and/or people belonging to the target audience may say whether or not they find purposes and contents clear, attractive and useful, and may give feedback on the appropriateness of the intended venue, time-scale, etc. In some cases activities can be tried out in advance on a small scale, and under more or less realistic conditions; when this happens in advance we speak of a 'pre-test'. Pre-testing is most easily done when tangible products such as leaflets, brochures, posters or computer programs are involved. When activities centre around group processes (e.g. in the context of conflict management) such pre-testing is often less feasible. It is important to note that soliciting feedback in itself requires some preparation. Apart from asking in an open way for 'general comments', it is often useful to explain the assumptions and considerations on which the preliminary design of an activity is based, and 'test' their validity in discussions with others. Thus, one needs to have an idea what one wants to know and from whom.

Questions for discussion

(1) Identify a problematic situation in which you would like to intervene. Identify an overall intervention objective, and then continue to make two sets of intermediate intervention objectives: one from an interactive and one from an instrumental mindset. What are the key differences between the two approaches? Which one do you like best and why? Would it be possible and/or necessary to combine the two sets of intermediate objectives?

(2) Identify a category of people that you are interested in (e.g. farmers, consumers, etc.) in view of a particular problematic situation. Discuss how diversity within this category is usually characterised (i.e. which segments are distinguished), and whether or not this is likely to be the most relevant characterisation of diversity in view of your intervention objectives. If applicable, discuss what you would like to do in order to arrive at a more relevant categorisation.

(3) Discuss and identify examples of 'intervention related' diversity and 'problematic context related' diversity in connection with a situation that you are familiar with.

PART 5
Organisational and interorganisational issues

Professionals can contribute more effectively to change and innovation when they work in an enabling organisational environment. Hence, it is important to discuss organisational issues in connection with communicative intervention. In Chapter 16 we discuss different ways of looking at organisations and their management, as well as several topical issues including the organisational implications of new ways of looking at 'extension', organisational learning and organisational research. At the same time, it is important to realise that organisations involved in communicative intervention do not operate in a vacuum. They work with problems, regions and communities that are also the concern of other organisations, albeit perhaps in a different manner. Hence, communicative intervention takes place in an environment where other organisations – e.g. local governments, health centres, marketing boards, unions, private firms, research stations, donors, environmental movements – may also play an active role. As change and innovation depend to a large extent on the effective mobilisation of (support) networks towards coherent practices (see section 8.2.2), we can say that it is crucial for communicative intervention organisations to establish formal or informal co-ordination and/or co-operation with other organisations. In this Part of the book, therefore, we also look at a number of interorganisational issues in connection with communicative intervention, focusing on the opportunities of and threats to co-operation. First, Chapter 17 elaborates further on the idea of agricultural knowledge and information systems that we touched on in Chapter 2, with special attention paid to the practical issue of networking. Subsequently, we discuss the emerging *knowledge markets* and their implications for communicative intervention and interorganisational co-operation (Chapter 18). We conclude this Part with a discussion of the relevance of, and obstacles to, cross-disciplinary co-operation (Chapter 19).

PART 5

Organisational and interorganisational issues

16 Organisational management, learning and research

As we have already indicated in section 1.1.2, there is increasing diversity in the organisational contexts in which change agents work. Professionals in communicative intervention work in public extension organisations, private extension organisations, non-governmental organisations, commercial businesses, government departments, project bureaus, consultancy firms, research institutes, etc. Moreover, they work in various political and cultural environments. Given this diversity, no recipe can be given as to how to organise and manage a group of communicative interventionists, and/or how to create an enabling organisational environment for them. Hence, that is not what we aspire to do in this chapter, and neither do we present a summary of what has been written in the literature on the management and organisation of public, profit and non-profit institutions (e.g. Clegg et al., 1999). Rather, we will discuss a few topical issues that relate to the organisation of communicative intervention. Before doing so, however, we discuss some conceptual modes of looking at organisations and their management (sections 16.1 and 16.2). Subsequently, we emphasise the importance of having an organisational mission (section 16.3), and raise a range of issues that relate to organisational learning (sections 16.4). We conclude in section 16.5 with some considerations regarding organisational research, including monitoring and evaluation. Issues related to the financing of communicative intervention are discussed in Chapter 18, because this issue has many implications that transcend the organisational level, and impinge on co-operation with other parties in a network.

16.1 Co-ordination in organisations: the significance of 'structure' and 'culture'

Developing an understanding of what organisations are, and how they can be 'managed', has a wider relevance for our field of study than just the management of an organisation with a mandate in communicative intervention. This is because we have argued that coherent innovations depend on co-ordinated action among human beings, and hence include new socio-organisational arrangements (see section 8.2.2). As we have seen in various parts of this book, the latter term refers to phenomena like the formal and informal rules, laws, agreements, procedures, facilities, contracts, markets, norms, values and/or shared understandings that may help stakeholders to co-ordinate their actions (e.g. Chapters 5 and 9). Formal organisations too can be seen as a particular social form through which human beings aspire to co-ordinate their actions, in this case towards a more or less clearly defined organisational purpose (adapted from Blau & Scott, 1962). To this end, formal organisations are usually characterised by an organisational framework or structure that may include a range of aspects. For example:

- a formal mission statement, clarifying the objectives for which the organisation exists and the core strategies for pursuing them;
- a description of the various tasks and sub-tasks to be performed by the organisation;
- a description of different areas of responsibility and authority in relation to these tasks;
- a distribution of sub-tasks, responsibilities and authority into several jobs/job descriptions;
- the clustering together of jobs into distinct organisational units and sub-units;
- a definition of the relationships between units (e.g. in terms of hierarchy, autonomy);
- lay-out of organisational procedures with regard to 'production', finance, monitoring, evaluation, communication, information management, staff development, etc.

It is crucial to recognise that such formal structures and procedures do not usually correspond with what happens in organisational practice. In many cases there are informal leaders, non-formal procedures and organisational routines, as well as un-official organisational goals. This is not necessarily bad, as it enables organisations to deal with the realities of everyday life, and frequently contributes to achieving organisational objectives. In fact, many things tend to go wrong when people start to work strictly 'according to the book', and 'work-to-rule' strikes have become a feared weapon in the hands of workers and/or trade unions. This reflects again our earlier assessment (see Chapters 4, 13 and 14) that social processes, including organisational processes, cannot be controlled or predicted in detail. Informal arrangements in an organisation can be seen as part of a wider organisational culture, that includes taken-for-granted mutual knowledge, norms, values, routines, artefacts, concepts, etc. (see sections 6.2 and 6.3 for further elaboration). In a book on organisational culture (Schein, 1992:12) the culture of a group is defined as:

> 'A pattern of shared basic assumptions that the group learned as it solved its problems of external adaptation and internal integration, that has worked well enough to be considered valid and, therefore, to be taught to new members as the correct way to perceive, think and feel in relation to those problems.'

While the idea of organisational culture is very relevant for understanding what happens in an organisation, it is equally important to realise that different (sub)cultures may exist in an organisation. Organisations are often not homogeneous wholes, and there may well be cultural differences, tensions and conflicts of interests between various individuals, organisational units or management levels (Morgan, 1998). In that sense, the difference between an organisation and a network of interconnected actors or stakeholders is *gradual*. In an organisation too there are different 'stakeholders' who represent and defend different organisational or personal realities, (sub)objectives and interests. An important difference with earlier discussed multi-stakeholder situations is that organisations tend to have a central management that has the formal authority to define common objectives, define priorities and make decisions in conflict situations, and often is in a position to get rid of employees who do not sufficiently conform to these. This tends to make it somewhat easier to forge co-ordinated action within an organisation, than it is to foster it among relatively autonomous stakeholders.

16.2   Images of organisation and the nature of management

Given the above complexities, the management of organisations remains a difficult task, even when a manager has considerable resources and formal authority. Over time different ideas have emerged as to how organisations work, and where management should focus. Morgan (1998) has analysed these ideas through discussing different *'images of organisation'*. In the previous section we have already touched on one of these: the image of 'organisations as cultures'. From this perspective, then, the role of management is mainly to foster shared meanings, norms and values. In this section we will draw on several additional images[1] that are particularly significant to our field of study. At the same time we will link them to different modes of thinking about organisations as *systems*. As the term 'systems' is widely used in the field of agriculture and rural resource management, the discussion below has a wider relevance than just organisational management. For example, the different 'systems perspectives' discussed below may also be applied usefully when thinking about 'agro-ecological systems', 'farming systems research' (e.g. Bawden, 1995) and/or 'agricultural knowledge systems' (see Chapter 17); i.e. such systems too can be approached as 'hard', 'soft', 'functionalist' or 'cognitive' systems.

16.2.1   Organisations as machines/hard systems thinking

A first image is that of 'organisations as machines'. In this mode of thinking, an organisation is seen as something that can be controlled and designed in order to attain certain pre-determined objectives in a rational manner. Thus, it builds on similar assumptions to the blueprint planning philosophy discussed in section 4.1.1. According to Harrington (1991) two strands of thinking about organisations and management can be distinguished: the scientific approach and the administrative principles approach. The main concern within the 'scientific management' approach, 'invented' by Taylor (1947), was to improve the productivity of labour. To this end, detailed time, movement and cost studies were made in relation to the tasks and sub-tasks in the production process, and the way their efficiency could be increased by means of specialisation, task division, mechanisation (e.g. the conveyer belt), etc. (Gilbreth, 1911, 1914; Taylor, 1947). The main preoccupation within the 'administrative principles' approach (Fayol, 1949) was not so much the execution of specific tasks, but administrative processes within organisations. Like Weber (1947), Fayol developed a list of administrative principles, in which he stressed the importance of authority, command lines, clear objectives, hierarchical decision-making procedures, etc.

---

[1]   Although we borrow Morgan's labelling of images, we sometimes deviate slightly from Morgan with respect to the approaches discussed under each heading.

The two approaches have in common that they start from the idea that organisations can be run as more or less predictable 'machines', if only managers pay sufficient attention to task division and the development of procedures. In terms of systems thinking, organisations are looked on as 'hard systems'; that is, as systems that can and should be engineered and optimised in a rational manner towards a known and previously defined goal (Checkland, 1985). Particularly when organisations aim to produce highly standardised products and services (e.g. cars or fast-food services) in a fairly stable environment, this way of looking at organisations can, to some extent, still be useful (Morgan, 1998). In order to be effective, however, many communication for innovation services (see section 2.2) need to be tailor-made and tuned to specific actors, stakeholders, contexts and emerging dynamics. Hence, it is rather doubtful whether organisations that provide such services can be looked at and managed in this way.

Nevertheless, a well known extension management approach that shows affinity with the idea of 'organisations as machines' is the Training and Visit (T&V) system which was developed by Benor and Harrison (1977) and promoted by the World Bank from the 1970s until the 1990s. This approach is characterised by a strong focus on producing high 'quantities' of purely technical agricultural production advice, using standardised, detailed and rigorously monitored schedules of contact farmer visits and staff training sessions. In short, it was proposed more or less to organise communicative intervention somewhat like a military operation. Like other extension systems of those times, T&V drew heavily on the adoption and diffusion of innovations approach, and has been criticised on similar grounds (see section 8.1.2; Howell, 1982; Purcell & Anderson, 1997). As an organisational management philosophy, however, its main problem was probably that it reduced the role of communication in change and innovation processes merely to 'giving technical advice' to individual farmers, thus turning communicative intervention into a *single standard product* (even if the advice itself could still be different for different people). But as we have argued throughout this book, change and innovation require much more than technical advice alone, and hence many communication workers and managers felt severely curtailed by the recipe of T&V (Hakutangwi, pers. comm.; Röling, 1988). As Howell (1982) reports, in India the system became informally called the 'Touch and Vanish' system of extension, partly owing to its inability to provide anything but advice, and the need for communication workers to move on elsewhere to meet the schedules and required numbers of visits.

16.2.2   Organisations as organisms/functionalist systems thinking

As a critique to the approaches discussed above, organisations became looked at as 'organisms'; that is, as 'complex living entities which do not have machine-like qualities' (Harrington, 1991:57). Social psychologists like Mayo (1933), for example, stressed the importance of healthy human relations and motivation, as influenced by formal and informal organisational arrangements, for the performance of an organisation. This followed his famous Hawthorne studies which showed that production increases did not so much depend on different technical work conditions (e.g. light availability in an assembly hall) but on social factors. Whatever the researchers manipulated in the context of their initial experiments on technical conditions, the productivity of labour increased. Hence, the researchers had to conclude that it was the *attention* of the research itself that had a positive influence. Subsequently, participation of employees in decision-making procedures, and co-operation between management and employees, came to be considered important ingredients of what became known as 'human relations management' (Roethlisberger & Dickson, 1961).

Although the priorities for managers changed (i.e. from dividing tasks and designing procedures to maintaining human relations and motivation), the idea of a fairly unambiguous organisational purpose remained. Like organisms, the key purpose of organisations was assumed to be survival. This way of looking at organisations coincides largely with structural-functionalist (sociological) theories of social systems (Parsons, 1951). Parsons uses several biological analogies and states that human activity systems consist of sub-systems that perform certain basic functions (adaptation, pattern maintenance, tension regulation, goal attainment, and integration), between which complex input–output relationships exist in order to maintain an equilibrium. In line with this, maintaining a balance between organisations and their external environment also became viewed as an important management task in so-called contingency theories (e.g. Mintzberg, 1979). A characteristic of this way of looking at organisations is that they are regarded as complex but essentially harmonious, integrated wholes of social behaviour. Within them, the behaviour of individuals is regarded as something that can be explained by looking at the larger system. Thus, in essence, human agents are looked upon as somewhat 'passive', and subordinate to the larger whole (for a critique see Long, 1990).

16.2.3   Organisations as flux and transformation/soft systems thinking

A next body of organisation theory can be metaphorically called 'organisations as flux and tranformation' or 'organisations as processes' (Harrington, 1991). Like Parsons, authors within this metaphor are influenced by 'open systems theory', and describe human activity systems as engaging in continuous exchanges with their environment. Unlike Parsons, however, the boundaries of systems are considered to be much more vague, and structures are not seen as having an important determining influence on behaviour. Instead, an organisation is seen as a system of interaction between individuals, without a definite form or hierarchy, that mainly exists because, as Harrington (1991:59) describes it, 'people are told it does by management teams, corporate identity and so on'. The processes and practices taking place in an organisation are seen as shaping and changing organisational structures. Thus, in this perspective organisations are seen as flexible, involved in dynamic processes, chaotic (see Marion, 1999) and subject to continuous change. Consequently, the main task of management is looked upon as dealing with complexity and dynamics processes, and facilitating interactions and organisational change.

Inspired by this way of looking at organisations (and also by the idea of 'organisations as cultures') are soft (or interpretative) systems thinkers, who at the same time reacted strongly against hard systems approaches (Ackoff, 1974; Churchman, 1979; Checkland, 1981; Vickers, 1983). In their view 'human activity systems' such as organisations must be looked upon as complex wholes, in which people have different world views (comparable to the concept of 'life-worlds', see section 6.2) or *Weltanschauungen* (Checkland & Davies, 1986) and therefore have different interpretations of the problems that exist, the goals to be achieved in relation to these, and the boundaries of the system itself. In order to manage this complexity, soft systems thinkers have developed methodologies (Ackoff, 1981; Checkland, 1981) aimed at reaching agreement and consensus on problems, ends and boundaries. For example, in Checkland's (1981) soft systems methodology, it is assumed that stakeholders in a particular problematic context (including experts, consultants, etc.) can develop a new 'systemic' shared perception (or model of reality) that can be compared with the original models that the different stakeholders had. Thereby the comparison of different interpretative models provides:

'. . . the structure of a dialectical debate, a debate which will change perceptions of the problem situation, suggest new ideas for relevant systems (leading to iteration), and concentrate thought on possible changes.' (Checkland, 1988:28)

'The aim of the debate is to find some possible changes which meet two criteria: systemically desirable *and* culturally feasible in the particular situation in question.' (Checkland, 1985:764)

As can be noted from the above, soft systems thinking has much in common with conventional ideas regarding interactive processes and participation (see sections 10.1 and 14.1.3) in that it is consensus oriented, and proposes that joint learning is the main route towards achieving this.

16.2.4   Organisations as political systems/critical systems thinking

The idea that an organisation could be understood as, or managed to be, a fairly harmonious system or collection of processes was met by the criticism that this ignored the significance of conflict and struggle within organisations. According to some, organisations are shaped primarily by internal and external politics; that is, by struggles over conflicting interests and/or the wish to generate or maintain power and influence over others (e.g. Pettigrew, 1973; Kolb & Bartunek, 1992). In this view, then, an important task of management is to find appropriate ways of 'distributing' power and dealing with organisational tension and conflict. In line with this, soft systems thinkers were criticised for their naive expectation that 'collective learning processes' towards organisational change would take place in a very open and eventually harmonious atmosphere, and for their failure to recognise that the conditions for an open debate are often lacking (Jackson, 1985; Ulrich, 1988). According to Jackson (1985:145), 'power structures' can considerably affect such a debate, and result in situations in which stakeholders have no equal say in discussion, and no equal access to relevant material and resources. Consequently, the application of soft systems methodologies can easily lead to a reinforcement of the status quo; as Jackson (1985:144) puts it:

'. . . soft systems thinking cannot pose a real threat to the social structures which support the Weltanschauung with which it works. It can tinker at the ideological level but it is likely simply to ensure the continued survival by adaptation, of existing social elites (Thomas & Lockett, 1979; Jackson, 1982). This is not at all what the designers of the soft systems methodologies intended. Nevertheless, there is some evidence that it is what is achieved by these approaches. Churchman, Ackoff and Checkland are baffled that their methodologies when applied to the real world tend to lead to conservative or, at best, reformist recommendations for change. Examples of such bafflement can be found in Churchman, 1971:228; Ackoff, 1979; and Checkland, 1981:15.

To overcome the shortcomings of the soft systems approaches, many critical systems thinkers have embraced Habermas' idea of communicative action and 'power free' communication (e.g. Jackson, 1985; Lyytinen & Klein, 1985; Ulrich, 1988; Fuenmayor & López-Garay, 1991; see section 14.1.4 for elaboration on Habermas). In the 'critically normative systems approach' proposed by Ulrich (1988), the focus is on the 'management of conflict by means of argumentatively secured mutual understanding' (1988:153). Similarly, according to Jackson (1982:25), systems methodologies should provide conditions for undistorted communication. In cases where communicative rationality cannot be secured, social groups may have to discontinue the process of dialogue and enlightenment, and temporarily engage in political struggle (Jackson, 1985:149). Thus, although he recognises the importance of power and conflict, he still assumes it can somehow be banned, at least temporarily, from processes of organisational change. Ulrich takes a slightly different view and suggests that it is not so much the task of systems methodologies to provide conditions for undistorted communication, but rather to critically deal with conditions of imperfect communicative rationality (Ulrich, 1988:158). This latter view comes close to our argument for 'integrative negotiations' in the context of interactive innovation processes (see sections 10.1 and 14.1.3).

16.2.5   Organisations as brains or psychic prisons/cognitive or autopoietic systems thinking

Lastly, we want to mention two related images of organisations which, like the image of 'organisms', draw heavily on biological analogies. The first image emphasises that the production and use of information, knowledge and intelligence are key processes in organisations, and draws an analogy between organisations and brains. Thus, organisations are seen primarily as a network in which signals, information and knowledge are exchanged and utilised to generate action. Hence, the management of knowledge, information, data (see section 6.1) and communication becomes the central concern for managers. A related view also focuses on the cognitive dimensions of organisations, and emphasises the limitations and barriers that exist to cognitive change (see sections 6.4 and 9.4). Here organisations are compared to 'psychic prisons', where 'groupthink' (Janis, 1972) and defensive mechanisms create blindness (Morgan, 1998).

This idea derives from two biologists who have developed an alternative theory of 'living'. Maturana and Varela (1984) propose that living systems are distinct in that they literally reproduce themselves continuously; in other words, they are autopoietically organised (the Greek word 'autos' can be translated with 'self', while 'poiein' means 'to make'). An important aspect of Maturana and Varela's conceptualisation of autopoietic systems is that in the process of (re)production the systems only refer to themselves; that is, even though a system may need material inputs from the environment, it is the system itself that determines the changes that take place as a result of the interaction between system and environment (Maturana & Varela, 1984). This implies that Maturana and Varela abandon the idea that living systems adapt to their environment, for the environment exists only to the extent that the system perceives and recognises it as its environment. Thus, autopoietic systems are organisationally closed systems of production relationships. Nevertheless, system and environment can interact through reciprocal perturbations which may even result in (but not determine or instruct) structural changes in either autopoietic systems or environment (Maturana & Varela, 1984). In the case of recurrent interactions between system and environment, Maturana and Varela speak of structural congruence or structural coupling. Maturana and Varela claim that their theory of living is also a biological theory of cognition; 'living' is seen as a process of cognition, and living (autopoietic) systems are therefore essentially cognitive systems:

> 'A cognitive system is a system whose organisation defines a domain of interactions in which it can act with relevance to the maintenance of itself, and the process of cognition is the actual (inductive) acting or behaving in this domain.'
> (Maturana, 1980:13 in Winograd & Flores, 1986:47)

The implication of this is that cognition too is an autopoietic process; that is, what autopoietic systems (e.g. human beings) 'know' about the environment needs to be understood in terms of their internal constitution, rather than as a representation of external 'facts' (Maturana & Varela, 1984). Thus, they propose that cognition is seen as a (neurophysiological) biological phenomenon. The fact, for example, that a frog can catch a fly is due to the fact that, as a result of historically developed structural coupling, the nervous system generates a specific pattern of activity which is triggered by specific perturbations (i.e. small moving dark spots) and not because it carries a *representation* of a fly (Maturana & Varela, 1984; Winograd & Flores, 1986). Several social scientists have proposed that *social systems* (e.g. organisations) can also be looked at as autopoietic systems (e.g. Luhmann, 1982; Van Twist & Schaap, 1991). Luhmann (1984:346 onwards) argues that human minds (psychic systems) are operationally closed in that every new reflection builds on previous ones. Likewise, in social systems every communication builds on previous communications, which is why they are essentially closed as well. Similarly, Van Twist and Schaap use the idea of autopoiesis to explain the limited capacity of organisations to adapt to changes in the environment. In our own field of study, Röling (2000, 2002) and others (see Leeuwis & Pyburn, 2002) have been inspired by cognitive systems thinking (see also sections 6.4. and Chapter 9). Clearly, the whole idea of autopoietically organised cognitive systems is closely related to both the earlier discussed connections between knowledge and ignorance (see section 6.4), as well as the various obstacles to learning that we have identified (see section 9.4). The key lesson for management here is that ways must be found to break through these barriers to learning and adaptation.

16.2.6   Conclusion: the implications of different images

As we have seen above, we can look at organisations in different ways. Depending on the spectacles one is wearing one sees different problems and solutions. And, as Morgan (1998:300) suggests, all spectacles can be relevant to a particular situation, which means that the very capacity to look in different ways is an important managerial competence. Depending on the situation, one image will be more illuminating and appropriate than another. We can also conclude that key areas that need to be considered when shaping and managing organisations include:

- formulation of goals and strategies;
- task division;
- organisational procedures;
- human relations;
- staff motivation;
- relationships with the environment;
- organisational meanings, norms and values;
- internal interactions and processes;
- resource distribution;
- conflict management;
- organisational change and adaptation;
- internal and external communication;
- management of explicit and implicit knowledge and information;
- organisational learning.

Many of the areas that require attention in fact resemble the issues that we have addressed when talking about the management of change and innovation by communicative interventionists. Hence, most of the theories and strategies discussed earlier in this book are relevant not only for understanding and dealing with 'outside' processes, but also for the internal affairs of an intervention organisation. This is not surprising, since we have identified the bringing about of new forms of coherence and co-ordination as one of the main tasks of communicative interventionists. In many ways, this can be seen as 'organising things' in complex situations that do not have organisation-like qualities and characteristics[2].

16.3   The importance of (re)formulating missions

It is difficult, if not impossible, to shape and manage any organisation without a fairly clear idea of why the organisation exists, what it tries to achieve, for whom and through what means. These are the kinds of aspects that are usually formulated in a mission statement. Such a statement can serve as a point of reference (not a

---

[2]   Similarly, it is interesting to note that some scientists in the field of organisation studies define organisations as 'the organising activities of its members' (Pepper, 1995:17), while communication is regarded as the main process involved in 'organising'. As Pepper puts it in a chapter entitled 'Organisations as communication events': 'To communicate is to organise' (1995:6).

recipe) for the members of an organisation when setting priorities and taking action. This is, of course, especially so when the members of the organisation are more or less in agreement with the mission, and feel that it is indeed *theirs*. Thus, imposing a mission from the top of an organisation is risky, as it may well be experienced as something from 'outside'. Such a strategy can also be seen as a missed opportunity for the development of shared meanings, and the mobilisation of creativity and implicit knowledge in the organisation. The latter is especially important since the environments in which organisations operate – and certainly those of conventional extension organisations – tend to change rapidly, and hence missions may need to be reformulated. Keeping track of relevant changes, then, requires one to explore the experiences of staff at various levels and corners of the organisation. Thus, it can make sense to employ interactive strategies and methods when (re)formulating missions (see Chapters 13 and 14).

16.3.1   Extension in crisis: the need for change and continuity

In the past, the mission of many (public) agricultural extension organisations was something like: 'to increase agricultural production and productivity through the transfer of relevant knowledge and information, and the offering of technical and economic advice'. As we argued in Chapters 1 and 2, we feel that today there are practical challenges and theoretical reasons for adapting and widening the mission of 'extension' into something like: 'bringing about new patterns of co-ordination (i.e. innovations) through the facilitation of learning and negotiation processes'; for example, aiming at societal goals like improved food security, poverty alleviation, ecological sustainability, food safety, increased market shares, multi-functional agriculture, etc. An implication of such a new mission would be that *process management* (see Chapter 14) becomes an important task and role for organisations aiming to stimulate rural innovation. We wish to emphasise that this is not just a nice conceptual idea, but in many situations a highly practical need. Several of the cases presented in section 10.1.1, for example, suggest not only that process management is sub-optimal, but even that it can be lacking completely; that is, none of the parties involved in a change trajectory consider that they are responsible for monitoring the process over a longer period, or for taking up a facilitating role with regard to learning and negotiation. In other words, we frequently witness a vacuum in process leadership (Mutimukuru & Leeuwis, 2003).

We do not wish to imply that process management is *the* new role and mission for extension, but we do feel that it is necessary for extensionists to think critically about changes in both societal goals and the role of communicative intervention itself. This is especially so since many have the feeling that extension is 'in crisis'. In many countries there is a crisis in funding and resources. Moreover, societal demands and contexts have altered significantly (see Chapter 1) so that old routines no longer suffice. In addition, the ways of thinking about extension have changed considerably (Ison & Russell, 2000; Röling, 2002). In fact, new popular ideas like 'participatory extension' (Agritex, 1998) and second order R&D (Ison & Russell, 2000) are easily presented as a radical breakaway of what extensionists have done in the past. In some respects we feel this is regrettable, in the sense that it may unneccessarily alienate communication workers from the work they identify with. Indeed, extensionists need to reconceptualise the idea of innovation, and throw out a lot of ideas that were ingrained during the 'adoption and diffusion of innovations' era. Moreover, on the basis of novel insights regarding innovation processes, classical extension organisations may well consider the provision of new services such as the facilitation of both interactive innovation processes and conflict management. Furthermore, their staff will need to improve their capacity to engage in joint exploration with clients and stakeholders in connection with the services offered.

However, this does *not* imply that there is no longer room for more 'traditional' extension services, and/or that communication workers need to unlearn everything they know, and be re-educated from scratch. On the basis of exploratory problem analysis, for example, it may still be concluded in many cases that conventional intervention goals and strategies are required and justified. Individual advice, horizontal exchange, organisation building and persuasion (see Table 2.1) and also functions like raising awareness, information provision and training (see Table 2.2) can still remain of utmost relevance within – or as an outcome and follow-up of – exploratory processes. The fact that many communicative intervention organisations need to improve their exploratory capacity does not justify that communication for innovation is more or less equated with exploratory processes only. Painting such a picture seems to unjustifiably deny or obscure its intervention character (Röling, 1988). Whether one likes it or not, communication for innovation – participatory or not – always takes place in an intervention context, and the way such processes unfold cannot be properly understood and/or influenced without taking this into account (see Chapters 2 and 14).

In all, we feel that many conventional extension organisations have every reason to rethink their missions, but at the same time we feel they should be careful not to 'throw away their babies with the bath water'. In rethinking, we hope that recent societal demands (see Chapter 1) and our distinction between different communication for innovation services, strategies and functions (see section 2.2) may offer inspiration.

16.3.2   Organisational implications of a novel mission

When the mission of an organisation changes, many things will have to change with it. When, for example, the mission of a development organisation becomes to 'bring about new patterns of co-ordination through the facilitation of learning and negotiation processes', this raises a lot of questions. Who is going to do this 'facilitation'? What kind of skills and training do these process managers need? How do they come to be in a position to facilitate? What kind of organisational and logistic support is required? What is the relationship between facilitators and regular staff and processes? How do we get support for this change from our bosses and clients? How do we communicate about our new mission? There are no standard answers to many such questions. We will just make a few general remarks about incorporating facilitators into a conventional extension organisation.

Functional requirements for process facilitators

In sections 10.2 and 11.2.3 we have already discussed some of the main tasks and roles of process facilitators. From this, several tentative conclusions can be derived with regard to what kind of people we need, and in what kind of organisational environment and position they may thrive.

- *Seniority*: We have already mentioned in section 10.3.5 that the process facilitator may often require a certain amount of leverage, status and credibility. Hence, such a person may need to be more 'senior' than the average field extension worker of yesterday and today.
- *Knowledge and skill*: A process facilitator needs to be equipped with adequate knowledge, skills and experience regarding the dynamics of learning and negotiation processes, as well as the possibilities of intervening in them in a sensible way.
- *Mobility*: Process facilitators need to be located 'where the action is'. And as 'the action' in a network of interrelated stakeholders is likely to take place in different places, process facilitators will often need to be mobile.
- *Autonomy and flexibility*: Facilitators need to be able to quickly respond to the capricious and unpredictable dynamics of innovation processes. This means that they must have a certain amount of autonomy and flexibility with regard to the allocation of their own time and other relevant resources (e.g. budgets).
- *Support and authority*: Especially in the case of relatively complex processes, facilitators may have to call upon the services of others, such as specialists and regular field extension workers. In order to effectively mobilise such support, they may need to have some say in what others in the organisation do.
- *Independence*: Although process facilitators cannot be neutral with regard to process matters (see section 10.3.5), they often cannot function in a multi-stakeholder situation if they have strong affiliations with particular societal interests. Hence, in such situations they must be able to operate 'at a distance' from substantive governmental, political and/or commercial interests.

Implications at the level of organisational structure

Conventionally, the formal structure of many extension organisations was rather hierarchical, and included the following 'line' and 'staff' functions (seen from the bottom upwards) (see Claar & Bentz, 1984):

- 'generalist' field extension workers (FEWs) operating individually in fixed geographical areas;
- several field extension workers (FEWs) are being supervised by agricultural extension officers (AEOs) at a district office (and/or by an intermediate level of 'supervisors');
- several agricultural extension officers (AEOs) are managed by a district agricultural extension officer (DAEO), who in turn is supervised by equivalents at provincial and/or national levels.

- the 'generalist' extension officers at various levels are supported by subject matter specialists (SMSs) who are located at district, provincial and/or national level.

To enhance knowledge sharing and an optimal provision of specialist services, there is a current trend towards the establishment of communicative intervention *teams* at the field level, whereby individual field workers are given space to develop their own specialism (e.g. crops, animal husbandry, etc.) and provide such specialist services in a larger area than before (see also section 16.4.2).

Positioning

How may process facilitators fit into such a picture? As we have indicated, process facilitators will often need to work at the field level (rather than in a distant office). Moreover, they frequently need to co-ordinate their activities with regular communication workers (i.e. those who perform 'traditional' communication for innovation services). Hence, a specialist position in a broader team could be an appropriate place for process facilitators to work from. Given the qualities that process facilitators need to have, and the many parallels between 'organisational management' and 'innovation management', it could make sense to combine the process management specialism with the role of team leader. Similarly, in the absence of teams, the task of process facilitation might fit best with those who currently supervise the field extension workers. However, this will often require a redefinition of the job description of these agricultural extension officers, with bureaucratic tasks reduced in favour of presence and process management work in the field. It must be noted that what constitutes 'the field' can vary considerably depending on the issues at stake and the stakeholders involved. Thus, 'the field' may well extend into the offices of district administrators, research institutes, non-governmental organisations and the like.

Areas of specialism

In terms of support, it is obvious that process facilitators may somehow need to be backed up by specialists in the 'subject matter' of process management. In general, we feel that there may be a need for many communicative intervention organisations to think critically about what 'specialisms' are relevant both at the level of field teams and higher up in the organisation. Commonly, specialisms in classical extension organisations have been defined along lines of agricultural disciplines (animal husbandry specialist, irrigation specialist, crop protection specialist, etc.) and/or basic communication media (mass media specialist, interpersonal communication specialist, etc.). However, from the perspective of new societal goals, missions and alternative ways of defining communication for innovation services and functions (see section 2.2), it seems pertinent to critically review current specialisms. In the future, it may make sense to formulate alternative technical or communicative intervention related areas of specialism, including perhaps food safety, ecological agriculture, multi-functional agriculture, environmental management, conflict management, exploration, horizontal exchange, organisation development, etc.

Public or private services

A key question that needs to be considered is what the opportunities and threats are for the provision of process management services by governmental, non-governmental and/or commercial 'extension' organisations. We are tempted to argue that no general conclusions can be drawn here, and that all will depend on the issues at stake, the specific context, and the extent to which these organisations can create conditions in which facilitators can function. However, it is unlikely that staff from highly politicised public extension services, typical one-issue NGOs, or organisations that have clear commercial interests can function as credible facilitators of multi-actor learning and negotiation processes. At the same time, there are public and private communicative intervention organisations that are in a position to act 'above the parties'. The ramifications of private and public service provision are discussed in more general terms in Chapter 18.

Decentralised priority setting

The identification of problematic situations in which process facilitators could play a positive role is something that cannot be done by a distant central government or management body. Finding out which problematic situations require attention, establishing whether or not the conditions for effective process facilitation are met, and defining what kind of outsider involvement may be needed and acceptable, typically requires understanding of locally specific initiatives, circumstances and social dynamics (see sections 14.2.1 and 14.2.2). This implies that the offering of process facilitation services requires sufficient autonomy and decision-making authority at lower organisational levels. From this perspective, the current trend towards decentralising governmental responsibilities and decision-making to lower administrative levels (Litvack et al., 1998) is indeed a positive one. At least it creates the possibility that decisions regarding priorities of communicative intervention, and the kinds of services that can be offered, are taken by people with a good under-standing of local situations. At the same time, it must be realised that decentral-isation, like privatisation (see Chapter 18), is not a magical formula that leads automatically to improved accountability and tailor-made service provision. Decen-tralisation will require the establishment of new interorganisational networks, pro-cedures and forms of co-operation (e.g. between local government and previously nationally steered extension organisations) that will take time to mature. Moreover, the involvement of local politics in setting agendas for communicative intervention may well be accompanied by the creation of new new biases in terms of who benefits and who does not.

16.4   The challenge of learning organisations: embracing tension

In order to meet the challenges outlined in Chapter 1 (section 1.1.2), both individual communication workers and their organisations as a whole will need to be adaptive and creative and anticipate diversity and continuous change (Van den Ban, 1997). A management style in which people are expected to do only what they are told from above is incompatible with these needs. Rather, an organisational atmosphere is required in which individuals take responsibility for their own actions, share experiences and play an active role in solving organisational problems (Argyris, 1994). This kind of organisation has frequently been called a 'learning organisation' (Senge, 1993; Easterby-Smith et al., 1999). Some key requirements and principles for such an organisation include (see Torres et al., 1996):

- the recognition that something may be learned from every experience (positive and negative) at all levels and corners of the organisation;
- that failures and problems are made explicit and treated as opportunities for learning and improvement;
- that relevant lessons, knowledge and information are shared with others in the organisation;
- that staff development policies (e.g. training facilities) offer opportunities for 'lifelong learning'.

Despite its relevance and beauty, the idea of a learning organisation is far from easy to realise in view of the problems mentioned under 'organisations as psychic prisons', section 16.2.5, the various additional obstacles that may exist to learning (see section 9.4), and the close associations between knowledge and power (see section 6.6). As a result, problems and failures are often denied or hidden instead of being recognised, and knowledge and information are strategically shielded off and/or manipulated rather than shared (e.g. Argyris, 1994; Dörner, 1996). Moreover, Weick and Westley (1999:190) point to the tension that:

> 'Organising and learning are essentially antithetical processes, which means the phrase "organisational learning" qualifies as an oxymoron. To learn is to disorganise and increase variety. To organise is to forget and reduce variety.'

Given all these problems, how can managers of communicative intervention organisations try to enhance organisational learning? In our view, there are two basic, interrelated strategies, which can be operationalised in a variety of ways. The first strategy is to *look for and/or create tensions between 'order' and 'disorder'* (see also Weick & Westley, 1999). The second is to *create opportunities and conducive conditions for critical debate* on the existence and practical implications of such tensions. Let us discuss in more detail what we mean. The idea of looking for tension is closely related to our assessment that *feedback*, and confrontational feedback in particular, is an important pre-condition for learning (see section 9.3). From that angle, organisational learning requires that we search for and/or solicit situations where there is a 'conflict' or 'confrontation' between the expected organisational 'order' and organisational practice. With the term 'order', we refer to the commonly accepted (formal *and* informal) organisational goals, values, routines, etc., while 'disorder' refers to situations where these are somehow 'challenged'. The second strategy mentioned stems from the experience that it is almost a default response of members of staff to hide and/or reason away such tensions, irregularities and problems. Here Argyris (1994) speaks of individual and organisational defensive routines which merely serve to 'avoid vulnerability, risk, embarrassment, and the appearance of incompetence' (1994:80). In view of these and other obstacles (see section 9.4) to learning, it is of utmost importance to try and create situations where these kinds of issues can be safely discussed. Below we give some practical hints in connection with these basic strategies.

16.4.1   Looking for and creating tension: some practical hints

Although most members of an organisation experience tensions between 'order' and 'disorder', these are often not explicit and not communicated with others. Hence, they must be actively *explored*. This means that many of the *exploratory methods* discussed in section 13.5 can be relevant to organisational contexts as well. The same holds for the *guidelines for exploration* presented in section 14.2.3. It is relevant to note that the exploration of tension is essentially an activity that involves *comparison* in order to find and analyse *differences*. Thus, some of what we have said about comparison in groups when discussing *horizontal exchange* can be relevant too (see section 13.3.2). Translating some of the ideas presented earlier to organisational contexts, the following strategies may be useful for the exploration of tension.

Systematic collection and comparison of information about the organisation and its environment

This may be information about clients, work activities, staff members, organisational units, types of issues and questions dealt with, client satisfaction, markets, goal achievement, etc. In essence we are talking about systematic monitoring and evaluation (see also section 16.5). On the basis of such information all sorts of comparisons may be made in order to find *differences* between, for example, expected and actual work activities, assumed and actual clients, the performance of different staff members and units, etc. As in the case of farm comparisons, identified differences require further qualitative debate and contextual analysis in order to contribute to learning (see section 13.3.2). Written and computer-based organisational memory and feedback applications (see sections 12.3.3 and 13.3.1) can in some countries play a useful role in all this.

Job rotation and/or regular visits to other units

These kinds of activities may enable staff to get to know different organisational realities, which can help considerably to challenge original assumptions (Garvin, 1993). This does not imply, however, that it is wise to shift staff around continuously. As a condition for being productive, communication workers usually need time to become familiar with a specific context (e.g. communities in a region) and 'grow' in their role. Thus, one may want to look for forms of rotation that do not render useless permanently the local contextual knowledge and social relationships embedded in staff.

Flexible and qualitative reporting on experiences

Many communication workers are obliged to write reports for their superiors, and frequently they do so by means of fixed forms and formats requiring mainly quantitative data. Although this may serve certain purposes of systematic data collection, such reporting systems can only capture a small part of the experiences of communication workers. Methodical lessons, identified problems, knowledge gaps, etc. are difficult to share through fixed formats, and may require different modes of reporting (e.g. writing contextual stories and case histories). Without more creative and qualitative forms of reporting, reporting systems easily become rather biased and artificial (see also our discussion of reward systems in section 16.4.2).

Experimentation

Members of organisations can be given space to put new ideas and knowledge to the test with the help of small experiments and/or pilot projects (Garvin, 1993). Experiments have the quality of exposing difference and tension between conventional practices and alternatives. Parallel to what we have discussed in section 13.5.8, organisational experiments too can take various forms, and can be expected to take place even without being formally created. Hence, organisational learning may not only come from inducing experimentation, but also from the ex-post reconstruction of 'experiments' and practices of improvisation (see also Weick & Westley, 1999).

Analysing humour and contested language

Tensions can also be identified through the analysis of everyday organisational talk (see also section 13.5.1). Particularly useful in this respect are jokes and other expressions of humour regarding the organisation (Weick & Westley, 1999). This is because jokes often make use of 'double meanings' (Freud, 1905) that may point to organisational contradictions and taboos. A lot can be learned, for example, from the fact that the Training and Visit system of extension became jokingly referred to as the 'Touch and Vanish' system (see section 16.2.1). Similarly, it can be interesting to look at language expressions that are, or become, contested in an organisation. Such expressions often point to wider tensions in an organisation regarding its goals, values, assumptions, methods, etc. For example, the fact that the word 'technology transfer' became and still is a controversial term in communication for innovation circles, is very telling about some of the problems and contradictions that characterise our field of study.

Analysing complaints

Spontaneously received complaints about the organisation clearly point to a tension between what people expect from the organisation and what they get. Careful reconstruction and analysis of the situations and dynamics that led to complaints can help to identify problems in relation to organisational procedures, the quality of services, internal communication, staff attitudes, etc. In connection with this, it can also be useful to actively solicit complaints, if only by having a clearly announced 'address' where they can be deposited.

Image research and external feedback

The deliberate organisation of feedback from outsiders with or for whom the organisation works can be very illuminating. Different categories of clients, funding agencies and co-operating institutions, for example, can be interviewed about how they feel the organisation is functioning, what the strong and weak points are, etc. This may expose a number of frictions between what an organisation would like their environment to think about them, and what they actually put forward. A specific way of doing this kind of investigation is *image research* (Van Riel, 1992; Van Woerkum, 1997). Here, the term 'image' refers to the multiple associations that people have with regard to various aspects (e.g. mission, activities, qualities, characteristics) of an organisation. In image research such associations are frequently studied with the help of bipolary scales, by which respondents can, for example, indicate to what extent they find an organisation friendly versus rude, dynamic versus static, open versus closed, knowledgeable versus stupid, co-operative versus competitive, effective versus ineffective, slow versus fast, etc. By submitting these kinds of questions to different people inside and outside an organisation it is possible to clarify:

- the image that management would like the organisation to have ('desired image');
- the image that members of the organisation have about themselves ('self-image');
- the image(s) that various outsiders have regarding the organisation ('external image').

Usually, such studies bring to light a number of discrepancies and help to identify areas where the image is seen to be undesirable or negative. This can lead to debate on whether negative images can be attributed to ignorance and incomplete images (see section 6.4) combined with sub-optimal presentation and communication, or to a more fundamental discrepancy between the actual functioning and identity of an organisation and the 'desired image' (Van Woerkum, 1997). In many cases, the latter is at least part of the problem, signalling that organisations need to improve their performance (or adopt a more realistic 'desired image').

Eliciting different situational and future perspectives

Different individuals or groups within the organisation, as well as different actors in the organisational environment, may have different views about the situation that the organisation is in, and about its future role; that is, they have different views regarding the organisation's relevant context, key problems, alternative solutions, obstacles, etc. For purposes of organisational learning and policy development, getting such perspectives on the table can be very enriching. Hence, internal and external stakeholders may be invited to explore a range of issues (e.g. Table 14.1 in section 14.2.3). Exploratory methods that may be of particular use here are future explorations and socio-technical problem tree analysis (see sections 13.5.10 and 13.5.7 respectively).

Some of the above strategies draw mainly on *internal communication* while others build on intensifying communication with the outside world (*external communication*)

(Van Woerkum, 1997). We can conclude that both modes of communication are important for the identification of tensions, and the enhancement of organisational learning.

16.4.2   Creating opportunities and conducive conditions: practical hints

In order for tensions to be brought into the open, and in order to capitalise on them, conducive conditions need to be strived for. Several issues and strategies are relevant in this respect.

Process management

It is crucial to realise that, in order to be feasible and useful, the exploration of tensions cannot be an isolated activity or event. In other words, it needs to be embedded in a wider process of organisational reflection and change. Hence, process management tasks other than just 'exploration' (section 14.2.3) may require careful attention. Many of the wider guidelines regarding the facilitation of interactive processes (see section 14.2) apply to organisational learning and change trajectories as well, and so do our suggestions concerning the planning of individual activities in such a process (Chapter 15). We will not repeat them here, but simply encourage managers of communicative intervention to translate the insights from Chapters 14 and 15 into the management of the organisation itself.

Facilitative management and leadership

As transpires from the above, an important task for the managers of learning organisations is the facilitation of internal learning, negotiation and change processes. As Senge (1996) has emphasised, this requires a different style of leadership from the classic 'authoritarian boss' or the 'bureaucratic administrator' (see also Schein, 1992). Using the metaphor of an ocean liner, Senge argues that learning organisations do not so much need a 'captain' who 'steers the ship', but rather a 'designer' who helps to build the qualities and properties of the ship in the first place. In his view, 'organisational design' is about creating shared purposes, values and meanings, fostering joint learning, and developing appropriate organisational policies, strategies and structures on the basis of it (Senge, 1996). Also, Senge emphasises that leadership in learning organisations requires an attitude of stewardship; that is of *serving* others to allow them to operate better in achieving organisational purposes. An important idea behind this kind of 'facilitative leadership' is that key responsibilities are left (or shared) with employees, and that this in itself is a stimulant to *take* responsibility, including the responsibility for signalling tensions and taking appropriate action.

Reward systems

As a complement to the above, it can be important to set up reward systems for practices like signalling tensions, suggesting solutions, sharing knowledge, taking responsibility, etc. If, for example, communication workers are only evaluated and rewarded on the basis of tangible 'outputs' like the number of clients served, this may not be conducive to taking up wider organisational responsibilities. Hence, it is important that other contributions are also made visible and rewarded. This can be done in various ways, for example through the simple expression of appreciation, the 'measurement' of organisational contributions, positive attention in an internal company magazine, organisational awards, financial incentives, promotion policies, etc. A precondition for all of this is that managers of organisations are in contact with staff at various levels of the organisation, and/or have adequate qualitative reporting systems (see section 16.4.1). Without 'eyes and ears' in 'the field' it is difficult to get a balanced idea of the qualities, contributions, problems and achievements of staff. In the absence of this, field-staff often paint a more positive picture of their work than is warranted. In hierarchical organisations especially, such 'polishing' practices can be repeated at several organisational levels. The result may be a totally distorted perception of on-the-ground organisational realities at the higher levels of the organisation. Clearly, this is not a very healthy situation from the perspective of organisational learning, and does not help the development of adequate organisational policies.

Working in teams

Transferring the responsibility for organisational performance from individuals to teams is a frequently advocated strategy for stimulating organisational learning (Garvin, 1993; Senge, 1993). The underlying idea is that people with different qualities and competences can learn a lot from each other while co-operating, and that as a group they are better equipped to deal with the tasks and problems they may encounter. Moreover, members of a group may stimulate each other to perform well, and compensate for each other's individual shortcomings. Hence, we see nowadays that even producers of fairly standardised products like cars delegate part of the production process to teams rather than to isolated individuals at a conveyer belt. As communication for innovation work typically requires a range of different technical and social competences, we feel that a team approach can, in principle, be useful. However, unlike staff in a car factory, rural communication workers often need to operate in a fairly large geographical area, and even more so when they work in a team. Therefore, a team approach in such a setting is only likely to work well when sufficient transport and communication facilities (e.g. cars, motorbikes, mobile phones) are available. In addition, adequate attention must be given to the stimulation of positive team dynamics (see sections 12.2.3 and 14.2.3). Similarly, we feel it may be important to maintain certain modes of registering individual staff performance, since it can be demotivating for people when individual contributions become invisible and are credited to the team only. Finally, a team approach may also (but not necessarily) imply that clients no longer have contact with a single communication worker, but get to deal with different staff on different issues (or at different times)[3].

---

[3]   Experience suggests that once a relationship of trust has been established between a client and a communication worker, both parties prefer each communication worker to have his or her own clientele.

When this is the case, it is crucial that a proper 'client information system' exists, in which relevant information about the client and previous contacts is documented.

Involving outsiders/creating safe havens

Frequently, staff find it easier and safer to communicate about tensions and problems in an organisation with outsiders. In this way they can remain relatively anonymous or 'behind the scenes', and avoid sanctions by their superiors. Argyris (1994), for example, describes a case where external consultants informed top management about problems and concerns that employees had known of for years. Thus, involving outsiders can be a strategy to enhance organisational learning at some point. However, as Argyris emphasises, one could also say that there is something fundamentally wrong with an organisation (in terms of, for example, management style, internal communication or organisational culture) when available insights and lessons cannot be brought into the open. Hence, deeper analysis is needed in such cases, to enhance regular organisational learning. Realistically speaking, however, it can be important to create 'relative outsiders' *within* an organisation; that is, to make provisions in an organisation for staff to express views, worries and problems to people on whom they do not immediately depend, and who have an authority to check or raise issues at a higher level, without necessarily revealing their sources. Moreover, similar arrangements may have to be in place in order to allow people from outside the organisation to file complaints and deposit problems.

Combining training and organisation development

Conventionally, training of communication workers has been approached as a largely individual affair, albeit in the interest of the organisation. That is, staff are sent on a specific course that is administered elsewhere and attended by similar staff from other regions or organisations. After participating in the course, the staff member returns to his or her work, and hopes that he or she can apply whatever has been learned. This kind of training may work where a particular individual lacks specific insights and skill in an otherwise smoothly running organisation. However, it frequently happens that individual training is provided for purposes of stimulating wider organisational changes, for example to make an organisation more 'participatory', 'client friendly' or 'gender sensitive'. A problem with this is that one cannot really tackle organisational problems through individual training alone. The chances are that after returning from the course, staff will find it very difficult to do anything with what they have learned, essentially because everybody else continues operating in the same old way. An alternative approach may be to bring all staff (including field-staff, managers, specialists, etc.) in a district or region together to learn about particular issues. Here 'training' can become more than providing insight, exchanging ideas and offering experiential learning experiences, but may also extend into forging agreement on how to change organisational routines and conditions. In essence, this means that training is combined with organisation development. An additional advantage of such an approach is that it can build on locally specific problems and training needs, as identified by organisational staff, clients and/or co-operating institutions.

Communication infrastructure

Organisational learning requires not only that lessons are drawn from experiences, but also that these are shared and communicated with others who may encounter similar situations. Following much experimentation, for example, communication workers and farmers may have developed a novel pest management strategy from which others may benefit. Similarly, organisations are often confronted with similar questions from different clients (as well as similar 'questions behind the question'), and in order to prevent a staff member having to 'reinvent the wheel', it would be helpful if they could make use of others' efforts. It can be important to document experiences, lessons, questions and answers in written or computer-based (organisational) search and access applications (see sections 12.3 and 13.6.1).

Clearly, this requires efforts in the collection and investigation of experiences, editorial activities, maintenance, etc. Apart from such internal communication requirements, lessons and their implications may well be relevant to people outside the organisation as well (if only to let them know that the organisation has specific expertise). In all, it is important that there are resources and people in an organisation who bear responsibility for internal and external communication, and for the 'research' activities needed to support this – both for the purpose of exploring tensions (see section 16.4.1), and for communicating about the lessons and changes that eventually arise from these.

Research capacity

As we have seen, learning in organisations requires activities like exploration, collection of information, image research, experimentation, critical analysis of language and experiences, etc. In essence, these are research-like activities. Hence, as a complement to communication staff and infrastructure, learning organisations may benefit from the availability of research capacity. The difference between 'communication' and 'research' is not always clear, especially where researchers engage in participatory forms of investigation. In the next section we will discuss organisational research in more general terms.

16.5   Organisational research, monitoring and evaluation

At various points in this book we have come across the need for communicative intervention organisations to engage in forms of 'research' in connection with their interventions. In section 2.3 we labelled such research 'decision-oriented research', and distinguished it from 'conceptual research'. In Table 16.1 we summarise and categorise the various forms of 'decision-oriented' research, and link them to conventionally used terminologies of 'monitoring and evaluation' (M&E).

Several of the forms of research in Table 16.1 (especially the first four) need to be undertaken by change agents as a regular part of their work, and may not even be

Table 16.1 Different forms of decision-oriented research connected with communicative intervention.

Our terminology	Essence and related (more conventional) terms
Preparatory research, e.g: • reconnaisance • context analysis (e.g. market, policy, technology, etc.) • stakeholder analysis • conflict assessment • assessing institutional space • targeting/analysing diversity • pre-testing (see sections 14.2.1, 15.2 and 15.6)	Generating relevant insights for improving the preliminary design (or 'formation') of interactive and instrumental forms of communicative intervention related term: *formative evaluation*
Process-monitoring, e.g: • communication patterns • task performance • critical events • cognitive change (see sections 11.2.3 and 14.2)	Analysing the dynamics of network building, social learning and negotiation processes and/or the effectiveness and efficiency of individual activities with the view of adapting interventions in the immediate future related term: *monitoring*
Joint exploration and situation analysis, e.g: • analysis of everyday talk • future explorations • socio-technical problem tree • visual diagramming • farm comparisons (see sections 13.3.2, 13.5 and 14.2.3)	Exploring a range of different issues (see Table 14.1) together with stakeholders, with the view of fostering greater mutual understanding, social learning and cognitive change, as elements in a wider strategy to arrive at novel forms of co-ordinated action related terms: *participatory appraisal, learning and/or action research*
Joint research and fact-finding, e.g: • laboratory research • literature research • on-farm experimentation • pilot studies (see sections 13.5 and 14.2.4)	Engaging in deliberate joint research activities in order to resolve identified knowledge gaps or conflicts, stimulate the creation of commonly agreed-upon insights, reduce uncertainty and/or enhance trust among stakeholders; this again as part of a wider strategy to arrive at novel forms of co-ordinated action related terms: *participatory appraisal, learning and/or action research*
Administrative evaluation	Assessing the overall impact and effectiveness (and sometimes efficiency) of intervention trajectories, campaigns or projects after they have been 'completed' related term: *summative evaluation*
Exploring organisational tensions, e.g: • analysing humour • image research • organising feedback (see section 16.4)	Looking for tensions between organisational 'order' and 'disorder' as an inspiration for wider organisational learning and change trajectories related terms: *organisational learning/organisational monitoring and evaluation*

thought of as 'research'. The reason for emphasising these research dimensions is that we feel many communication for innovation efforts could benefit from more deliberate and critical investigation activities. At the same time, this may require more support of regular staff by research specialists and/or a research unit.

16.5.1   Why deviate from the conventional M&E terminology?

The reason for distancing from conventional M&E terminologies is that these have become associated with somewhat unfortunate definitions and meanings. The FAO (1985; in Groot et al., 1995), for example, defines monitoring and (summative) evaluation as follows:

> 'Monitoring is a continuous process of periodic surveillance over the implementation of a project to ensure that input deliveries, work schedules, target outputs and other required actions are proceeding according to what has been planned.'

> 'Evaluation is a systematic process that attempts to assess as objectively as possible the relevance, effectiveness and impact of a project in the context of the project objectives.'

We have several difficulties with decision-oriented research carried out in line with such still influential definitions:

- *Blueprints fostering ignorance*: The definitions clearly start from the assumption that change and innovation can be planned in advance. Hence the preoccupation with detailed work plans and predefined project objectives. As we have indicated at various points, the idea of planning change is highly illusionary (see sections 4.2 and 13.2). For innovation trajectories to be productive, continuous adaptation of goals and plans is necessary. Hence monitoring and evaluation activities that assume the desirability of predefined goals and plans may at best foster 'single loop' learning (see section 9.2), but at the same time are likely to create considerable ignorance with regard to alternative directions and routes. And in the process, such activities tend to reproduce and reinforce outdated ideas about innovation, and come to look at truly adaptive projects as 'failures' because they do not meet original objectives. Or perhaps even worse, they help to define projects as 'successful' that from stakeholders' points of view would not qualify for continuation or scaling-up.

- *Simplification fostering ignorance*: By taking the objectives and work plans of the organisation or project as the central starting point for monitoring and evaluation, one tends to ignore that projects and intervention trajectories are arenas in which different stakeholders and target audiences have their own interests, plans and objectives (Crehan & Von Oppen, 1988; Long & Van der Ploeg, 1989). Of course it is legitimate for organisations to have their own objectives and evaluate these. But by not taking into account the diverse concerns of others, one is likely to miss out on crucial insights that are relevant to the intervention (from the perspective of both goal achievement and goal adaptation). In addition, the focus on plans and objectives may lead to insight into measurable outputs, but does not help much to illuminate the underlying dynamics and reasons through which these take shape. Again, investigating such less tangible issues can provide important insight for developing further intervention strategies.

- *Measuring the unmeasurable*: Conventionally, summative evaluation in our field focuses on assessing the contribution of communicative intervention to changes in awareness, perception, attitude and, mostly, practices and behaviour. This is often in the context of justifying further investments in communication for innovation activities. Quantitative research methods (e.g. surveys) are often dominant in summative evaluations (Mohr, 1992; De Jong & Schellens, 2000), as these are regarded by many as more 'credible' to funding agencies. However, we have seen in Chapter 5 that many factors and actors can play a role in shaping changes in practice and perception, and, in Chapter 8, that innovation depends on co-ordinated changes in a network of actors. It is, methodologically speaking, extremely difficult if not impossible to isolate the contribution of communicative intervention. In other words, whenever there is change, innovation and problem solving, other (f)actors than change agencies may have been crucial in inducing it; and neither can one automatically blame communicative interventionists when innovation is lacking. In order to achieve the near impossible, rather complex and elaborate research designs are advocated (Rossi & Freeman, 1993; De Jong & Schellens, 2000), which may at times become larger than the intervention under study.

  Leaving the wishes of funding agencies aside for a moment, we feel that this kind of summative evaluation is itself often inefficient from a learning perspective. Plausible insights into the contribution, strengths and weaknesses of communicative intervention can often be generated with the help of less expensive and elaborate research strategies such as regular process-monitoring, case histories, in-depth interviews, participant observation and forms of interactive exploration (see section 13.5).

The above criticised conceptions of monitoring and evaluation are quite typical of mainstream thinking in the 1970s and 1980s. Since then, there has been an upsurge of alternative thinking on monitoring and evaluation, emphasising its learning dimensions and the need to approach it in a more participatory way (e.g. Feuerstein, 1986; Uphoff, 1989; Horton et al., 1993; Groot et al., 1995; Abbot & Guijt, 1998; Guijt, 1999; Estrella et al., 2000). Key elements in such approaches include:

- an emphasis on monitoring and evaluation for purposes of learning, interpretative debate and goal-seeking, rather than for control and justification;
- a greater emphasis on internal monitoring and evaluation rather than external monitoring and evaluation;
- paying more attention to continuous monitoring of activities than to evaluation at programme level;
- evaluation and monitoring on the basis of emergent rather than pre-determined objectives;
- taking on board different stakeholder perspectives, objectives, criteria and indicators in the monitoring and evaluation process;

- organising monitoring and innovation processes interactively;
- paying more attention to the monitoring of learning and negotiation process dynamics and process objectives (instead of focusing mainly on measuring concrete outputs in terms of, for example, behaviour change);
- a more elaborate use of qualitative research methods in addition to quantitative methods.

Clearly, such an approach is much more compatible with both our theoretical perspective on change and innovation processes (see Chapters 8, 9 and 10), and our related plea to focus on activities or 'micro-episodes' (see Chapter 15). Despite these alternative ways of looking at monitoring and evaluation, the ideas developed in the 1970s and 1980s are still prominent in practice. In addition to accountants' declarations, many donors, ministries and funding agencies still demand conventional modes of effect and impact evaluation as a condition for getting access to their funds. This is usually as a strategy to enhance control over the projects they fund, in combination with a wish to further ease the worries of taxpayers and citizens that public or donated money may not be well spent. In connection with this, popular but highly illusionary tools for planning and linking goal hierarchies (see section 15.1.1) with criteria and indicators for monitoring and evaluation, are still being advocated widely. A well known case is the 'logical framework' tool (see GTZ, 1987; Poate, 1993; Shields, 1993 and for a critique Hersoug, 1996). Thus, engaging in conventional monitoring and evaluation is in many instances a political and financial necessity. But for purposes of learning and improving interventions, we feel it is more often than not a sub-optimal, inefficient, and ignorance inducing effort. This is why we speak of 'administrative evaluation' in Table 16.1.

In all, we wish to stress that forms of research and critical investigation remain of utmost importance in connection with communicative intervention (see also Singhal & Rogers, 1999). However, we would like to advocate different purposes, areas, methods and modes from those suggested in conventional modes of monitoring and evaluation. Hence the introduction of a slightly different terminology (Table 16.1).

16.5.2   Key questions to ask in decision-oriented research

Although every situation requiring decision-oriented research is different, there are several questions that can be addressed usefully in each case. Here we indicate some relevant issues and questions, and for more detailed discussion of the pros and cons of various options we refer readers to specialist literature on monitoring and evaluation (e.g. Marsden & Oakley, 1990; Rossi & Freeman, 1993; Groot et al., 1995) and to general publications on research methodology (Denzin & Lincoln, 1994).

Relevant questions in relation to decision-oriented research have been grouped by Groot et al. (1995) under headings like 'Why?', 'Who?', 'What?', 'For whom'?, 'How?' and 'When?'. In Table 16.2 we have followed their approach and related these headings to basic questions and options. Parallel to what we have argued in connection with the different elements and aspects of developing regular communication for innovation activity (see Chapter 15), the answers to the questions posed must be carefully tuned to each other in order to arrive at coherent research design. Each answer to each question may have implications for other issues.

Table 16.2 Questions that need to be addressed in connection with decision-oriented research.

Relevant questions	Options (incomplete)
Why? <ul><li>For what purpose do we want to engage in research?</li><li>What purposes may others have for co-operating (or not) with such research?</li><li>Can these different purposes be viably combined?</li></ul>	<ul><li>preparation of intervention?</li><li>enhancing motivation by showing results?</li><li>assessment of project results for outsiders?</li><li>gaining understanding about reasons and dynamics underlying results?</li><li>fostering learning and adaptation of efforts?</li><li>improving the project/future projects?</li><li>enhancing organisational control?</li><li>staff evaluation?</li><li>justification (or 'killing') of further efforts?</li><li>networking or image building?</li><li>gaining access to resources?</li></ul>
What? <ul><li>In relation to what level(s) of objectives do we want to do research?</li><li>Whose objectives do we take into account?</li><li>What kind of social dynamics are we interested in?</li><li>What kind of reasons are we interested in?</li></ul>	<ul><li>ultimate objectives, intervention objectives, specific aims, actor aims and/or activities?* (see section 15.1)</li><li>objectives of intervention managers, field-staff, facilitators, donors, government, stakeholders, etc.?</li><li>learning, negotiation, problem solving, communication, network building, diffusion, exclusion, politics, none, etc.?</li><li>related to 'know', 'want', 'be able to', 'be allowed to' or none?</li></ul>
For whom? <ul><li>Whom do we expect/want to do something with the results?</li></ul>	<ul><li>intervention managers, field-staff, facilitators, donors, government, stakeholders, etc.?</li></ul>
Who? <ul><li>Who decides about the design of this research?</li><li>Who conducts/facilitates the research?</li><li>Who interprets the findings?</li><li>Who presents the findings?</li></ul>	<ul><li>donors?</li><li>internal staff? (internal research)</li><li>stakeholders? (participatory research)</li><li>independent external experts? (external research)</li><li>jointly? (collaborative research)</li></ul>
How? <ul><li>What concepts/variables relate to our research interest? (see 'What?')</li><li>How can we operationalise these (e.g. what indicators can we use)?</li><li>What research methods are we going to use?</li><li>How do we create a safe environment?</li><li>How are we going to analyse results?</li><li>How are we going to report about it?</li></ul>	(too context specific to present options) (too context specific to present options) <ul><li>e.g. quantitative, qualitative, exploratory (see section 13.5)</li><li>process agreements, anonymity arrangements (see also section 14.2.2)</li><li>e.g. statistical analysis, qualitative analysis, interpretative debate?</li><li>e.g. a written report, oral presentation, seminar, stakeholder meeting</li></ul>
When? <ul><li>When do we do research?</li></ul>	<ul><li>preceding prospective interventions?</li><li>at the end of each activity?</li><li>at regular intervals?</li><li>at the end of an intervention project?</li></ul>

* In conventional monitoring and evaluation terminology studying 'impact' involves looking at the level of ultimate objectives, while 'effectiveness' can be studied in connection with goals at various levels. The latter holds for the study of 'efficiency' as well; here one essentially looks at the balance between the resources used and the 'effects' obtained at various levels in a goal hierarchy.

Questions for discussion

(1) Identify an organisation that you are familiar with, and discuss how the organisation is managed. What are the central concerns of those who manage the organisation? What are the underlying views or 'images' that the managers have about their organisation? Do you feel these concerns and 'images' are appropriate in view of the tasks and activities that the organisation is involved in?

(2) On the basis of your own experience, describe an occasion where an organisation failed to learn and adapt to changing circumstances. Was learning absent completely, or did it remain restricted to certain groups or levels in the organisation? Can you explain why organisational learning was sub-optimal in this case?

(3) Conventional modes of monitoring and evaluation are rather persistent in the development sector. Discuss why this may be the case. What are possible consequences?

(4) Identify some jokes that people tell about an organisation that you are familiar with. What could the organisation learn from these jokes?

17 Agricultural knowledge and information systems

For a considerable time, scholars in extension science were primarily interested in issues relating to the adoption and diffusion of innovations (see Chapter 8), and the prospects for particular media and methods getting certain messages and/or technological packages across. As Röling and Engel (1990:7) point out, the question of 'how do we get the message across?' was soon followed by the question 'why don't they do what we want them to do?' At first attempts were made to discover the socio-psychological causes of what was termed individuals' 'resistance to change', 'lack of innovativeness' and/or 'traditionalism'. At a later stage the theoretical models developed in relation to these themes were complemented by marketing approaches (Kotler, 1985) and models for communication planning (see Chapter 15; Van Woerkum, 1982; Windahl et al., 1992) in an effort to induce better targeting, greater client-orientation and higher effectiveness of communicative intervention practice. All these approaches suffered from, among other factors, pro-innovation bias and a focus on individuals (see also section 8.1.2), which led them to focus mainly on processes happening at the interface between extension agents and individual farmers.

The recognition that the messages and technologies promoted by extension were frequently inadequate and inapplicable implied that focusing only on farmers and extension agents was too limited. This is because many others (e.g. university staff from different disciplines, applied researchers, politicians, policy-makers, agro-industry, bureaucrats, etc.) played a role in bringing about such offerings. Consequently, coherent innovations could only be expected to emerge when the multiple actors (including farmers), who could influence the bringing about of adequate knowledge and technology, co-operated to improve collective performance. In order to realise this, it was argued that social scientists should study and help to improve the functioning of Agricultural Knowledge and Information Systems (AKIS). In section 17.1 we further clarify the 'knowledge and information systems' perspective. Subsequently, we discuss the Rapid Appraisal of Agricultural Knowledge Systems (RAAKS) methodology, which was developed to analyse and enhance systems in a participatory manner (section 17.2). We conclude with a brief discussion of networking within knowledge systems (section 17.3).

17.1 Knowledge and information systems thinking

The notion of 'knowledge systems' was first developed by Nagel (1980) in relation to the Indian context, and inspired by the American Land Grant Colleges which brought agricultural research, education and extension together in one institution (Lionberger & Chang, 1970; Swanson & Claar, 1984; Havelock, 1986). Later, the idea of Agricultural Knowledge and Information Systems was theorised and operationalised in more

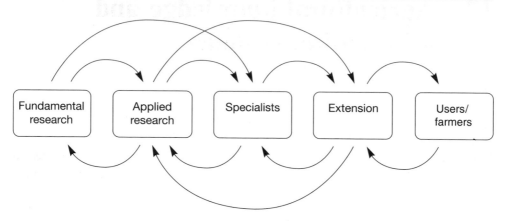

Figure 17.1 Early model of actors and (two-way) linkages in a fictitious knowledge system (Havelock, 1986; McDermott, 1987; Röling, 1988).

detail by the Wageningen scholars Röling and Engel (e.g. Röling & Engel, 1990). In an early definition Röling describes an AKIS as:

> 'a set of agricultural organisations and/or persons, and the links and interactions between them, engaged in such processes as the generation, transformation, transmission, storage, retrieval, integration, diffusion and utilisation of knowledge and information, with the purpose of working synergically to support decision making, problem solving and innovation in a given country's agriculture or domain thereof.' (1989:1)

The idea of 'synergy' in the above definition stems from general systems thinking, and expresses the idea that a system as a whole (e.g. a car) has properties that transcend those of the individual parts (e.g. engine, gearbox, wheels, etc.); that is, a car can drive and transport passengers, whereas the individual parts cannot. Similarly, an AKIS would be able to achieve more than an incoherent set of actors in contributing to societal problem-solving. In the early days of knowledge systems thinking, the idea of AKIS was often visualised as knowledge institutions and effective communicative linkages between them (see Figure 17.1).

In later definitions, it is emphasised that the actors in such an AKIS do not necessarily work synergically, but have the *potential* to do so. This follows critiques (e.g. Leeuwis et al., 1990) that it was misleading to attribute a clear purpose to a 'system' that consisted of so many actors with possibly diverging interests and perspectives. Similarly, it was argued by, for example, Van Woerkum (1990b) that institutions other than formal knowledge instutions (e.g. policy institutions) were relevant as well. Moreover, under the influence of, among others, Checkland (1981; see also section 16.2.3), it was realised later that the notion of AKIS might not only be useful for agriculture, but also for other 'human activity systems' (e.g. natural resource management, organisations, etc.). Incorporating such insights, a later definition speaks of KIS instead of AKIS, which is defined as:

> 'the articulated set of actors, networks and/or organisations, expected or managed to work synergically to support knowledge processes which improve

the correspondence between knowledge and environment, and/or the control provided through technology use in a given domain of human activity.' (Röling, 1992:48)

As shown in section 16.2, there are many branches of systems thinking. The ideas concerning (A)KIS that we present in this section have been inspired mainly by soft systems thinking and critical systems thinking. Accordingly, it is recognised that an AKIS or other 'human activity system' cannot be literally considered to be 'a system' with fairly clear objectives and a life of its own. Instead, the idea is not that an AKIS *is* a system, but rather that it can be useful to *look at it* as if it were a system. Analysing and talking about a complex multi-actor setting with the help of concepts from systems theory, is expected to enhance joint learning and negotiation towards better co-ordination among actors, and more effective adaptation to the environment. In this line of thinking 'a system is a construct with arbitrary boundaries for discourse about complex phenomena to emphasise wholeness, interrelationships and emergent properties' (Röling, pers. comm.; see also Röling & Wagemakers, 1998). For more detail on this kind of systems thinking, we refer readers to our discussion of soft and critical systems thinking in sections 16.2.3 and 16.2.4, and to the overview provided by Engel (1995).

Notions like (A)KIS have become popular in international policy institutions such as the International Service for National Agricultural Research (ISNAR), the World Bank and the Food and Agriculture Organisation of the United Nations (FAO). In a joint publication, the latter two speak of AKIS/RD which they describe in the following way:

> 'An Agricultural Knowledge and Information System for Rural Development links people and institutions to promote mutual learning and generate, share and utilize agriculture-related technology, knowledge and information. The system integrates farmers, agricultural educators, researchers and extensionists to harness knowledge and information from various sources for better farming and improved livelihoods.' (FAO & World Bank, 2000:2)

Note that in this description, the AKIS is composed mainly of farmers, researchers, educators and extensionists, whereas authors like Checkland (1981), Röling (1992) and Engel (1995) would argue that it is not useful to work with such pre-fixed boundaries, and that, depending on the problematic context, it may make sense to include other actors (e.g. consumers, agro-industry, policy-makers, agri-business, etc.) in one's description of a KIS.

Lastly, we wish to mention at this point that the notion of Knowledge and Information Systems has been subject to heated theoretical debates, resulting in many changes in the perspective over time (e.g. Röling & Leeuwis, 2001). Remaining critiques centre around its rather Habermassian orientation towards dealing with conflict, and its tendency to focus on cognition and knowledge exchange in isolation from politics and the exchange of other resources (see sections 14.1.4 and 16.2.4). In connection with the latter, an AKIS is often reduced from being a *social* system to an *aspect* system. But although actors in an AKIS are frequently dealing with knowledge products (instead of material goods), we cannot usefully understand their practices without looking also at other issues such as reward systems, politics, resource distribution, etc. However, this does not render 'knowledge' useless as an entry point, provided that the influence of other aspects on knowledge processes is taken into account.

17.2   Rapid Appraisal of Agricultural Knowledge Systems (RAAKS)

The abstract ideas presented above are clearly not sufficient to improve the synergy in (A)KISs. Hence, a range of analytical and diagnostic concepts has been developed over time (e.g. Röling, 1989; Kaimowitz, 1990; Van Beek, 1991; Engel, 1995). At the same time it was realised that having outsiders look at KISs from a distance with the help of such concepts, would not be likely to lead to much change in system performance. What was needed instead was some kind of participatory methodology through which actors in a KIS could analyse their own system, and eventually agree on co-ordinated courses of action. To this end Engel and others have developed a methodology called Rapid Appraisal of Agricultural Knowledge Systems (RAAKS) (Engel, 1995; Engel & Salomon, 1997). Below, we briefly introduce this methodology and touch on the key analytical and diagnostic concepts that have been incorporated into it. Before doing so, however, we would suggest that much of Chapter 16 is relevant to the study and management of KISs as well. This is because, if it comes to co-ordination, the differences between organisations and an inter-organisational phenomenon like an AKIS are gradual (see section 16.1).

While in abstract terms the RAAKS methodology is designed to foster 'systemic[1] learning' (King & Jiggins, 2002) according to the principles of soft systems thinking (see section 16.2.3), its strategic and operational objectives include, according to Engel (1995:188–90):

- to make a strategic diagnosis of the system's performance, constraints and opportunities;
- to identify opportunities for intervention aimed at improving collective innovative capacity;
- to create awareness among relevant actors about existing constraints and opportunities;
- to identify people who may act effectively to remove impairments and make use of opportunities.

When it is decided to use the RAAKS methodology, the idea is first to form and train a 'research team' that is going to organise and facilitate the process. Forming such a team can, of course, be a sensitive issue. The idea is that at least some of the team members are part of the (A)KIS under study. After a preparatory phase – including team building, getting acquainted with RAAKS and building relationships

---

[1]   The term 'systemic' carries not only the connotation of deliberate and systematically organised learning, but also the idea that one is learning about a system of which one is part, with the view or possibility of changing that system. Along similar lines Alrøe and Kristensen (2002:3) speak of a 'systemic science' as 'a science that influences its own subject area'.

Box 17.1 Overview of RAAKS phases and windows, incorporating key analytical concepts and questions. Source: Engel 1995:265, Box 7: RAAKS windows: appreciating a situation and the social organisation of innovation from different angles. Copyright Wageningen University.

Phase A: Problem definition and system identification

A1   Redefining the objective of the appraisal: Whose problem is it anyway? What is it about?

A2   Identifying relevant actors: Who is involved, or should be? What is it about in their eyes?

A3   Tracing diversity in mission statements: Who pursues what, why? Who perceives what 'problem'?

A4   Environmental diagnosis: Natural, economical and socio-cultural factors to be taken into account.

A5   A first approximation: Clarifying the problem situation; who is relevant, why, how?

Phase B: Constraint and opportunity analysis

B1   Impact analysis: Volitions cause assessments to differ; what is the outcome in practice?

B2   Actor analysis: Not all actors are equally relevant for, or interested in each type of innovation.

B3   Knowledge network analysis: Studying interactive communication for innovation.

B4   Integration analysis: Studying linkages and resource coalitions.

B5   Task analysis: What should be done to innovate and who does it?

B6   Co-ordination analysis: Studying leadership and orchestration.

B7   Communication analysis: Cultural barriers in the way of effective communication for innovation.

B8   Understanding the social organisation of innovation: How does it work? Or, does it?

Phase C: Policy articulation/intervention planning

C1   Knowledge management: What can be done to enhance innovative performance?

C2   Actor potential analysis: Who can, and is willing to do what?

C3   Strategic committments: Who will do what? Who will participate in carrying out the activities?

---

with relevant stakeholders – the RAAKS trajectory consists of three phases (Engel, 1995). For each phase several so-called 'windows' have been identified. Each 'window' represents a different way of looking at (i.e. exploring) the system, using different analytical concepts and related questions for investigation. According to Engel (1995:264) such windows help to quickly focus attention, and stimulate the construction of, and 'juggling with', different images about the system. Phases and windows are summarised in Box 17.1.

The RAAKS windows can be used by the RAAKS team during the in-depth interviews and/or regular workshops with stakeholders that are part of a RAAKS exercise. For each RAAKS window, specific exercises, procedures and forms of visualisation have been developed (see Engel & Salomon, 1997). Depending on the context, the RAAKS team may select several of these. In Table 17.1 we provide some examples of RAAKS tools for visualisation.

Given the qualities and benefits of visualisation (see section 13.5.5), the tools mentioned in Table 17.1 can help us to gain more insight into the system under study, and can stimulate critical dialogue among the participants during the process. As with other participatory methodologies that have been described in great detail,

Table 17.1   A selection of visual tools used in RAAKS (as elaborated in more detail by Engel & Salomon, 1997).

Type of diagram or matrix (relevant to window 'X')	Gives insight into what (different) groups perceive to be
Prime mover septagram (A5/B6)	the relative influence and/or leadership of key actors (e.g. donors, farmers, consumers, etc.) on the way the system functions; those actors exerting strong leadership are labelled 'prime movers'
Info-Source-Use matrix 1 (B3)	the relative importance of different sources (journals, extension, colleagues, etc.) for acquiring different kinds of information (e.g. strategic, operational, technical, market and policy-related)
Info-Source-Use matrix 2 (B3)	the relative importance of different sources (journals, extension, colleagues, etc.) with respect to the fulfilment of different decision-related functions (image formation, determination of needs, comparing alternatives, implementing solutions)
Communication network sheet (B3)	intensity of contact between an actor (written in the centre) and other actors/sources (written in concentric circles, whereby more peripheral circles indicate lower levels of contact)
Linkage matrix (B4)	the intensity of contact/linkage between pairs of system actors, whereby different matrixes may be made for different kinds of linkage (e.g. communication, finance, formal/informal, etc.)
Task analysis matrix (B5)	the main innovation-related tasks (e.g. policy formulation, research, programme management, facilitation, evaluation, etc.) of different actors in the system
Basic configurations diagram (B6)	the way different actors are configured vis-à-vis each other in terms of dominant, important or weak relationships (drawn as bold, normal or no lines between actors in a mental map). Maps may indicate whether a system is 'user-driven', 'donor-driven', 'industry-driven', 'policy-driven', etc.

there is a considerable risk that practitioners use RAAKS windows and tools as a recipe and/or as an end in itself (see sections 13.2 and 13.5.11). Hence, we want to emphasise again that each context requires its own methodology, in which RAAKS may serve usefully as *one* out of many sources of inspiration (see Chapter 13) in a wider effort of process management (see Chapter 14).

17.3   Networking in knowledge systems

When talking about interorganisational issues, many authors (and actor-oriented sociologists especially; e.g. Leeuwis et al., 1990) prefer the concept of 'networks' over 'systems'. This is mainly because the term does not have in-built connotations of a common purpose and clear boundaries, and hence serves better to describe what happens in most situations. As we discuss in section 20.4, the idea of multi-actor 'networks' has additional conceptual and methodological merits. However, in knowledge systems thinking too, the concept of networks plays an important role, not only as a key analytical concept in RAAKS (e.g. Table 17.1), but also as an important practical activity (i.e. 'networking') in the context of knowledge systems, communicative intervention and innovation (Engel, 1995). In this section we focus on the practical side and significance of networking and network building.

In many ways, contributing to innovation can be equalled to establishing novel, effective relationships between multiple human and non-human entities. In other words, innovation is about network building and/or reconfiguring existing networks (see section 8.2.2). Key tasks and activities to that effect are social learning and negotiation, as well as process management (see Chapters 9, 10 and 14). However, such processes cannot start from a vacuum, and require that relevant stakeholders know each other and recognise each other as relevant partners in an innovation process. A term that is frequently coined in connection with this is *networking*. In our vocabulary, this term refers to activities that result in, and/or are directed to, the establishment of, at first, relatively non-committal contacts between actors, as a foundation for possibly more committed relationships in the future. Thus, networking contributes to what Van Woerkum and Aarts (2002) call 'relationship capital'; i.e. the quality and quantity of relationships on which an organisation can draw in order to play a meaningful role and secure its existence[2]. Today, many private and public institutions invest in 'networking' (making it literally a kind of *work*; see Engel, 1995), to create opportunities to become involved in often only vaguely defined activities and projects in the future. In summary, we look upon networking as an activity that widens the options and/or increases the chances for actors to become involved in network *building*, and which revolves around the creation of new social and technical arrangements – through learning and negotiation – along with further specification of network relationships in a specific innovation context. Important aspects of networking include:

- *Establishing personal contact*: It is important for organisations (including those involved in communicative intervention) to make sure that their staff have informal interpersonal contact with relevant staff or people inside and outside the organisation. Typical mechanisms here include attending, or organising, workshops and seminars, making introductory visits, participation in social events, etc.

- *Making oneself known*: It is important that those with whom informal linkages exist have an adequate idea about what an organisation stands for. In other words, one needs to somehow communicate about what the organisation aspires to do, the roles it can and is willing to play, the views to which it subscribes, past achievements, and the skills and expertise that it has available. Insights regarding communication planning and image management are relevant in connection with this (see Chapter 15 and section 16.4).

- *Maintaining contacts and relationships*: Contacts must not only be established but also maintained. In this context, it can be important that new and existing contacts are reminded regularly about an organisation's existence, and that up-to-date information is provided, e.g. through newsletters, etc. Another aspect

---

[2]   Such 'relationship capital' can be seen as part of 'social capital' (see Putnam, 1995; Dasgupta & Serageldin, 2000).

of 'maintenance' is that efforts are made to restore relationships which have somehow become 'disturbed'.

- *Gathering information about other actors' networks*: It can be very important not only to keep track of contacts and relationships that exist in one's own organisation, but also to document relationships that others are known to have. Several studies have demonstrated what Granovetter has labelled 'the strength of weak ties'. This refers to the phenomenon that significant developments and new ideas frequently arise from network contacts that only existed indirectly and/or distantly (Granovetter, 1973). Similarly, Boissevain (1974) shows the significance of 'friends of friends' in getting things accomplished.

As mentioned earlier, the networks that take shape through networking activities can be seen as the foundation from which more specific and meaningful network building and alignment efforts can arise, involving phenomena like learning, negotiation, enrolment, and resulting potentially in new forms of co-ordinated action on the basis of, for example, new agreements, institutions and cognitive change (see Chapter 9) that define and/or impinge upon future relationships.

Questions for discussion

(1) Identify a network of organisations that, in your view, should better co-ordinate their activities. Discuss whether or not a RAAKS-like process might improve collective performance in this situation. What do you consider the strengths, weaknesses and risks of embarking on such a trajectory? Who would be in the position to initiate it?

(2) In order to enhance co-ordination among organisations, it is often suggested that the 'linkages' between them must be improved. Identify a network of organisations that you are familiar with and discuss the different kinds of linkages that exist between them. What kind of linkages do you consider most important and why? What forms of linkage would you like to add?

(3) Discuss the difference between 'networking' and 'network building'. What risks do you foresee when the focus is on only one of these activities rather than on both?

18 Privatisation and the emergence of 'knowledge markets'

Throughout the world the actors and dynamics in agrarian knowledge networks or systems have changed considerably in the last decades (e.g. Rivera & Gustafson, 1991). Important changes include the tendency to treat knowledge as a commodity in knowledge networks, along with the legal privatisation of conventional extension and research institutions (Rivera & Zijp, 2002). According to some (e.g. FAO & World Bank, 2000), the idea of effective 'knowledge systems' and the introduction of 'knowledge markets'[1] are compatible or even identical, but in this section we also point to some practical and theoretical contradictions. In order to better understand the idea of knowledge markets, we first discuss some basic ideas in economic theory (section 18.1). We then take a closer look at developments in the financing of communicative intervention and research (sections 18.2, 18.3). In section 18.4, we discuss some initial experiences and observations with regard to the functioning of commoditised knowledge networks. We round off with a conceptual reflection on the sense or otherwise of promoting commoditisation of knowledge.

18.1 Economic theory and the commoditisation of knowledge

From the viewpoint of economic theory, it is possible to identify four basic types of goods along two dimensions (Umali & Schwarz, 1994). The first dimension is 'subtractability' (or 'rivalry'), which refers to the extent to which one person's access to an item or service *reduces* its availability for others. When, for example, there is only a certain amount of fish in a lake, the catches of one fisherman limit the availability of fish for others. Hence, fish in a lake constitutes a 'high subtractability' item. The second dimension is 'excludability', which refers to the extent to which mechanisms are in place that reserve access to goods to some, excluding others. When, for example, social mechanisms are introduced that effectively prevent the number of fishermen or nets from increasing, we can say that fish becomes more excludable. The four basic types of goods and services that are associated with these dimensions are presented in Box 18.1. As shown in the Box, the various goods tend to be associated with different ownership arrangements as well.

---

[1] Theoretically speaking the concept of 'knowledge markets' is somewhat misleading, as a market is not the only institutional arrangement through which knowledge can be exchanged as a commodity in a privatised setting. Contracts too are a frequently used exchange mechanism, which strictly speaking is a different institutional arrangement from a market (e.g. Ménard, 1995). Hence, it is perhaps better to speak of a 'commoditised knowledge network'. However, for the purpose of this book we use the term 'knowledge market' and 'commoditised knowledge network' interchangeably, and consider both terms to include contract arrangements as well.

Box 18.1 Four basic types of goods and services according to economic theory (based on Umali & Schwarz, 1994; see also Beynon, 1998; Katz, 2002).

<table>
<tr><td colspan="2" rowspan="2"></td><td colspan="2" align="center">Excludability</td></tr>
<tr><td align="center">Low</td><td align="center">High</td></tr>
<tr>
<td rowspan="4">Subtractability</td>
<td>Low</td>
<td>Public goods
Collective or non-property which is abundantly available for all, for example:

• air for breathing
• daylight
• a public radio station
• a national animal disease prevention strategy</td>
<td>Toll goods
An individual or group property that can be a used by others who are granted access and/or pay a fee, for example

• a toll road
• an agricultural exhibition
• a website accessible for a fee</td>
</tr>
<tr>
<td>High</td>
<td>Common pool goods
Collective or non-property that is available to all until supply runs out, for example:

• fish in an unprotected lake
• grazing area in a communal land</td>
<td>Private goods
A property that can be 'owned' by individuals, often after paying a price, for example:

• an individual fishing right for a demarcated fishing ground
• private grazing land
• a bag of fertiliser
• tailor-made veterinary advice</td>
</tr>
</table>

It is important to keep in mind that the same material goods can take on a different nature, depending on the specific context (e.g. scarcity) and social arrangements. Agricultural land, for example, is nowadays a 'private' good in the Netherlands, but in other settings it may be a 'common pool' or 'toll' good. Similarly, agricultural advice can at one time be organised as a 'public' or 'common pool' good, and later can be reorganised into a 'private' good.

From the perspective of this classification, it can be noted that 'knowledge' (in its mental capacity) is inherently characterised by a low subtractability, since its 'quantity' does not diminish once it is shared with others (if anything, it increases). To a lesser extent, the same is true of 'information', which we defined in section 6.1 as 'knowledge expressed in a tangible form' (e.g. in a book, leaflet, website, simulation model, etc.). Clearly, information can be more subtractable than knowledge in its mental capacity; the number of books or leaflets available, for example, can be limited. However, with the help of arrangements like copying machines, libraries and the internet, much can be done to keep subtractability low, and/or frustrate attemps to make information subtractable.

In relation to excludability, we can assess that knowledge and information are not necessarily excludable, but that mechanisms to this effect can be relatively easily invented. People can, for example, be denied certain services (e.g. expert advice, an agricultural journal, up-to-date market information, and/or access to a television channel or website) unless they are prepared to 'pay' in money or kind. From the above we can conclude that turning these classic forms of knowledge and information into 'commodities' is often most feasible by increasing their 'excludability', thus turning them into toll goods. In addition, we have seen in Chapter 6 that knowledge can be incorporated and applied in physical products like seeds, machines, pesticides, etc. In that form, 'knowledge' becomes both easily excludable *and* subtractable, and hence can be easily converted into private goods.

As implied by the above, the nature of goods (public, private, toll or common pool) is, to a considerable extent, a matter of societal choice. Thus, privatisation of agricultural extension and the stimulation of knowledge markets reflects a political decision to organise excludability; that is, to no longer perceive and treat agricultural knowledge as public or common pool goods, but to define it as toll or private goods to be exchanged against a price on a market and/or through contracts (see also Rivera, 2000; Rivera & Zijp, 2002). This implies simultaneously that governments shift a greater part of the responsibility for agricultural knowledge and development to the private sector. The underlying rationale is discussed in section 18.2.

The pros and cons of markets

According to mainstream neo-classical economic theory, a 'perfect market' is in principle the most efficient way of exchanging goods and services. 'Efficient' here means that the supply and demand of goods is optimally balanced, so that users can obtain the best quality product given the price that they are willing to pay. Important conditions for this to happen, however, are sufficient competing suppliers of the same goods (perfect competition), and users with adequate information about the prices and qualities of goods and services (perfect information). In many instances, such conditions are not met; i.e. markets are 'imperfect'. Even so, it frequently happens that development policies (including agricultural and resource management policies) are based on the assumption that markets *are* perfect, which leads to all sorts of problems with regard to their effectiveness (Baland & Platteau, 1996; Stiglitz, 2002). Typically, policy-oriented economists have responded in two ways to the widespread occurrence of such 'market failures'. Many maintain the normative ideal of a perfect market, and argue that in the case of 'failure' the obstacles to perfect competition and perfect information should be removed. An increasing number of economists, however, argue that the assumptions underlying neo-classical economic theory (e.g. that people make rational calculations of costs and benefits) are far too simplistic, and that new, less normative theories are needed to explain economic behaviour. In this line of thinking, deviations from neo-classical theory are no longer rejected as 'imperfect' or a 'failure', but rather seen as phenomena in their own right that may also have positive qualities. In particular, such economists argue that a much better understanding is needed of the functioning of economic institutions (Ménard, 1995; Baland & Platteau, 1996; Williamson, 1998), including 'non-market' arrangements for exchanging goods and services.

This is not the place for an in-depth discussion of economic theory. Suffice it to say that, even from the perspective of economic theory, 'markets' are not a panacea, and that there may be good reasons for exchanging goods and services through other arrangements such as, for example, organised bodies, family networks, distributive rules, etc. As Van der Hamsvoort et al. (1999) point out, the transaction costs necessary for organising excludability and/or subtractability of goods, as preconditions for a market to operate, may be excessively high in comparison with the benefits incurred. In a given context, for example, the technical and/or administrative costs for making people pay for listening to the radio can exceed the expected benefits. Also, the societal costs can be high, as particular groups may be excluded from the product, and this may be deemed undesirable especially where there is a public benefit for widespread provision of the product (i.e. the merit good argument). According to Van der Hamsvoort et al. (1999), the key advantages of introducing market arrangements for goods that used to be provided by the state (in their case 'nature and landscape') are typically that (a) the government can reduce costs; (b) one can expect a better connection between supply and demand; and (c) providers of goods can diversify their sources of income, and reduce risk. As risks they mention that (a) the provision of certain goods may be endangered as they will be substituted by goods that are easier to market (substitution risk); (b) clients will obtain goods elsewhere where no market has been organised (relocation risk); (c) certain groups will be excluded (exclusion risk); and (d) providers may incur losses and go bankrupt (market/continuity risk).

Let us now turn to more empirical observations concerning the functioning of 'knowledge markets'/commoditised knowledge networks.

18.2   Public and private forms of financing communicative intervention

The provision of communication for innovation services is becoming more pluriform. As noted in Chapter 1, forms of communicative intervention are provided by government extension organisations, private (or privatised) extension organisations, non-governmental organisations, commercial businesses, government departments, project bureaus, consultancy firms, research institutes, etc. The rationales for providing communication for innovation services usually vary considerably among such organisations, and the same is true of the financial arrangements through which the services are paid for (e.g. Van den Ban, 2000). In Table 18.1 we have summarised some basic rationales and funding mechanisms.

The types of communicative intervention organisations listed in Table 18.1 have co-existed for a long time in many countries. In some situations, this co-existence could be described as 'peaceful' in the sense that the services provided were largely complementary, whereas in other contexts they were competing for clients and/or supporting contradictory messages or policies. In the last decades especially, we have witnessed that classic governmental agricultural extension services are in decline or being abolished altogether. At the same time, the private extension sector is expanding (Marsh & Pannell, 1998; Rivera, 2000). As indicated earlier, this is not just an autonomous trend, as it is frequently encouraged by governments and international institutions (e.g. the World Bank). On the following page we list a number of arguments commonly used to justify privatisation policies. As we will see, they partly differ from context to context (e.g. industrial and developing countries).

Table 18.1 Typical rationale and funding arrangements for providing communication for innovation services associated with different types of organisations.

Type of communicative intervention organisation	Typical rationale for providing communication for innovation services to farmers and/or communities	Typical modes of funding
Conventional government extension service	• developing/realising policies that are considered to be in the public interest (e.g. economic growth) • controlling rural/farming populations	• taxpayers' money • product levies • (sometimes) direct fees for service • donor subsidies/contracts ·
Commercial input providers or output buyers	• securing customers • selling products • guaranteeing certain qualities of products bought • increasing customer satisfaction with products bought	• incorporating costs of services in selling/buying prices • making advisory services part of contract farming
Non-profit non-governmental organisations	• realising certain political/developmental aspirations • maintaining jobs for staff • access to resources*	• donor subsidies/contracts • donations from individuals • membership fees • government subsidies/contracts
Farmers' associations	• providing support to members • influencing policy	• membership fees • donor subsidies/contracts • government subsidies/contracts
Private extension services/consultancy firms/publishers	• satisfying an apparent need • maintaining jobs for staff • making profit	• direct fees for service • contracts with governments • contracts with commercial organisations • journal subscriptions

* In many countries, non-profit NGOs have (also) become part of a strategy of well-educated people to get access to external funds, earn a decent living, maintain linkages with foreign networks, do interesting work, etc. Especially where few other opportunities exist, such private aims may be an important reason for mobilising funds for continuing development work and/or providing communication for innovation services.

Influences and arguments in support of privatisation

As indicated above, many governments have partly or wholly 'privatised' their state agricultural extension services. That is, these organisations were reorganised into profit or non-profit organisations that were independent to some degree and had to earn an important – and often gradually increasing – share of their income through 'the market'. Important influences and arguments that play a role in bringing about privatisation include (e.g. Verkaik & Dijkveld-Stol, 1989; Le Gouis, 1991; Wilson, 1991; Umali & Schwarz, 1994; Rivera, 2000):

• generally increased trust in the efficiency of market forces, combined with the perception that many public extension organisations operate in an inefficient, bureaucratic and inflexible manner, and are vulnerable to corruption, nepotism and the like;

- sufficiently high incomes in commercial agriculture for farmers to pay for agricultural advice that leads to increased profit (in industrial countries especially);
- increased specialisation and less common interest among farmers (in industrial countries especially);
- agricultural overproduction and reduced public support for subsidising the agricultural sector (in industrial countries especially);
- sufficiently high incomes among consumers, and therefore a reduced need to keep the prices of agricultural products low (in industrial countries especially);
- reduced electoral and economic importance of the agricultural sector (in industrial countries especially);
- a wish to make agricultural service provision more demand driven and client-friendly, and less top–down and paternalistic;
- a wish to resolve the long-standing friction that extensionists experience between 'being a policy implementor, acting in the government interest' or 'being a consultant, acting in the client interest';
- a wish to 'open up' agricultural knowledge networks (i.e. reduce the influence of agricultural lobbies of primary producers in setting extension and research agendas) in order to create more space for 'new' concerns such as environmental issues, natural resource management, consumer concerns and chain management;
- a wish or need to reduce government spending in view of deficits, structural adjustment policies, etc;
- a wish to reduce government responsibility and liability in the case of 'bad quality' advice.

As can be noted from the above, there are a number of arguments in favour of privatising extension that are specific to industrial countries, implying at the same time that there is perhaps more reason to look critically at efforts to create a 'knowledge market' in developing countries. Moreover, it is important to realise that it cannot be taken for granted that the privatisation of extension will automatically help to solve pertinent problems such as lack of client orientation, and bureaucracy. Important problems and possible pitfalls associated with the privatisation of extension are discussed in section 18.4.

Privatisation in extension can take many forms

In its most extreme form, privatisation implies that governments withdraw completely from the extension arena, which means that all communication for innovation services become funded by private persons, and are provided by private organisations (this is called *private funding and private delivery*). It is important to note that this is *not* what most advocates of privatisation propose. It is widely recognised that there are always important issues of public interest (e.g. environmental and developmental issues) for which it is not easy to acquire private communicative intervention funding. Hence, it is argued that the provision of several forms of agricultural knowledge and information has to remain a public responsibility (Verkaik & Dijkveld-Stol, 1989; Umali & Schwarz, 1994; LNV, 1998). Rather, it is proposed to *separate funding from delivery* (e.g. Zijp, 1998). This can be operationalised in many different ways.

In some countries, governmental bodies still pay for communicative intervention (e.g. on issues of public interest), but subcontract the delivery of services to one or more private companies who compete with each other. In this mode of *'public funding and private delivery'* (Zijp, 1998), central and/or local governments act as clients, and can hire the private company that provides the best 'value for money'. Similarly, a government can give *vouchers* to farmers for obtaining certain services, and farmers can choose an organisation that in their view provides the best quality service. The (public or private) organisation then gets paid by the government after handing in the voucher.

Another way of separating funding from delivery is through *private funding and public delivery* (Zijp, 1998). Here, the communicative intervention organisation remains a government body which, for the provision of specific services, becomes co-financed by farmers and/or private companies through direct and indirect modes of cost-recovery. We speak of *indirect* modes of cost-recovery when a private or public organisation gets paid for providing communication for innovation services by *others* than those to whom the services are directed. A donor or commercial company, for example, can pay an advisory organisation to provide free advice to farmers on pest control. In the case of direct cost-recovery, those with whom the communication workers interact (e.g. farmers) pay for the services themselves. Several forms of *direct* cost-recovery have been developed, both by fully private and (semi)public communicative intervention organisations. These include:

- charging a fee for each specific service provided to an individual or group;
- offering packages of services to which individual farmers (or groups) can subscribe for a certain period and for a fixed subscription fee;
- providing advice and/or inputs to farmers in return for a share of the product ('share crop extension').

Given the different basic models of privatisation and various practical arrangements that are possible, there are myriad ways in which 'privatisation' of communicative intervention can take shape in practice (see Rivera & Zijp, 2002).

Governments can intervene in 'knowledge markets' in several ways

We have seen that many governments feel it is important to keep paying for specific communication for innovation services. In essence, this is to correct for knowledge market 'failures' of various kinds. On specific (e.g. environmental) issues there may not be an autonomous interest and demand from farmers, or there may be an active interest but no financial resources or 'buying power' to make it effective. Moreover, in the absence of effective demand, the supply side may well be underdeveloped (or vice versa).

In addition to 'correcting' for a lack of demand and supply of specific services (both in substantive and economic terms), there are also other ways in which governments may enhance the functioning of knowledge markets. Governments (or other institutions) may try to enhance the transparency of the knowledge market by collecting and distributing information about the properties, qualities and prices of services offered by different providers. Thus, clients can become better informed about where they can get 'value for money'. Similarly, governments may develop certification systems in order to get a better grip on the quality of the services that are delivered with public funds. It must be noted, however, that it is not very easy to 'measure' and control the quality of a service. Moreover, it appears that farmers may not easily switch from one service provider to another, even if they have information that the other provider could be cheaper or 'better' (De Grip & Leeuwis, 2003). Over the years, farmers often develop relationships of trust with their advisor, and the advisor develops increasing insight into the history, characteristics and specific circumstances of each farmer. Thus, changing from one communication worker to another may involve considerable 'costs' in that new relationships and mutual understanding need to be developed, while old investments are rendered useless. However, improved transparency in the knowledge market may also be beneficial to existing relationships in that it may enhance the 'negotiation' power of a client, and provide new impulses.

Finally, governments may want to correct certain imbalances in the market. Demands from less vocal or smaller categories of clients, for example, may easily be overlooked when service providers have plenty of work. Government intervention may help to put specific categories of clients, or particular types of services, 'back on the map'.

18.3   Privatisation in research

Like private extension, private research has existed for a long time. Commercial companies (e.g. plant breeders, chemical industry) invest substantial sums of money in research for the development of new products (e.g. new varieties, genetically engineered seeds, agro-chemicals, machinery). Commercial companies obviously do this in the expectation of eventually making a profit. This would be severely threatened if the findings of commercial research were freely available to all. Competitors could then manufacture a similar product with far less investment cost. As it is in many cases impossible to keep knowledge 'a secret', other mechanisms have been created to organise excludability. The most significant is the patenting system. Through this, new knowledge – in the form of novel production procedures or products – is registered and thereby becomes the 'private property' of an individual or company for a limited, but considerable, period (e.g. 15 years). At the same time, the production procedure or product is described in detail and is made accessible to anybody (including competitors) who asks for it. However, although the underlying knowledge becomes more or less available for free, it cannot be used freely; that is, others are not allowed to produce the same product or use the same production technique unless they pay a price to the owner of the patent. Thus, the owner of the patent can make a profit by producing and selling products for a price that incorporates the research costs, and/or allows others to produce for a fee per product. A topical issue with particular relevance to agriculture is whether or not forms of life can be patented, and if so by whom and under what conditions. While forms of protection have already been in place, for example for conventional plant breeders, this discussion is widening with the upsurge of modern genetic engineering techniques. A relevant question, for example, is whether or not individual genes from existing organisms can be patented when isolated and/or built into other organisms or even species.

More recently, privatisation is increasingly becoming an issue in previously public organisations carrying out applied and strategic agricultural research. In the Netherlands, for example, since 1998 former public research institutes have become legally independent organisations that have to compete for research contracts through a tendering system in a 'market'. Thus, they no longer get a 'lump-sum' budget from the state, but receive money for specific projects commissioned by the state and/or other parties (e.g. product boards, farmer groups, etc.). In the official terminology, 'input' financing is replaced by 'output' financing. In this set-up, the state still funds research deemed to be in the public interest, but increasingly uses an open tendering procedure for selecting the research institute that offers the 'best value for money' in delivering specified outputs. This form of privatising agricultural research is advocated with arguments similar to those used for agricultural extension (see section 18.2).

18.4 Initial observations regarding emerging knowledge markets

From the discussions in sections 18.2 and 18.3 it is clear that privatisation of research and extension is likely to have significant consequences not only for farmers, but also for research and extension organisations. It is not only the eventual knowledge for farmers that may become a 'commodity', but also, and perhaps even more significantly, several research and extension organisations are becoming competitors in a 'market' for often still publicly funded research and extension contracts[2]. Thus, the 'knowledge market' has different segments and layers. In view of the diverse ways in which privatisation takes shape, it is not easy to draw general conclusions or list general experiences. Besides, there has not been much systematic qualitative research on how the dynamics in commoditised knowledge networks have changed in different contexts. Drawing mainly on the Dutch experience, we present some provisional observations that we expect to have a wider relevance. In addition to some positive developments, we will also point to several risks that may occur in a research and extension system that is privatised to some extent. Many of our observations relate to situations of 'public funding, private delivery' (see section 18.2).

Communicative intervention organisations can survive with private funding mechanisms only

The first observation to be made is that it is possible for a former state agricultural extension service to survive as a private organisation, albeit often with significantly less staff. Experiences in Europe and elsewhere (e.g. Umali & Schwarz, 1994; Rivera & Zijp, 2002) show that certain farmers are willing and able to pay for specific services that used to be provided free of charge by public extension organisations.

---

[2] Note that, in contrast to eventual 'knowledge' and 'information' produced, such contracts for performing research and extension activities are characterised not only by a high degree of 'excludability' but also by a high degree of 'subtractability' (see section 18.1).

This is particularly so when communicative intervention organisations provide access to knowledge, information and innovations that have already proved themselves to be relevant and/or adaptable to individual farmers' situations. Frequently, such knowledge and information originates at least partly from other farmers, but is selected, translated, combined and adapted by communication workers into tailor-made technical and/or economic advice. In addition to classical advisory services, we see that private organisations frequently offer related services, for which farmers appear willing to pay; these include administrative services (e.g. bookkeeping and other paperwork), up-to-date marketing information, detailed local weather forecasts, real estate services, etc. Apart from providing new services, we also see that new farmer and other audiences (see below) are catered for. The Dutch privatised extension service, for example, now operates internationally and provides services and advice to institutions and farmers in a wide range of countries, to the regret, by the way, of several Dutch farmers, as they feel that 'their' knowledge and experiences are now exported to their international competitors.

Although it is still early to draw firm conclusions in relation to privatised research, the indications are, at least in the Netherlands, that privatised agricultural research institutions do manage to attract sufficient clients (including at times groups of growers/farmers) to guarantee continuity.

Increased clarity about mutual expectations

An advantage frequently mentioned by both clients and service providers is that, in a knowledge market, it is often much more clear and well-defined what they can expect from one another, at least when something has been commissioned (e.g. Van Deursen, 2000). This is because there is usually a contract that stipulates, for example, what 'product' is to be delivered, when, and how much time and money is available for it. Before privatisation such things were often less clear; in fact, expectations could change at any time, so that at times certain activities were 'never' finished and kept dragging on. Although the increased clarity can in many ways be regarded as a positive development, we will argue later that it also has negative consequences.

Shifting client perceptions as a default

When institutions need to survive in a market, it is only logical that they will look for clients who have money to spend. Hence, a default tendency for privatised extension and research organisations is to look for resource-rich clients and/or commissioners of contracts (Kidd et al., 2000; Hanson & Just, 2001; Katz, 2002). When talking about farmers, then, there is a good chance that services become oriented towards relatively well-off farmers (if they weren't already), unless this is counteracted by governments or donors who are willing to pay for services (and/or patented products) delivered to less resource-rich farmers (i.e. through indirect cost-recovery). Alternatively, resource-poor farmers may remain in the picture when they organise themselves to pay for joint services. However, it is important to realise that farmers are just one of the possible clients of privatised institutions, and that other parties in the agricultural production chain (e.g. input providers, processing industries, governments) tend to be much more resource-rich than even wealthy farmers. Not surprisingly, therefore, privatised extension and research organisations become relatively more oriented towards providing services to (or on behalf of) these other clientele. The problems of such clients, therefore, can easily become the points of reference for communicative intervention and research organisations, as these clients are the ones who define and pay for the expected outputs. At the same time, the orientation towards farmers is likely to suffer, which is somewhat paradoxical since an important objective of privatisation strategies was to make institutions more client – 'farmer' – oriented.

Enhanced steering capacity of the government regarding public funds

The paradox mentioned above relates to the fact that, for fully privatised extension and research organisations, the government is no longer only the relatively 'distant'[3] financier of activities but also, in many instances, the commissioner and client. In such a client role, a government will usually have particular ideas about the outputs to be realised, and may go elsewhere (i.e. to a competing organisation) if dissatisfied. It is evident, at least in the Netherlands, that a financing system where the government is both financier *and* client tends to increase its control and steering capacity over collectively funded research and communicative intervention. The government is not only in a better position to decide on the specific themes and topics that need to be tackled, but also to stimulate (and/or enforce) co-operation between different actors in the knowledge network. This is because such co-operation (along with other demands) can easily be made a condition in a tendering procedure. However, there are strong indications that a knowledge market can also pose several threats to co-operative activity (see below).

In other situations (e.g. in Uganda) the government delegates the client role to separate bodies or institutions at local level. However, it has been observed that even in such situations government bodies can still play an important role in setting the agenda, for example by defining all sorts of criteria that contracts should meet (Kayanja, 2003). Similarly, one can frequently witness limited and superficial attention paid to the identification of relevant questions and demands from beneficiaries (see De Grip & Leeuwis, 2003; Kayanja, 2003). Thus, in situations of 'public funding and private delivery', greater client-orientation (in the sense of farmer-orientation) is far from automatic. This is likely to be rather different in cases of direct 'private funding' (see section 18.2) by farmers, provided that farmers can choose between different service providers. Here communication workers and researchers are likely to feel greater pressures to be 'client-oriented' since their clients may go elsewhere if they are dissatisfied.

---

[3] For example, although the Dutch government always had a say in defining agricultural research and extension policies, farmers' organisations, research institutes and the public extension service also had a significant influence. Moreover, research and extension efforts tended to be evaluated ex-post, rather than be defined in advance, thereby giving research and extension organisations a certain amount of autonomy.

Non-co-operation and parallel networks as a default

In a situation where various private organisations need to survive in 'the market', there is bound to be competition. In terms of economic theory, this is a necessary condition for a knowledge market to operate, and healthy as it allows the client to choose the best product (see section 18.1). At the same time, however, a competitive atmosphere induces relevant knowledge and insight to become strategic resources for organisations, that cannot be shared freely and especially not with (potential) competitors. We frequently witness that previously public organisations which are in the 'knowledge business' are now growing apart. In the Netherlands, for example, we see that communication workers and applied research staff who formerly worked together for the same boss now work for different independent companies which find it difficult to agree on who should pay whom for what, and why. Should communication workers pay to get detailed insight into the results of collectively financed research, carried out by an independent research organisation? Should applied research pay communication workers for the identification of relevant questions and problems from the field? Can the now legally independent applied researchers set up their own 'advisory branch'? Do their clients pay communication workers for the knowledge they supply or only for the transmission and translation of that knowledge[4]? The tensions that have arisen seem to be leading to an effort by the applied researchers to set up their own advisory channels, while the privatised extension services take initiatives in the research market.

This example shows that privatisation may induce the formation of largely separate and competing knowledge networks. Such forms of competition might act as a stimulant to innovative and client-friendly service provision and research, but could also lead to research duplication and a decrease in research capacity on specific issues. Perhaps most significantly, the example shows that, in a fully privatised context co-operation and exchange of experience are no longer easy and the 'normal thing to do', unless one is paid to co-operate by a third party and/or when the parties expect that pooling resources might help them to win a tendering procedure (i.e. attract additional funds). As we discuss below, this may hamper the solving of everyday problems that demand co-operation between communicative intervention, research and/or others, but for which no easy money is yet available.

Risks with regard to pro-activeness of the network on public (and private?) issues

Our observations with regard to co-operation are underlined by regularly heard, more general complaints by farmers, fully privatised communication workers and research

---

[4]   Regarding the relationships between communication workers and farmers, another pertinent question is whether or not communication workers should pay farmers for the knowledge and insights they extract from them and pass on (i.e. sell) to others.

staff alike, that 'nothing happens unless somebody pays for it', or '. . . unless a bag of money is floating in the air'. Such complaints apparently refer to situations, issues and problems that demand immediate attention, but for which no adequate funding is yet available. Due to the necessity in commercial organisations to allocate working hours to paid projects, nothing much can be done. Funds may not be unavailable because the issue is deemed unimportant, but rather because it is not yet on a resource-rich institution's agenda, or because joint payment is difficult to organise, and/or because there are disagreements as to who should pay for it. In the case of environmental issues, for example, farmers and governments may disagree about whether the environment is a 'public' or 'private' responsibility.

Farmers, for example, may argue that solving environmental problems is in the public interest, and that the government should fund research and communicative intervention. In any case, they may be reluctant to pay because investments in sustainability usually lead to increased costs and not necessarily to greater profits. At the same time a government might argue that 'the polluter should pay', and hence that farmers should be carrying the costs. Moreover, even if farmers agree that they need to pay, creating an appropriate mechanism through which costs can be shared may well cause problems. Also, when a government decides eventually to fund particular research and communication for innovation services collectively, it may take a while before the funding materialises owing to a certain level of (market induced) 'bureaucratisation'. In view of national and international laws that regulate free competition, for example, the Dutch government is obliged to organise increasingly formal tendering procedures before any major research and communicative intervention effort can be commissioned. This usually takes a considerable amount of time (not least since fairly precise outputs need to be defined in advance to make a tendering procedure possible), and requires much investment of time from people who eventually will not get the job. Similarly, when researchers, communication workers and/or farmers themselves identify new areas for research and/or service provision, they will have to estimate whether there may be funds available somewhere, and will often need to put much effort into lobbying and writing proposals, again with considerable risk of being turned down eventually. All of this probably causes the transaction costs of research and communication for innovation to increase (and sometimes even to become prohibitive) when compared to a situation where institutions receive a 'lump sum' and need to decide only on where to put priorities, and justify their activities 'ex-post'. The possibility of using the 'gut feelings' and creativity of researchers and others may also be limited in view of the above; according to an exploratory study by Van Deursen (2000) this is a real concern. It cannot be taken for granted that people with good ideas will also be good at obtaining funds.

Under fully privatised conditions it may take considerable time to get new research and communication for innovation activities off the ground, especially when government funding is involved, and/or when many individuals need to organise themselves to obtain sufficient funds. This poses risks with regard to the capacity of the network to respond quickly to newly emerging needs and insights. More generally, it has been assessed earlier that innovation and bureaucracy do not usually mix well (Leeuwis, 1995).

Risks regarding learning capacity, discontinuity and coherence

The fact that the 'clarity' of activities tends to increase, also has a disadvantage. In a fully privatised 'output financing' system, such 'clarity' is created through the ex-ante formulation of fairly detailed tenders, contracts and/or proposals. In these, things such as research questions, objectives, activities, target audiences, contributing parties, participants, time schedules and budget allocations tend to be specified in some detail. The chances are that these are taken so seriously by the parties involved that they effectively become a 'blueprint'. Especially in the case of longer term projects, this may bring about many of the problems that we have discussed in connection with blueprint planning (see section 4.1). Rigidities and inflexibility may again be built into research and communication for innovation trajectories, causing reduced learning and adaptive capacity. Another obstacle towards learning, referred to by Bartstra (2001), is that for commercial organisations especially, it may be difficult to treat mistakes as learning experiences for all parties involved. This is because admitting mistakes might increase their legal liability, and encourage clients not to pay for 'substandard' services. A related risk is that 'output financing' can easily lead to discontinuity. At a certain point a project period and/or budget can run out, causing activities to come to a premature standstill. Similarly, it can easily occur that a government commissions a number of distinct projects with different (sub)contractors, but on interrelated themes. In the Netherlands, for example, there are currently numerous research and innovation projects on mineral management, each with their own target audience, research objectives, communicative intervention approach, and participating institutions. Even though these projects could learn a lot from each other, experience shows that it is not easy to foster coherence and co-ordination among these independent projects and implementing institutions (Leeuwis, 2002b).

Undermining of mutual knowledge exchange and shifting communicative intervention methods

In a situation where access to outside knowledge and advice becomes more expensive, one might expect that farmers intensify mutual knowledge exchange. However, although this may be happening in several contexts, there is also evidence of the opposite. In the Dutch context, there are indications that farmers and horticulturists are less inclined to share advice and research findings for which they have paid. The Association of Dutch Horticulture Study Groups (NTS) had to give up on its long-defended ideal of open, mutual, knowledge exchange (Oerlemans et al., 1997). Under pressure from horticulturists it has been decided that study groups will be allowed to withhold and shield new insights from other members for a certain time. Although this cannot be explained only with reference to commoditisation of knowledge (e.g. there is also increased specialisation and competition for survival), it reinforces our earlier observations of the possible emergence of parallel, separated and less 'open' knowledge networks.

In the Netherlands, mutual knowledge exchange is further undermined by the fact that the privatised extension service is less inclined by itself to support study groups or engage in other group activities. This is not surprising as both the support of horizontal knowledge exchange (see section 13.3.2) and the supply of knowledge to a group can be regarded as 'spoiling' one's own market. From a commercial point of view, it often makes more sense to sell one's ideas and experiences 15 times to 15 individual farmers, than once to a group of 15 farmers. Hence, the tendency seems to be that group activities take place only when insisted upon by indirect cost-recovery clients. Interestingly, Marsh & Pannell (1998) report the opposite in relation to the Australian situation, where more intensive use is made of group methods. However, they are referring to a situation where a public extension service is facing severe cutbacks in resources, and has decided to work with groups mainly in order to reach as many farmers as possible. A similarity, then, is that here too individual farm visits and advice become largely the domain of private organisations, while the government takes responsibility for organising and/or paying for group activities.

Remaining dilemmas with regard to different clients

One of the aspects that communication workers tended to like about privatisation was the prospect of becoming independent from the government, in the expectation that this would resolve the dilemma between 'serving the client' and 'serving the government'. However, many privatised extension organisations cannot afford to reject resource-rich clients (such as the government) who are willing to commission indirect cost-recovery communicative intervention work. This means effectively that in many situations the dilemma has, at best, only partly been resolved. After all, communication workers may still end up approaching farmers on behalf of not only the goverment, but also other resource-rich clients (e.g. agro-industry). Thus, communication workers may even be 'wearing more hats' than before. Little is known at present of how communication workers deal with such 'multiple bosses'. The impression is that communicative intervention organisations tend to make a strict division (as regards content, methodology and time) between services paid for directly by farmers, and services paid for by third parties. In particular, communication workers seem careful not to jeopardise their relations of trust with individual farmers, and hence prefer to reserve individual contacts for farmer-paid work. Typically, then, in a group meeting or leaflet paid for by the government, a communication worker is likely to put forward the government's point of view on, for example, sustainable mineral management, whereas for an individual consultation a rather different message may be presented. From the point of view of a market-oriented organisation this is logical. However, from a public policy perspective, one can question whether this leads to a coherent and integrated approach towards achieving certain societal ends.

Potential government isolation and inability to give advice

With reference to the Dutch situation, Van der Ploeg (1999) argues that the central government has effectively lost its 'eyes and ears' in agrarian communities after the privatisation of the extension service. And consequently, he argues, the government is much less able to formulate well-adapted policies, including policies for research and communicative intervention. This seems a valid observation, which is underscored by recent attempts by the ministry to improve its communication with the agricultural sector. What we see in essence is that government departments will always require mechanisms for intensive communication with the outside world, and that certain communicative functions cannot be privatised. Communication is required not only in the context of policy formation, but also in relation to policy implementation. Hence, something like a 'communication division' remains necessary, if only to inform citizens about policies. When such a division exists alongside a privatised extension service, we see that it can sometimes lead to friction. In recent years, for example, the Dutch government has tried to streamline its communication about newly established, and fairly complex, laws and regulations concerning mineral management. There is a debate as to how far the government can go in informing farmers on this issue. The dominant view seems to be that the government can only inform people about the rules and regulations that apply to their specific situation. However, communication staff of the ministry are frequently confronted with farmers who do not just want to know what the rules are, but also want some guidance as to which options provided by the rules would be best for them. While providing such advice was normal before privatisation, the policy is now to refer farmers to private consultants to sort out such questions, because the government does not want to engage in 'false competition' with private organisations, and also does not want to take responsibility for advice given to individuals. Again, this is consistent with privatisation policies, but it may not exactly enhance the image of the government as a helpful organisation.

Transition period

As transpires from the above, the more rigorous forms of privatisation may be accompanied by a change process towards new organisational rationales, clientele, services, patterns of co-operation, etc. Experience teaches that such change processes are far from smooth. Perhaps most importantly, being successful as a commercial organisation requires communication workers with different attitudes, skills and aspirations. Staff must, for example, have the capacity to 'sell themselves' both to outside clients and to internal managers who distribute 'declarable hours' from larger projects. Not surprisingly, in the Netherlands for example, many pre-privatisation staff have left or been forced to leave the organisation. In any case, one can expect it to take a number of years before a former public extension service can operate successfully as a commercial organisation. Much time is required, for example, for reorganisation, staff training, the development of new routines, lay-offs, recruitment, the design of new products and packages, attracting new clients, etc. Also, the outside world may need considerable time to adapt and get used to a privatised situation. It may take a while for farmers to realise that it may at times make sense to pay for external advice and/or other services. Similarly, government staff may need time to discover what it means to be a client. Formulating clearly defined outputs that can be put in a tender or contract, for example, is not as easy as it seems. It requires a fairly clear policy and internal agreement, which are not always present. And neither is it simple to organise a workable and fair procedure for commissioning contracts, or to avoid duplication and inconsistencies in contracts commissioned over time. Moreover, it takes time to get used to the idea that, in principle, one gets only what is in the contract – no more, and no less. Adapting a contract on the basis of changed insights and goals usually requires renegotiation, and will often be at a price.

18.5 Reflections on the commoditisation of knowledge

Although the above analysis is based partly on provisional impressions and refers mainly to the Dutch situation, there seem to be sufficient reasons for anxiety and critical reflection on policy assumptions and underlying economic theories.

The linear connotations of 'supply and demand'

The idea that applied agricultural knowledge can be treated as private and/or toll goods that can be exchanged effectively and efficiently through a 'knowledge market', may only be valid where proven and easily adaptable innovations are already available. Where this is not the case, however, the overall impression is that 'knowledge markets' may well complicate the very innovation processes that are necessary to arrive at such proven innovations. This is because commoditisation of knowledge can easily lead to interaction patterns that hinder the flexible co-operation and creativity necessary in an innovation process. In conceptual terms, we can say that the notions of 'supply' and 'demand' are in many ways not applicable to innovation processes. In discourses about the 'knowledge market', the 'demand' side is mostly associated with *users* of knowledge (farmers) while the suppliers are thought of as *developers* (researchers) and *transmitters* of knowledge (communication workers). 'Supply and demand' therefore carries with it the idea of a clear division of tasks between the three parties. In this sense, it draws implicitly on a linear model of innovation[5]. But as we have seen in Chapters 8 and 17, innovation processes usually benefit greatly from non-linear and non-exclusive task-sharing. In other words, in an innovation process it tends to be rather unclear who 'supplies' and who 'demands' knowledge and information, as successful innovation requires the integration of relevant (but often implicit; see sections 6.2 and 8.2.6) knowledge and information from several parties. This means that it is inherently unclear who should be paying who in a multi-party innovation process. Consequently, it is difficult in a privatised setting to initiate an innovation process unless one manages to mobilise a third party who pays for it all. This may be far from easy and can cause significant delays, and/or may mean considerable redirection of the ideas formulated by the original initiators. Overall, our fear is that the overall innovative capacity of agrarian knowledge networks may suffer from knowledge commoditisation. Additional research, however, is needed in this area.

---

[5] Here we are confronted with another paradox of privatisation: on the one hand it is promoted as giving greater client-orientation, and on the other hand it reinforces linear thinking.

Focusing on knowledge or cognitive change?

In connection with the above, the idea of a knowledge market still seems to start from the assumption that innovation and change are primarily about 'knowledge'. In the Dutch context, where privatisation has progressed significantly, many policy documents on transition and change are still full of terms like 'knowledge generation', 'knowledge transfer' and 'knowledge diffusion'. In Chapters 5 and 9, however, we have seen that what is generally understood as 'knowledge' (see Chapter 6) constitutes only a specific type of human cognition, and that other kinds of cognition and perception are equally significant when talking about innovation and change. However, these kinds of cognition are hardly mentioned in dominant discourses on the privatisation of research and extension, not least because they are not very easily captured in terms of exchange in a market. There is a risk that discussions on privatisation will stimulate both commissioners and providers to define a relatively narrow set of communication for innovation services that are clearly knowledge related (e.g. technical advice and horizontal knowledge exchange). In other words, communication for innovation services and functions (see Tables 2.1, 2.2) that depend on process facilitation and broader forms of cognitive change (e.g. supporting organisation development, conflict management, exploration) may be less easily funded through a knowledge market, whereas they are clearly very relevant from an innovation perspective. Thus, in a context of privatisation it is all the more important to be aware of the need to rethink agricultural extension (see sections 1.1.2, 2.1.2 and 16.3).

The need to differentiate between economic and substantive demand

The wish to make publicly funded research and extension more 'demand driven' is an important element in policy discussions about privatisation. We have seen that, as far as 'farmer demands' are concerned, such objectives are realised far from automatically. Apart from the need to clarify *whose* demands are envisaged to be the driving force, the concept of 'demand' itself needs further clarification. In discussions on privatisation two meanings of 'demand' are easily confused. The first is 'demand' in the economic/financial sense, which refers to whether or not there is sufficient economic buying power to pay for certain services required, as a condition for creating an interaction between market parties. The second meaning is 'demand' in the substantive sense, referring to the interest that clients have in certain services and contents, and the questions clients pose. Although policy discourses often suggest that the 'substantive demand' of farmers must be the driving force, we regularly see that policy measures are primarily about stimulating 'economic demand'. The distribution of vouchers to clients, for example, and also the handing over of funds to decentralised bodies or local communities, in themselves have little to do with making research and extension more responsive to 'substantive demand'. In fact, such vouchers and funds often come with strings attached, which may have arisen from government policies that do not coincide at all with farmers' immediate substantive interests and demands. The Dutch government, for example, has handed out vouchers that could only be spent on advice regarding environmental issues, but farmers had little autonomous substantive interest in these (De Grip & Leeuwis, 2003). Similarly, the Ugandan government stipulated that contracts with private extension companies needed to be oriented towards particular cash crops, rather than to the priority problems of farmers (Kayanja, 2003).

Although such policies may be quite legitimate from a government point of view, they have little to do with making research and extension responsive to farmers' substantive demands. In many situations, however, the creation of an economic demand can be a *precondition* to making research and service provision more responsive to substantive demands. However, the latter will require additional effort, exploration and skill. Much of what we have said about information and other 'needs' (see sections 13.6.2 and 14.2.3) applies to the notion of 'demand' as well. This implies that we cannot assume that demands are articulate, stable and uniform. They need to be discovered and explored in a process of problem analysis (see section 14.2.3), which may also benefit from the views and insights of outsiders and 'suppliers' of services (see section 11.2.3). Thus, one cannot usefully look at substantive 'demand' and 'supply' as totally independent categories. Good facilitation on the part of communication workers is an important condition for articulating demand adequately (see sections 12.2.3 and 16.3.2). And whether or not substantive demands do become a driving force depends not only on the individual qualities of a communication worker, but also on the organisational space to accommodate them (see sections 10.3.4 and 16.3).

The special vulnerability of sustainable agriculture

Elsewhere we have argued that the risks and constraints elaborated in section 18.4 are perhaps most significant in connection with the development of ecologically sustainable forms of agriculture (Leeuwis, 2000b). This is because this area is relatively complex and knowledge intensive, and relatively little valid knowledge and information is available as yet. Moreover, ecological agriculture is least amenable to standardised recipes and uniform advice and typically requires the generation of locally specific innovations on the basis of intensive co-operation between farmers, researchers and communication workers (see section 1.1.1). This is at odds with the observation that the conditions for this kind of co-operation seem to be deteriorating (due, for example, to knowledge protection, individualisation, 'bureaucratisation' and discontinuity), and also with the likelihood that client perceptions shift towards well-resourced commissioners of contracts, and away from local-level primary producers. Moreover, tensions about who should be paying for research and communication for innovation services can easily emerge in relation to ecological issues in particular, in view of discussions regarding their public and private benefits. Finally, there is the threat of the ever-rising costs of knowledge acquisition for farmers who wish to use sustainable production practices. This is because of the eventual devolution of the various payments within knowledge networks to primary producers, decreases in 'free' knowledge exchange, and the knowledge-intensive character of sustainable agriculture.

The dependence on government functioning

We have seen that in the case of 'public funding, private delivery' arrangements, the influence of central or local governments in setting research and extension agendas is likely to increase. This is because governmental bodies need to formulate fairly specific outputs in order to make tendering procedures possible. This means that the functioning and quality of activities in the privatised knowledge network still depend to a considerable extent on the quality, integrity, accountability and proactiveness of governmental institutions. Across the world, many governments have a poor reputation in this respect. In connection with other societal domains, Stiglitz (2002) signals that instead of reducing corruption and nepotism, privatisation may have increased it, causing many to speak of 'briberisation'. In all, privatised research and extension systems may remain vulnerable to badly functioning government institutions.

Arguments for privatisation revisited

When we return to the economic and other arguments in support of privatisation (see section 18.2), we are tempted to argue that the assumed advantages must be treated with reservations. Although private organisations may be more efficient and client-friendly in providing proven and tailor-made advice and other services to individual farmers, there are indications that it is much easier for public extension organisations to contribute to collective change and innovation trajectories. Hence, improved matching of 'supply' and 'demand' is likely to depend on the issues at stake.

Although there is little doubt that a government can reduce costs by privatising extension and applied research, other costs may increase. Governments will probably have to expand their communication division, and will have to put considerable effort into organising proper tendering procedures and becoming a well-prepared client for the commissioning of research and communicative intervention that is in the public interest. In addition, a government will have to pay commercial rates for any activity to be carried out. At the same time, the return on investments in research and communicative intervention may reduce in view of decreased continuity, flexibility, coherence and proactiveness. An area where privatisation is most likely to be effective is in opening up the agricultural knowledge networks for new non-traditional and relatively well-resourced clients. But at the same time this may imply less orientation towards primary producers and public issues, which may again have to be corrected by the government.

Furthermore, we have seen that it is rather unlikely that privatisation resolves the role conflict that communication workers find themselves in, because they still end up having mixed loyalties. When phrased in terms of the risks and advantages predicted by economic theory (see section 18.1), we can easily recognise some of the indicated risks, in particular the exclusion risk (some farmers will be excluded from relevant knowledge), the substitution risk (research and communicative intervention will focus on those issues and/or methods for which money is easily available, that is on well-resourced clients), and possibly high transaction costs ('bureaucratisation'). At the same time, some of the expected advantages (cost reduction and improved matching of supply and demand) are dubious.

It may be that the above observations and conclusions are biased by the transition problems mentioned in section 18.4. All the same, we feel that the idea of a 'knowledge market' starts from a much too simplistic notion of knowledge and information. Looking at knowledge in terms of 'supply' and 'demand' reinforces linear thinking, and encourages us to look at knowledge and communication in terms of clear-cut products and well-articulated demands. This contrasts with the idea that much relevant knowledge arises from a process of experiential learning, which is fuzzy and much more difficult to capture and 'sell'. There is also a tension between the focus on 'knowledge' and the idea that other areas of cognition are equally significant in a context of change. In all, we doubt that 'a market' is an adequate arrangement for exchanging knowledge and encouraging innovation in agriculture and rural resource management. Other co-ordination mechanisms such as forms of organisation and/or co-operation contracts with an open agenda and long duration may well prove to be more effective eventually.

Questions for discussion

(1) Make a list of knowledge and information 'products' in a sector and country that you are familiar with, and determine whether they must be considered public goods, toll goods, common pool goods or private goods. Choose another sector or country and do the same. Are there any differences? How can these be explained?

(2) Discuss to what extent the observations made in section 18.4 resemble or differ from the functioning of a knowledge network that you are familiar with. What circumstances may explain the differences?

(3) Discuss whether or not you agree with the following statement: 'Publicly funded agricultural research and extension is not likely to be farmer-driven regardless of whether it is publicly or privately delivered, because a paying government will always have different interests from farmers.'

(4) Discuss the potential of 'voucher' systems for improving demand-driven research and extension. What are the limitations of such a system?

19 Co-operation across scientific disciplines and epistemic communities

Although working towards effective innovation involves much more than 'doing scientific research', we have concluded that academic researchers may at times provide useful contributions (see section 11.2). Due to the socio-technical and multi-dimensional nature of innovations, we see frequently that *several* natural and social science disciplines (e.g. soil science, entomology, plant breeding, economics, sociology, psychology, etc.) can be relevant. Experience has taught that co-operation among such distinct scientific communities is not easy, not least because members of these communities are likely to have different problem perceptions, and use different analytical concepts, theories and methodologies, and may even constitute totally different epistemic cultures (see section 6.5). The latter is especially the case when we compare social scientists with natural scientists. In view of this complexity, it is relevant to pay attention here to topical issues like multi-disciplinary, inter-disciplinary and/or trans-disciplinary co-operation. In order to properly understand the challenges, we first discuss some key differences between the social sciences and the natural sciences (section 19.1), and then continue to discuss what 'cross-disciplinary co-operation' might entail (section 19.2). We will end in section 19.3 with a discussion of some of the obstacles to this kind of co-operation.

19.1 Methodological differences between the social and the natural sciences

The differences between the social and natural sciences have everything to do with their object of study. Typically, social scientists study and try to explain (patterns of) human (inter)action and behaviour. Different social sciences have a different focus and emphasis; for example, economists study how human beings deal with scarce goods, clinical psychologists investigate psychological disorders, their causes and remedies, and sociologists may study how scientists go about producing 'facts'. Conventionally, natural scientists study natural phenomena, including manipulations and applications of these in the form of technical practices and/or devices. Thus, entomologists may study insect behaviour as well as the response of insects to pest-management strategies, physicists may study different kinds of radiation, including ways of applying these in medical treatments or food conservation, etc. In relation to this, it is important to note that scientists who seek to explain human (inter)action face rather different methodological challenges from scientists who seek to understand natural and/or technical phenomena. Such differences include the following:

- *Options for controlled experimentation*: While it is relatively easy for natural scientists to study natural and technical phenomena in a controlled environment, this is not the case for the social sciences. In a laboratory, for example, it is easy to study genetically identical potatoes with almost the same history, and expose them to different treatments while keeping all other known conditions the same. This is impossible with human beings, as they tend to have very different backgrounds and often cannot be isolated for experimental purposes for a prolonged period.

- *Societal relevance of controlled experimentation*: Some social science disciplines (e.g. social psychology and economics) make use of controlled experiments. This is usually by creating an artificial context in which respondents, often students, are exposed to different conditions. For example, respondents can be given a role (e.g. a fisherman or car-buyer) and a different amount of information (e.g. regarding the behaviour of other fishermen or the quality of a car). Subsequently, they can be asked to answer questions about what they would do in such a situation, or can be observed during a role play. Often questions can be raised on whether the results obtained in such experiments are valid for *real* fisherman or car-buyers, who are operating in a much more complex and multi-faceted context than the experiment. This of course may also be true for technical experiments in a laboratory or experimental station; after all, we have argued in section 6.5 that they have, in principle, a *local* relevance only. However, in relation to natural and technical phenomena, it is often much easier to *copy* laboratory conditions in society, and hence *make* the experimental results relevant more widely (Van der Ploeg, 1987). One can, for example, make a factory production line, design a glasshouse, and/or manipulate the hydrological qualities of an agricultural field in such a way that it *resembles* the experimental situation. Again, this is less feasible with social circumstances.

- *Influence of measurement*: In most situations, measurements of natural phenomena do not themselves influence the value of what is being measured. One can measure the growth of a plant in a particular period, without actually changing it by the fact that it is measured[1]. However, many questions regarding human conditions, reasons and motives may themselves change these. By measuring a person's level of 'awareness of' or 'interest in' something by means of an interview question, one may effectively increase it. And by confronting people with a list of reasons for doing something or not doing it, one may influence a person's reasoning about his or her behaviour.

- *Likelihood of strategic response*: In contrast to natural research objects, human respondents may strategically try to influence the outcome of a measurement

---

[1] Note that, even in the natural sciences, this is not always the case. In quantum physics, for example, the detection of small particles by means of some device may actually change the very particle being detected (e.g. Knorr-Cetina, 1992). Also, several measurements on a plant (e.g. measuring the weight of its roots) may cause us to effectively destroy the plant. In this situation, however, measurements can be easily extrapolated to other plants that remain untouched.

and/or study; that is, they may give answers that suit them, or that they expect to make them look better in the eyes of others (including the researcher). Similarly, researchers can easily influence (wittingly or unwittingly) the outcomes of responses, e.g. by asking leading questions. Plants, animals and other natural or technical phenomena do not interact with those who investigate them in such a strategic manner (even if researchers may withhold or manipulate unfavourable outcomes).

- *Single and double hermeneutics*: In the natural sciences, scientists are the only ones who are trying to explain what is happening; that is, a plant does not try to explain why it responds to fertiliser in a particular manner, and neither is the 'behaviour' of the plant changed by what the scientist says about it. In the social sciences, this is totally different. Social scientists are confronted with respondents who themselves may give an explanation about their behaviour and/or the reasons for a particular situation. Moreover, in giving explanations, respondents may well make use of scientific terminology and be influenced by scientists. In explaining a stagnant situation, for example, respondents may make use of concepts like 'stress', 'alienation', 'dependency syndrome', 'exploitation', etc., which all originate from some academic discourse. In view of these 'double hermeneutics' (Giddens, 1976:156), a social scientist always has to distinguish between his or her own explanation of a situation, and that of the respondents.

- *Levels of observability*: Many natural phenomena can be monitored and observed almost continuously, e.g. by means of automated measurement, camera surveillance and other mechanisms. This is far less feasible when human (inter)action is concerned. Many things happen 'behind closed doors', within a person's mind, and/or in contexts that cannot be monitored continuously. As a consequence, social scientists often cannot use direct observation of (inter)action, but need to depend to a large extent on what respondents say about it, even if they know there may be discrepancies between words and practice.

- *Stability*: When measuring natural phenomena, the outcomes tend to be fairly stable; that is, when one repeats a measurement or experiment, the outcome can be expected to be more or less the same. However, human repondents may describe and interpret a situation quite differently at different times, due to changes in context, mood, experience, interviewer effects, etc.

- *Consistency*: Human beings tend to have multiple identities, and hence can be approached in different capacities. Depending on the 'active' identity, a person's response to interview questions may differ; that is, a person may act differently and/or have different views in relation to something (e.g. animal welfare) in his or her role as a consumer, member of a religious community, family member, or employee. This kind of complexity exists to a lesser extent in connection with natural phenomena. Even if, for example, 'water' may play different roles at the same time (i.e. drinking water, ecological resource, health hazard, agricultural resource, physical barrier, etc.) its physiological qualities (e.g. temperature, quantity, pollution level) do not tend to differ according to the role under consideration. One can argue, however, that the *relevance* of specific qualities may vary.

As can be seen from the above, social and natural sciences are rather different in that they require significantly different methodological principles and rules. Not surprisingly, therefore, they can be associated with different epistemic beliefs, methodological traditions, etc. Many natural scientists, for example, believe that unequivocal truths exist, and that universal laws can be discovered through reductionist experimental designs (see section 6.5). In contrast, many social scientists have come to accept that this is certainly not the case for the social sciences, and moreover that this is questionable for the natural sciences as well (see sections 6.4. and 6.5). In line with this, the social sciences are characterised by a much greater prominence of qualitative (instead of quantitative) research methods. Such methods allow for a much more context-sensitive analysis of phenomena, but clearly the inferences made are far less verifiable than in quantitative approaches and cannot usually be generalised. Another consequence of these circumstances is that the social sciences have a much greater diversity of conceptual (or theoretical) approaches than the natural sciences. Depending on the concepts and theories used, one situation can be studied, described, interpreted and analysed in radically different ways. Thus, a situation in which the uptake of a new technology by farmers is limited, can be analysed rather convincingly in different ways:

- drawing on adoption and diffusion of innovation theory (see section 8.1), one may conclude, for example, that the *compatibility* of the innovation has been the limiting factor;
- drawing on knowledge systems theory (see Chapter 17), one might sensibly argue that insufficient *knowledge integration* and *co-ordination* within the agricultural knowledge system has led to the emergence of an inadequate technology;
- drawing on the perspective of interactive (socio-technical) innovation processes (see section 8.2), one could argue that *network building* and *alignment* efforts have failed, e.g. due to sub-optimal *facilitation* of *social learning* and *negotiation*, and/or because the conditions for negotiation were not met (see Chapters 9, 10 and 14).

Note that the explanations do not necessarily contradict each other, and that all three may be supported by plausible evidence. In the social sciences this is a normal situation. And although members of different 'schools of thought' may 'fight' about which perspective is more useful and credible, this kind of theoretical and interpretative diversity may well continue to exist forever. In the natural sciences, then, interpretative controversies are often more contradictory (i.e. they cannot be true at the same time), and are more likely to be resolved by some kind of decisive evidence.

A biased and simplified representation?

It is quite possible that our representation of the differences between the natural and the social sciences is somewhat biased by the dominant epistemological views in the natural sciences and/or by our own position within the social sciences. Indeed, we have spoken so far about natural scientists who prefer to work in controlled environments, and about social scientists who do not. Of course, there are a considerable number of natural scientists who are trying to say useful things about complex phenomena in the 'real world', such as water systems, climate change and ecological processes. Such scientists too are confronted with uncontrollable situations, multiple interactions among variables, a shortage of measurement opportunities, and the like. Not surprisingly, therefore, natural scientists with an interest in ecosystem management (e.g. Holling, 1995; Berkes & Folke, 1998) emphasise fundamental limitations for understanding and predicting the dynamics in such systems (see also Prigogine & Stengers, 1990). The differences between these natural scientists and social scientists are perhaps less pronounced than depicted above. Nevertheless, we can safely conclude that in interactive innovation processes we are usually confronted not only with epistemic differences between scientists ('outsiders') and stakeholders ('insiders') (see sections 6.5 and 13.5.8), but also between different groups of 'outsiders'. As we elaborate later, these differences may complicate interactive and cross-disciplinary co-operation.

19.2   What does cross-disciplinary co-operation entail?

In the vocabulary of this book, the term 'cross-disciplinary' includes related terms such as 'inter-disciplinary' (between two or more disciplines), 'multi-disciplinary' (involving multiple disciplines) and 'trans-disciplinary' co-operation. The latter term refers to a situation where disciplinary boundaries are transgressed, e.g. through the development and use of novel concepts and methods of investigation, that are not part of the original disciplines (see our discussion of the 'spiderweb metaphor' below). In view of the idea of socio-technical innovation (see Chapters 1, 2 and 8), it is clear that we are especially interested in cross-disciplinary co-operation between the natural/ technical sciences and the social sciences. It is important to note, however, that much cross-disciplinary co-operation may take place (and/or be desirable) *within* the natural sciences (e.g. between soil scientists and scholars in crop protection) and within the social sciences (e.g. between economists and sociologists).

Different metaphors of cross-disciplinary co-operation

On the basis of discussions and workshops with (natural) scientists, Jiggins and Gibbon (1998) concluded that there are various ideas about what cross-disciplinary co-operation entails. Here we introduce some of the mental models that Jiggins and Gibbon (1998) identified and summarised as different 'metaphors':

● *The 'tree' metaphor: cross-disciplinarity as a historical phenomenon*: In this view, cross-disciplinarity is seen as something embodied in the history and structure of universities. The 'roots' of the agrarian sciences are formed by the basic disciplines (e.g. mathematics, physics, biology, philosophy, chemistry, etc.) from which different agricultural disciplines have emerged (i.e. the branches of the tree; e.g. soil science, irrigation, crop protection, etc.). Between the branches, new sub-disciplines ('twigs') can emerge over time, as a result of co-operation, specialisation and/or specific developments in science (e.g. molecular biology, bio-informatics, etc.). From this perspective, most 'disciplines' in agrarian science are already 'cross-disciplinary'. This kind of cross-disciplinarity is indeed widespread (in fact Communication and Innovation Studies represent a good example in the social sciences; see section 2.3), but rarely involves natural and social sciences.

- *The 'hedgehog' metaphor: cross-disciplinarity as problem-solving teamwork*: This way of looking at cross-disciplinary co-operation starts from the idea that complex societal problems exist (the body of the hedgehog). Such problems are to be studied by scientists from different disciplines (the spines of the hedgehog). It is proposed that only when the scientists somehow co-operate while looking at the problem (while remaining within their discipline), is it possible to solve the problem (the hedgehog walks in a desirable direction).

- *The 'plant' and 'umbrella' metaphors: cross-disciplinarity as a synergetic addition*: In this perspective on cross-disciplinary co-operation the starting point is not a problem, but rather a specific phenomenon (a plant). In order to understand the phenomenon, different disciplines have to look at different constituent parts (leaves, flowers, roots, stems, etc.). The disciplines do not co-operate in a team, but do exchange insights and results. This exchange supposedly leads to an improved understanding of the plant (compare the idea of 'synergy'; see section 17.1). The 'umbrella' metaphor represents a closely related mode of thinking. Here the different disciplines (ribs) are subject to central steering (the central rod of the umbrella), and the results of the different disciplinary studies are centrally collected and integrated.

- *The metaphor of the 'spiderweb': trans-disciplinary network formation*: In this view, cross-disciplinarity requires a team of scientists (the spider) from different disciplinary backgrounds (the legs of the spider). The team only progresses if there is sufficient mutual understanding and co-ordination (co-ordination between the legs), for example on the basis of a shared language and common concepts (e.g. systems thinking). The team's task is to establish relationships between seemingly distinct phenomena and problems (i.e. weave a web consisting of different threads). The network of threads may be flexible and dynamic, but retains its basic form (a web). The team (the spider) looks at the network of relationships (the threads) from different angles (the corners of the web). Thus, new ideas, insights and concepts are generated (caught in the web).

We feel that the 'hedgehog' and 'spiderweb' modes of thinking about cross-disciplinary co-operation are perhaps most relevant to societal (socio-technical) innovation and problem solving. The 'tree' metaphor does not offer much practical guidance on how to foster cross-disciplinary co-operation in areas where this has not grown 'organically'. Within the 'plant' and 'umbrella' metaphors, scientists from different disciplines remain relatively isolated from each other, which in our view reduces the possibility of in-depth understanding and design of the interrelations between the social and the technical (see below). Moreover, such isolation seems less conducive to embedding cross-disciplinary co-operation in an interactive innovation process with stakeholders. It is important to remember that science and scientific research have only a relatively minor role to play in such interactive design processes, and need to be carefully tuned to stakeholder dynamics (see section 11.2).

The trans-disciplinary (spiderweb) and the problem-oriented (hedgehog) model both propose the formation of a multi-disciplinary team that jointly studies a complex phenomon, and the members of which need to find a way to co-operate with each other (and, we would add, with societal stakeholders). Whether or not this eventually results in (or happens with the help of) 'trans-disciplinary' concepts and methods, is secondary. And in any case it is important that the language and procedures used are transparent to stakeholders.

The significance of (re)framing questions and problems

Our suggestion that forms of teamwork and interaction between scientists (and stakeholders) are important does not imply that scientists cannot usefully 'withdraw temporarily' to their own discipline, libraries and/or laboratories during an interactive and cross-disciplinary process (see also our discussion on the role of scientists in section 11.2.3). In our view, the essence and purpose of cross-disciplinary co-operation is not to remove disciplinary boundaries and become all 'the same', but rather to combine and make use of disciplinary expertise *in a more useful manner*. The key to achieving this, in our view, is to generate more relevant *questions* and *problem definitions*, and to use these as entry points into several disciplines. In other words, the essence of interactive and cross-disciplinary co-operation is to *reformulate*, *sharpen* and *co-ordinate* the questions that scientists ask themselves. By engaging in debate and discussion around the making of a socio-technical problem tree (see section 13.5.7), for example, new interrelations and problems can be identified, leading to novel questions and corresponding solutions. Taking the problem tree presented in Figure 13.1 (section 13.5.7) as an example, one can easily imagine that a soil scientist may at first have asked the question 'What composting techniques are technically sound for the purpose of restoring soil fertility?'. After the completion of the tree, however, he or she may be tempted to rephrase the question into something like 'What *low-labour intensive* composting techniques are technically sound for the purpose of restoring soil fertility?' or 'What composting techniques can be used effectively *during the period in which off-farm activity is at a low*?'. Similarly, the technical insight that low soil fertility plays a significant role in explaining food shortages, may eventually lead social scientists to further explore the internal conflicts regarding land tenure arrangements and the fenced community grazing area (that would facilitate manure collection). If the technical problems had been in the area of water availability the questions asked by the social scientist would probably have been very different. As the example shows, efforts to combine, confront and integrate social and technical analyses of a problematic situation are likely to lead to non-trivial changes in questioning and research effort on both sides, and hence may result in even more radically different answers and solutions.

19.3   Obstacles for cross-disciplinary co-operation

While many agree that societal problem-solving can benefit greatly from cross-disciplinary co-operation and interactive approaches towards innovation and research, we often see that the members of knowledge institutions (universities, research institutes, etc.) find it difficult to realise this kind of co-operation, and remain working along disciplinary lines. In conceptual terms, much of what we have said in section 9.4 about 'pre-conditions and obstacles to learning' may help to explain why 'learning' by researchers on cross-disciplinary co-operation is slow. Specific practical and motivational problems are given below.

Disciplinary institutions

In most countries, scientific institutions have come to be separated along disciplinary lines. Researchers are enrolled and paid by organisational bodies (e.g. faculties, institutes, departments) that represent a particular discipline (e.g. soil science) rather than a societal problem field (e.g. food security). Thus, researchers with diverging disciplinary backgrounds function in distinct organisational hierarchies, and often also at a different spatial location. Similarly, many academic and professional journals, societies and conferences are characterised by a disciplinary identity as well. As a result, formal and informal contact between people belonging to different disciplines is far from self-evident. Hence, researchers frequently have little affinity with what others can do. Moreover, the first loyalty of researchers lies frequently with their disciplinary unit, resulting often in an inclination to reserve maximum resources (e.g. research budgets) for the own unit, rather than dividing them between several units. In a wider sense too, their belonging to largely separate networks is likely to be accompanied by differential societal problem perceptions, political agendas and social pressures, which means that there is often no autonomous tendency to join forces.

Assessment procedures

Commonly, academic status and promotion opportunities depend to a considerable extent on the number and quality of publications that a researcher produces. Getting quick results that are publishable in high status journals, therefore, is imperative if one wants to move ahead in one's career. However, cross-disciplinary (and interactive) co-operation can be a slow and difficult process with uncertain outcomes, and it is not very easy to get cross-disciplinary articles published in high status, often disciplinary, periodicals. Thus, assessment rituals can make researchers wary of engaging in cross-disciplinary and interactive research.

Cultural diversity on epistemics and linearity

As we have seen, different researchers may hold on to rather different epistemological and methodological beliefs (see section 19.1). Hence, tensions can emerge easily on whether insights are 'true', 'valid' and 'relevant' or not (see also section 6.7) Similarly, many scientists are still influenced heavily by linear thinking (see section 8.1.2), whereas others strongly believe in the blessings of interactivity. Thus disagreement may arise regarding the appropriate role and attitude of researchers in an innovation process (see section 11.2). Such different cultural convictions on the nature and role of science and scientific knowledge may well complicate cross-disciplinary and interactive processes, especially in an atmosphere of disrespect for each other's views.

Hiding behind 'fundamental' science

In relation to the above, scientists frequently argue that it is only 'applied' science that should be cross-disciplinary and interactive, but that 'fundamental' or 'curiosity driven' science should have little to do with this, and would even be under threat if it did. As Bouma (1999:220) argues, however, making a strong separation between fundamental, basic, strategic and applied research, to be carried out by distinct and relatively isolated groups of researchers, is rather unhelpful. The solving of societal *and* academic problems can benefit much from cross-fertilisation and mutual inspiration between knowledge and research at different analytical scales (e.g. molecular, cell, organ, organism, field, farm, ecological zone, etc.) that may derive from different disciplinary units or institutes. Thus, Bouma makes a plea for interaction and co-ordination in 'research chains' (Bouma, 1999). While in such a 'chain' it may be legitimate for, for example, university scientists to focus on 'basic underlying processes', the precise choice and framing (see section 19.3) of the topic on which an academic focuses can have a major effect on its applicability. It makes quite a difference, for example, whether the question is asked of how the migration behaviour of a useful insect can be modelled in an undisturbed environment, or of how this behaviour is influenced by various types of interim work (spraying, manual weeding, mechanical weeding, etc.). Both questions raise 'fundamental' issues, and one might obtain a PhD by making a computer model in answer to either of these. But it is immediately clear that the latter question – which is in fact more complex and academically challenging than the first – holds more promise for practical application. As the example shows, agrarian scientists at agricultural universities and institutes for strategic research can, in principle, allow themselves to be inspired by applied concerns and research questions, without devaluing scientific endeavour.

Worries about 'visionary' solutions

Some scientists are concerned that an interactive and cross-disciplinary approach may be incompatible with developing visionary innovations and a radical breakaway from existing practice, even if the circumstances may call for this (Vereijken, pers. comm.). Involvement of stakeholders in particular would reduce the bandwidth in which alternatives are sought, and lead to risk-avoiding solutions[2]. In view of such concerns, it is proposed that scientists may need to remain 'in charge' of innovation design processes. As we have argued in section 11.2.3, however, scientists may usefully bring in

---

[2]   In conceptual terms, it is feared that interactive approaches lead to 'regular innovations' within the rule set provided by the dominant 'technological regime' (Kemp et al., 2001), and not to 'architectural innovations' (Abernathy & Clark, 1985) that require a fundamental reorganisation of existing relationships and rules (see section 8.2.4). At the level of learning, then, it is implied that an interactive approach fosters 'single loop learning' (i.e. learning without challenging the underlying assumptions) rather than 'double loop learning' (Argyris & Schön, 1996) (see section 9.2).

visionary and revolutionary technological and organisational solutions as *feedback* in an interactive innovation process. In addition, we feel it is a mistake to assume that interaction with stakeholders would be less useful for the *development* of such visionary horizons and architectural innovations. In order to induce a learning process, radically different solutions must somehow be appealing and speak to the imagination. While thinking along totally new lines (e.g. for the development of a radically new farming system), it is inevitable that a number of options emerge, and numerous decisions will have to be taken in order to operationalise the ideas. To arrive at a credible package that speaks to the imagination, therefore, it is important that stakeholders help to make choices and give direction in this creative process. As we have elaborated in section 14.2.1, this is not to say that one should bring in all the diverging interests; rather it is important to involve a coalition of interdependent actors who show an interest and are willing and able to think along particular lines. Moreover, the development of a credible revolutionary package also requires iteration between 'thinking' and 'doing' (i.e. experiential learning) in order to arrive at a coherent and well-adapted innovation. This iterative application and adaptation of radically new solutions requires, at least temporarily, a 'protected space' (see section 8.2.3). Without proper safeguards stakeholders will indeed be more inclined to experiment only with less radical alternatives within the boundaries of the existing technological regime (see section 8.2.4). In all, we can conclude that there is no inherent tension between radical thinking and an interactive approach, as long as (a) a safe space for experimentation is provided, and (b) sufficient attention is paid to the selection of an appropriate coalition of stakeholders. To put it more strongly: a radical idea still requires an interactive design process in order to become credible, appealing and applicable.

Lack of methods and process management

Cross-disciplinary and interactive processes involve many actors and hence require process management in order to evolve constructively. However, universities and research institutes do not usually employ professionally trained process facilitators, and frequently assume that this will be taken care of by senior scientists with limited experience and skill regarding process issues. As a result, cross-disciplinary co-operation may become even more difficult and complex than it is already. In addition, generally accepted and/or applicable methods for coherent cross-disciplinary analysis and synthesis seem to be lacking, which further complicates the process. As we have hinted in section 19.2, we feel that several of the exploratory methods outlined in section 13.5 can be helpful for arriving at a coherent and relevant set of (at times disciplinary) questions in connection with a complex problematic context. As mentioned, scientific institutes often lack credible and independent process managers who can assist in translating complex problems into the most relevant research activity, and in dividing available resources accordingly.

Imposition of perspectives and methods

Several of the elaborate approaches for cross-disciplinary co-operation draw on 'soft' or 'hard' systems thinking. Typically, these approaches accompany the development of a special language and set of methodological devices. The 'soft' side is illustrated by the work of Checkland (1981) and Bawden (1994). 'Hard' systems approaches for cross-disciplinary work frequently make use of computer languages and modelling techniques in order to capture and integrate knowledge from different disciplines (see Van der Werf et al., 1999). In itself the idea of developing a common language and methodology to facilitate cross-disciplinary work is logical and commendable. But the more rigorous and precise such 'hard' or 'soft' approaches become, the greater the chances that scientists from specific disciplines are alienated. Many social scientists, for example, feel extremely uncomfortable with exact modelling techniques, and find it impossible to incorporate into them insights and expertise that they find relevant (see sections 13.3.1 and 13.5.10). Similarly, there are many natural and social scientists who have doubts and hesitations with regard to 'soft systems' frameworks and modes of operation (see section 16.2.4). Yet, at times systems concepts and methodologies are effectively made into doctrines that are experienced by some as rigid and externally imposed. In such cases, the proposed 'common language' becomes an obstacle rather than a stimulant for cross-disciplinary work. Paradoxically, therefore, we can say that cross-disciplinary work requires, to some extent, a common 'language', but one that is flexible enough to allow for the co-existence and bringing in of dialects and 'foreign' discourses.

Funding arrangements and rigidities

The dominant modes of funding research can frustrate cross-disciplinary co-operation in several ways. As mentioned at the outset, much funding of research takes place along disciplinary lines and through distinct disciplinary networks (e.g. separate councils for social and natural science research). Clearly, this does not encourage researchers to look over the fence. Moreover we have seen that commod-itisation of knowledge and output financing can result in mechanisms and rigidities that hamper the co-operation and flexibility needed in an interactive and cross-disciplinary process (see Chapter 18). Instead of dividing the budget according to how a learning process unfolds, budgets are frequently allocated from the outset and earmarked for pre-defined purposes, activities and disciplines, not least because, in a commercial context, parties usually want to know in advance how much money and resources can be expected to come in, before they commit themselves. Another pervasive characteristic with regard to funding is that – at the level of governments and international development agencies – the funding of 'research' is often separate from the funding of related activities such as 'extension communication' or 'rural development'. The implicit message here is that these are somehow completely different spheres of activity, which clearly they cannot be when looked at from our perspective on innovations and innovation processes (see sections 5.3.5 and 8.2). The existence of largely disconnected streams of resources can easily reinforce incompatibilities and lack of co-operation and co-ordination between research and various efforts geared towards societal problem solving. Not surprisingly, then, scientific research often integrates better – in terms of exact questioning and/or research set-up – with previous research than with practical problems and real-life innovation efforts.

Defending separate organisational mandates

In many countries it is difficult to overcome strict divisions between, on the one hand, fundamental, strategic and applied research, and on the other, research, extension and rural development activity. This is because such activities have become the specific mandate of specialised organisations with their own separate budget allocations (see above). Particularly when resource allocations are under threat, organisations tend to emphasise their uniqueness and guard carefully against others trespassing on their territory. In other words, organisations tend to go into 'defensive mode'. Such strategies can be effective for organisational survival, but can also generate certain risks (see sections 16.2.5 and 16.4). In any case, maintaining strong divisions is not likely to be in the interest of societal problem solving and innovation.

We can conclude that the conventional organisational infrastructures, incentives and reward systems in and around research institutions can easily obstruct cross-disciplinary (and interactive) co-operation. The 'costs' in terms of time and effort needed to overcome inherent complexities and adverse conditions are likely to be high, while the academic recognition to be gained is dubious. At the same time this means that much progress can be made through the development of new organisational set-ups, facilities and routines.

Questions for discussion

(1) Discuss whether or not cross-disciplinary co-operation is important in all problem situations, or only in specific types of problem contexts.

(2) In your experience, which epistemological and methodological differences between natural and social scientists form the greatest obstacle to cross-disciplinary co-operation? What strategies would you like to propose to overcome these difficulties?

(3) Which metaphor of cross-disciplinary co-operation do you find most appealing and why?

(4) Select a problematic context that you are familiar with and try to arrive at a coherent set of natural and social science questions for research by using the socio-technical problem tree method outlined in section 13.5.7.

PART 6
Epilogue

As a way of concluding this book, we present some ideas about doing research in Communication and Innovation Studies/Extension Science. At this point we are not primarily thinking about the various forms of *decision-oriented* research discussed in section 16.5, but rather about *conceptual* research (see section 2.3) that may take place in the context of an MSc or PhD programme. We hope that scholars are inspired by the issues and approaches suggested, and will help us to further develop the field of study. In view of our hopefully common aspiration to help meet challenges as outlined in Chapter 1, a deepening of our theoretical and practical understanding of the relationships between communication and innovation remains of eminent importance.

20   Approaches and issues for further conceptual research

The range of issues that can be studied under the banner of Communication and Innovation Studies is enormous. There are several reasons for this. The first is that, even within the domain of agriculture and resource management, there are numerous innovations to be looked at. They can, for example, relate to different farming domains or levels (see section 5.1), centre around various natural resources (e.g. water, land, biodiversity, etc.) and involve different spatial contexts (village, catchment, region, country, world) as well as different human institutions (culture, government, markets, knowledge systems, law, organisations, etc.). Moreover, when talking about the conceptual development of our field of study, there is every reason to gather experiences and studies on the relationships between communication and innovation in other societal domains such as the health sector, industry or the environmental branch. Second, our field of study is inspired by various disciplines (see section 2.3), which means that, in each innovation context, issues can be identified from the perspective of multiple theoretical and methodological traditions. And finally, we are not just an academic discipline, but also a specific societal practice that is institutionalised in extension divisions, communication units, projects, etc. Such organisational forms can be studied by more distant disciplines like management science, economics and law, and still yield highly valuable insights.

Research questions in Communication and Innovation Studies, then, may be framed from numerous contextual and theoretical angles, and can be studied with the help of the full range of available social science methods and techniques. In view of this diversity and complexity, we want to clarify what we think are important questions and relevant approaches to research in Communication and Innovation Studies. Clearly, this is mainly a matter of personal academic taste and preference, and should not be seen as a denouncement of other lines and frameworks of investigation. Finally, we want to emphasise that, even though we make some specific remarks on research design and methodology, the purpose of this chapter is not to provide a detailed overview or guide on how to do social science research. That is a field of study in itself, and we strongly encourage scholars to study relevant literature (e.g. Nooij, 1990; Strauss & Corbin, 1990; Denzin & Lincoln, 1994).

20.1   Overall focus: communication and processes of socio-technical design

Historically, the interest in our field of study is not so much in the 'autonomous' *emergence* of social-technical innovations (as defined in section 1.1.2), but in the ways in which this can be *deliberately* stimulated and supported through communicative means. This interest in issues of communicative intervention remains of essence to our field. At the same time, there is a lot we can learn from innovation processes that occur without the involvement of professional communication specialists and interventionists. Similarly, it is important to recognise that, in any case, deliberate communicative intervention is likely to constitute only a fraction of the communicative processes that take place in change and innovation contexts (see also section 5.3.2). Thus, we feel that our research should not be limited to 'communicative intervention' in a narrow sense, but should focus on 'communication' in and around situations where social actors (including perhaps communication professionals) engage in a deliberate effort to resolve problematic situations. In other words, we are looking at communication in situations were certain actors strive to *design* innovative solutions consisting of social and technical arrangements. As we have stressed throughout this book, such design processes should not be thought of as predictable and controllable, but should be approached as multi-faceted and inherently capricious sets of social learning and negotiation processes (see section 5.3.5 and Part 3 of this book).

Drawing on the above, we propose that research in Communication and Innovation Studies should first and foremost centre around *processes* (see also Aarts, 1998). After all, if one aspires to say useful things about processes, one needs to study them. Evident as this may seem, much research in our field does not focus on improving understanding about processes, but rather on making assessments about specific points in time, without taking into account what happens in between. Many studies in our field still measure 'adoption rates', 'acceptance levels', 'awareness', 'opinions' or 'attitudes' at discrete points in time (e.g. at yearly intervals), and seek to assess how differences between measurements have been influenced by different forms of communicative intervention. Similarly, studies may seek to assess the impact of participatory technology development projects, and/or compare this to the impact of more conventional technology transfer efforts. Often such studies treat what happens between different points of measurement as a black box, and simply assume or make plausible statistically that registered differences between points in time can be attributed partly to this, that or the other. While this may be sufficient for purposes of administrative evaluation and justification (see section 16.5), we feel this kind of research does not help much to advance our field, because such studies make largely invisible the very communication, learning and negotiation processes through which outcomes are forged. In our view, having detailed insight into the complex dynamics of, and interrelations between, such processes is essential for both the development of theory and the improvement of professional practice. As we discuss below, the proposed general orientation towards 'processes' still leaves considerable space for defining the precise phenomena, specific kinds of processes and/or variables that a researcher may focus on.

20.2   The role of theory in formulating specific areas and questions for research

When setting out for conceptual research, one needs to have a conceptual question in mind; that is, a question expressed in terms of conceptual notions and possible interrelations between them. Although posing such conceptual questions makes a lot of sense, there is reason for caution as well. When encountering a complex situation with specific questions and concepts in mind, one can easily become blind to other issues, interrelations and concepts that may be appropriate for improving one's understanding of it. If, for example, one draws solely on the theories presented in sections 9.2 and 9.3, one could be tempted to ask a question like:

> How does the (non)availability of various forms and qualities of feedback influence different levels of social learning in problematic situations X and Y?

However, by posing and limiting the question in this specific way, one may easily overlook other more contextual factors that can have a bearing on learning as well, such as the existence of different kinds of obstacles to learning (see section 9.4) or non-conducive conditions for integrative negotiation (see section 10.3). Thus, the framing of specific questions can cause considerable blindness (see sections 6.4 and 19.2). Some social science researchers respond to this risk by arguing that researchers should study social processes with as little pre-prepared conceptual notions and specific research questions in mind as possible. The idea, then, is that by looking with an open mind, the situation itself can inspire the selection and/or development of relevant theory (e.g. Strauss & Corbin, 1990). Although we are sympathetic to the latter idea, we feel that this requires researchers to bring in a *maximum* rather than a minimum of possibly relevant theories and research questions. This is because we feel a researcher can only *select* relevant concepts and questions when he or she is conversant with different theories and possibly relevant academic questions. When empirical research is not preceded by forms of explicit conceptual exploration and thinking about possibly relevant research questions, and starts with an 'empty' mind, the chances of overlooking relevant phenomena may be even greater then when one particular perspective is adopted. Moreover, it is fundamentally impossible to look at situations with an 'empty' mind, as human beings (including researchers) always apply language and interpretative frameworks to make sense of them (see Chapter 6). In academia, therefore, it makes more sense to try and be explicit about the interpretative frameworks one may apply, than to look at a situation with all sorts of implicit notions that are probably equally theory-laden. In all, we feel that setting out for conceptual research requires researchers to seek the inspiration of a wide range of theories and conceptual ideas, and formulate an equally wide range of possibly relevant research questions. These notions, then, can serve as *sensitising concepts* (Nooij, 1990) during empirical exploration. At a later stage in the research they may be narrowed down to a more specific focus, resulting also in the further identification and operationalisation of specific issues and variables.

In terms of the above, this book provides a selection of concepts and theory that we find potentially relevant to the study of our overall theme 'communication and processes of socio-technical design'. Although the book brings together quite a few different theoretical perspectives, it is by no means complete and exhaustive. Thus, we wish to encourage scholars not to use only this book as a source of inspiration, but also to explore other frameworks. While recognising the limitations, we want to suggest some research areas and questions that derive from the conceptual frameworks presented in this book.

(1)   Diverse administrative and political environments and their influence on processes of innovation design

At various points in this book, we have concluded that innovation processes require an institutional environment that is conducive to, among others, network building, social learning and negotiation. At the same time we have argued that current administrative control infrastructures (e.g. hierarchical planning systems) and dynamics in political arenas (e.g. the need for politicians to make their mark) may pose threats to this in various ways. In relation to this there is a need to gain a better understanding of possibly constraining or enabling circumstances to innovation design in the administrative/political sphere. Specific questions that might be addressed to further explore this include:

● How and why is communication in different kinds of innovation trajectories influenced by interactions with diverse administrative/political environments, and how does this shape the course of such trajectories?
● How and why do different forms of interaction with the administrative/political environment impinge on social learning and negotiation processes in the context of innovation design?
● What are the conditions that shape these interactions, and to what extent can they be regarded as 'enabling' or 'constraining' from the point of view of innovation?
● How and why do process facilitators/communication workers deal with and/or relate to constraints from the administrative/political environment, and to what extent do their interventions contribute to innovation design processes?
● What practical implications and guidelines derive from this?

(2)   Different forms of privatisation and knowledge commoditisation and their influence on processes of innovation design

As we have seen in Chapter 18, many agricultural knowledge systems and networks are currently in a process of transition, with the introduction of 'market-mechanisms' as a leading philosophy. Although we have pointed to some preliminary experiences in the Netherlands, there is still great scope and need for studying how such changes may shape and/or alter the dynamics of innovation processes. Relevant questions in this area may include:

● How and why do different forms of privatisation and knowledge commoditisation shape and/or alter communication in knowledge networks and the way in which agricultural innovation trajectories (and social learning and negotiation processes within them) evolve?
● To what extent do different modes of privatisation affect, positively or negatively, the overall innovative capacity of agricultural knowledge and information networks for conventional, ecological and/or low external input forms of agriculture?
● To what extent are emerging patterns of communication, interaction and exchange consistent with economic and other theories and/or with the policy assumptions that underlie privatisation policies? For example, do knowledge systems indeed become more 'demand driven' as a result of privatisation?

- How and why do process facilitators/communication workers deal with and/or relate to knowledge commoditisation, and to what extent do their interventions contribute to innovation design processes?
- What practical implications and guidelines derive from this?

(3)   The relationships between different areas of cognitive change in innovation design

We have argued that what is conventionally termed 'knowledge' is in fact a specific area of human cognition (see Chapter 6), and that different kinds of cognition and cognitive change are important in the context of innovation as well (see Chapter 9). At the same time, we have seen in Chapter 5 that different kinds of perception and cognition are closely intertwined. However, there is considerable scope for improving our insight into how precisely such different 'learning fronts' (see Chapter 9) may affect each other. In other words: How do changes in beliefs about the functioning of the biophysical and social world (i.e. knowledge change) influence other forms of cognitive change, such as different kinds of aspirations, feelings of responsibility, trust, identity, efficacy, etc., and vice versa? Such questions are of interest both at an individual level, and at the level of interactions between different social actors. When studied in the context of an emerging innovation process relevant questions may include:

- What is the order in which different kinds of cognitive change come about?
- How is a specific kind of cognitive change (or stability) connected with earlier and/or later changes (or stability) in cognition of a different kind?
- How does the nature and quality of social relationships among actors affect such patterns of cognitive change?
- What kinds of tensions do actors experience between different (or similar) kinds of cognition, and how are such tensions resolved (or not)?
- What is the significance of such tensions for processes of innovation design?
- How and why do process facilitators/communication workers influence different kinds of cognitive change (or stability) and the relations between them, and to what extent do their interventions contribute to innovation design processes?
- What practical implications and guidelines derive from this?

(4)   The social shaping of relevant knowledge constructs or 'frames' in innovation design

Several areas of knowledge are likely to play an important role in processes of innovation design (see section 11.2.1). More specifically such knowledge can be expressed or 'wrapped' in various forms such as the stating of a 'fact', the definition of a 'problem', 'crisis' or 'research question', the formulation of 'conclusions' or 'lessons', the identification of 'solutions' or 'stakeholders', the giving of 'feedback', the proposing of a 'research methodology' or 'way forward', etc. As we have argued in Chapter 6, knowledge is socially constructed, and this is equally true for these kinds of knowledge expressions, which hence might better be termed 'knowledge constructs' or 'frames' (Gray, 1997). An interesting area of investigation, then, is how specific knowledge constructs are selected and made to count in design processes (or not) (e.g. Te Molder, 1995; Aarts & Te Molder, 1998). In connection with this, it can also be instructive to look at how specific forms of societal problematisation (e.g. hot issues or media hypes) come about and impinge on innovation trajectories. Relevant questions in connection with this could be:

- What communication and other strategies do participants apply to make their knowledge constructs count and/or become accepted by others as 'relevant', 'opportune' and/or 'credible'?
- How do such strategies relate to unequal access to power resources?
- What are the consequences of different strategies and access to resources for innovation design processes at large?
- How and why do process facilitators/communication workers influence these processes of social construction and power generation, and to what extent do their interventions contribute to innovation design processes?
- What practical implications and guidelines derive from this?

(5)   The interaction between network building, social learning and negotiation in innovation design

One of the central arguments of this book is that the productiveness of change and innovation trajectories depends on the quality of network building, social learning and negotation, and hence that communication and communicative intervention must be geared primarily towards enhancing these three processes. However, we do not have much insight yet into the different ways in which the three sub-processes may interact, alternate or interfere positively or negatively with each other. Questions that may need to be addressed therefore include:

- How do different moments and modes of network building, social learning and negotiation influence each other, and how does this affect innovation trajectories?
- Are there any patterns of alternation or interaction between network building, social learning and negotiation that are more 'productive' than others?
- How do processes and patterns of communication and communicative intervention impinge on the interactions between network building, social learning and negotiation?
- How and why do process facilitators/communication workers influence such patterns, and to what extent do their interventions contribute to innovation design processes?
- What practical implications and guidelines derive from this?

(6)   The role of uncertainty and trust in processes of innovation design

As Aarts and Van Woerkum (2002) have suggested, different kinds and sources of uncertainty may affect processes of change and innovation design. One cannot, for example, predict how the wider agro-ecological world and the social world will 'respond' to the various intermediate options and/or experimental interventions that may emerge in the context of innovation design. In other words, one does not know in advance whether or not proposed changes in practice will 'work' in technical and social terms (e.g. be acceptable to constituents and/or administrators) Clearly, this implies that, when embarking on an innovation trajectory, neither stakeholders nor facilitators can predict the overall outcomes of the design process. In addition, stakeholders may face numerous 'process' uncertainties of a more interactional nature. What do other stakeholders really want? Are they participating with sincere intentions? What is their space for manoeuvre? How will they respond to our proposals? What can we expect from the facilitator? In the absence of certainty, then, trust is an important issue as well (Lagerspetz, 1998; Sorrentino & Roney, 2000). This is because different forms of 'trust' may contribute to keeping actors 'together' despite uncertainty. For purposes of research, the following questions may be pertinent:

- What are the different kinds and sources of uncertainty and trust that stakeholders and facilitators experience in innovation trajectories?
- How do participants cope with various forms of uncertainty, why, and how does this affect processes of innovation design negatively or positively?
- How do processes and patterns of communication and communicative intervention in and around innovation design trajectories affect feelings of uncertainty and trust among participants?
- How and why do process facilitators/communication workers deal with various forms of uncertainty and trust, and to what extent do their interventions contribute to innovation design processes?
- What practical implications and guidelines derive from this?

(7)   Articulation between 'top–down' and 'bottom–up' moments and interventions in innovation design

We have argued in sections 11.1 and 14.1.3 that innovation processes may, in principle, benefit from both instrumental/persuasive and interactive/participatory forms of intervention, as well as from communicative and non-communicative policy instruments (see section 4.1.2). And although we have made some initial suggestions regarding productive forms of alternation and sequencing, there is still much scope for further investigation. Further clarification is needed on questions like:

- What are the specific patterns and modes in which 'top–down' and 'bottom–up' interventions interact in practice, and how do such patterns shape and influence different innovation design trajectories?
- Is there a relationship between the kinds of problematic situations and innovations (e.g. regular or architectural), and the likely 'costs and benefits' of various forms and patterns of alternation?
- How and why do process facilitators/communication workers use different modes and patterns of alternation, and to what extent do their interventions contribute to innovation design processes?
- What practical implications and guidelines derive from this?

(8)   Integration of knowledge from different epistemic and disciplinary communities in innovation design.

We have argued that innovations are multi-dimensional, and are usually composed of different biotic and abiotic artefacts, human practices and social arrangements. Innovations require the integration, or at least connecting, of knowledge and insight from different social actors, including diverse epistemic and/or disciplinary communities (e.g. stakeholders, social scientists, natural scientists). The perspectives and specialist expertise of such actors may not only be shaped by different social interests, backgrounds and experiences, but may also refer to different levels of aggregation, varying from, for example, molecule, cell, organ, organism, field, farm, ecological zone, and from individual, group, community, society, etc. Clearly, connecting or integrating such highly diverse perspectives and knowledge into a coherent innovation is a major challenge. An interesting area of study, then, could be the communication tactics and methods of knowledge representation (including visualisation, modelling, socio-technical problem trees, etc.) through which effective integration of knowledge is being forged (or not). Relevant questions may include:

- What communication tactics and modes of knowledge representation do actors belonging to different social groups and/or epistemic/disciplinary communities use during different innovation design trajectories and why?
- How did knowledge integration come about in apparently productive innovation processes?
- How, to what extent and/or under what conditions do such tactics and methods contribute to, or constrain, knowledge integration?
- How and why do process facilitators/communication workers go about fostering knowledge integration, and to what extent do their interventions contribute to innovation design processes?
- What practical implications and guidelines derive from this?

(9)   The organisation of communicative intervention and the implications for innovation design

It is clear that the current landscape of organisations involved in communicative intervention is changing rapidly (see sections 1.1.2 and 18.2). Different organisational missions, structures, cultures, management and processes are emerging around 'extension', research and development intervention, and these are bound to have an impact on the way in which communicative interventions in innovation processes unfold.

- How and why is communicative intervention and communication in different kinds of innovation trajectories influenced by (inter)organisational changes of various kinds, and how does this shape the course of such trajectories?
- How and why do different (inter)organisational set-ups and/or changes impinge on network building, social learning and negotiation processes in the context of innovation design? And how do they affect and shape processes of organisational learning?

- To what extent can different (inter)organisational set-ups and changes be regarded as 'enabling' or 'constraining' from the point of view of innovation?
- How and why do process facilitators/communication workers deal with different (inter)organisational set-ups, and to what extent does this contribute to innovation design processes?
- What practical implications and guidelines derive from this?

As is usual in research oriented towards conceptual progress, the above areas for research are formulated in fairly abstract terms, and deviate considerably from conventional themes in our field of study such as 'research/extension linkages' , 'knowledge and information management', 'participation', 'knowledge transfer', etc. It is not that such themes are no longer important or worth looking at. On the contrary. What we propose, however, is to look at the 'old' themes with new concepts in mind – concepts that derive from more recent insights into innovation, are less 'mechanical' and control-oriented, and serve better to study and understand dynamic processes. Such reframing of classical questions, then, may help to advance our field of study. Finally, we want to stress that the above areas of research are just examples of the kinds of questions that may be relevant. There are many additional interesting questions that can be posed with concepts from this book or other literature in mind. In any case, specific research questions will need to be inspired by and tuned to unique research contexts.

20.3   A note on research design and methodology: towards 'comparative process ethnography'

The above questions are geared towards gaining an in-depth understanding of the dynamics in innovation design trajectories. This requires us to look at the everyday practices of, and interactions between, social actors, and interpret these in relation to their relevant context. An obvious research tradition that allows such contextual analysis of human (inter)action is a case-study approach, and particularly what Van Velsen (1967) calls 'the extended-case method' or 'situational analysis'. In connection with innovation processes, this would mean the close following (or ex-post reconstruction) of events and interactions in and around a particular innovation trajectory, as well as the gathering of participants' reflections and rationalisations in connection with these. Paraphrasing Pottier (1993:30), who calls for the making of 'project ethnographies', we refer to 'process ethnographies'. This is to underline that the processes we are interested in (network building, social learning and negotiation in the context of innovation) may well exceed the arbitrary boundaries of a formal project. A particular challenge in this context is to develop qualitative and quantitative approaches for studying and documenting different forms of perception and cognition, and changes in them over time. Clearly, in order to answer the kind of questions raised above, one cannot usually just analyse a single case-study (i.e. one process ethnography); rather one needs to purposively select and *compare* several of these. Thus, we make a plea for 'comparative process ethnography' as an important approach to research in our field.

   Much more can be said about the different ways in which process ethnographies can be conducted and analysed. However, this is largely beyond the scope of this book, and we therefore refer to specialist literature in connection with this (Van Velsen, 1967; Long, 1989; Strauss & Corbin, 1990; Yin, 1994). Nevertheless, there are two issues that we would like to touch upon.

Case-study methods

First, it is important to note that adopting a case-study or process ethnography approach does not automatically imply the inclusion or exclusion of specific research methods and techniques. All sorts of methods and techniques, both qualitative (e.g. participant observation, in-depth interviewing, qualitative content analysis, analysis of narratives, life histories, etc.) and quantitative (e.g. survey techniques, quantitative content analysis, etc.) can be used and/or combined in a case-study. However, we propose to grant primacy to qualitative research methods, or, as Knorr-Cetina (1981a:17–20) puts it, to 'sensitive' rather than 'frigid' methodologies. It is only through such methodologies that complex dynamics and interrelations can be understood and interpreted in their relevant context. Qualitative case-studies have a significance of their own, which Elias and Scotson (1965) have coined 'sociological significance', and which obviously is quite different from 'statistical significance'. In many ways a rich story or case description can be more telling, insightful, credible and convincing than an abstract and anonymous statistical association, especially when the researcher grounds his or her analysis in the perspectives and reasoning of a variety of stakeholders, uses multiple sources, and reflects critically on his or her personal assumptions and biases. At the same time, quantitative methods can be useful in order to further assess the significance and relevance of qualitative findings in a wider perspective. Moreover, additional quantitative studies can serve to enhance the credibility of qualitative findings for specific audiences, most notably the many scientists and policy-makers who still have a greater appreciation of statistical significance than sociological significance. Whenever quantitative methods are used, however, we feel that the interpretation of statistical findings must be grounded in in-depth qualitative case-studies.

Defining a case-study

Second, we want to emphasise that the nature and set-up of a case-study or process ethnography may still vary considerably, not only in terms of methodology, but also in terms of the choice of what constitutes a 'case-study'. Depending on one's research interest the 'cases' that are chosen and compared can be specific social actors, projects, technologies, innovative ideas, communities, intervention methods, organisations, problems, etc. One may, for example, study and compare processes of organisational learning in different types of organisations. Here different organisations constitute the case-studies through which one hopes to obtain a greater understanding of certain aspects of organisational learning. Alternatively, one may be more interested in how organisational learning evolves around different types of organisational problems, and opt to select the latter as case-studies. Similarly, one can analyse how and why specific research findings (or knowledge gaps) are constructed and communicated to various audiences in privatised or non-privatised knowledge networks. Here different segments of knowledge networks (e.g. the networks of researchers and communication workers working on pest management in rice in privatised context A and non-privatised context B) may constitute different case-studies, through which one aspires to gain insight into the relationships between social construction processes, communication and different modes of knowledge commoditisation. Although the boundaries and entry points of case-studies may thus vary considerably, the idea of process ethnography still demands that we take an in-depth look at interrelated interactions and events that are relevant to a case-study. The idea of defining what constitutes a case-study, therefore, should not be taken to mean that rigid boundaries need to be imposed on networks of actors and interactions across time and space. Where a case-study begins and ends is something that needs to be inspired by the findings themselves, and against the background of a specific conceptual interest.

20.4   Process ethnography as network analysis

In the context of process ethnographic case-studies, one is typically looking at interconnected interactions among social actors, across time and space. We propose that the notion of 'networks' is relevant to this kind of study; after all one could say we are looking at networks of actors, networks of interrelated events and interactions, etc. Moreover, there are several additional reasons to pay attention to the concept of networks in our field of study. As we have hinted in section 17.3, the concept of networks has advantages over the concept of 'systems' in that it has fewer connotations of a pre-assumed common purpose and clear boundaries when looking at multi-actor situations. Also, the idea of 'networks' is attractive in that it makes clear that individual agents are in fact part of, or even constituted by, a wider web of relationships that impinge on the practices and actions that 'individuals' engage in (see also section 5.2). In connection with this, Long (1990:9) proposes to take the *social actor* rather than the individual as a unit of analysis, while others prefer the term '*actor-network*' (Callon et al., 1986). More pragmatically, some of the challenges discussed in Chapter 1 (e.g. market liberalisation, globalisation, chain management, knowledge society, etc), for example, are closely intertwined with changes in the functioning of networks, and so are other important developments in the world (e.g. acceleration of interaction patterns, the changing meaning of time, creolisation of cultures, etc.) (e.g. Hannerz, 1996). Thus, zooming in on networks may help to identify qualitative and quantitative changes that may be relevant to communicative intervention. In line with this, we feel that in-depth network studies can – like the tools and procedures in RAAKS; see section 17.2 – provide critical feedback to a set of interdependent actors about existing dynamics and interaction patterns, and hence can contribute to interorganisational learning and change.

Finally, since we have concluded that building networks is a key component of innovation processes (see section 8.2.2), network studies can also help to improve our conceptual understanding. Many of our conceptual and practical ideas regarding participation and process management, for example, have been inspired by detailed qualitative studies of interrelated (i.e. networks of) events in the context of innovation trajectories (e.g. Leeuwis, 1993; Pijnenburg, 2003). Hence, we feel that focusing on networks can be an important strategy in conceptual research in Communication and Innovation Studies, and can also inform practitioners when dealing with multi-actor situations. In the next sections we discuss several modes in which process-ethnographies can be combined with, or approached as, the study of networks.

20.4.1 Classical network analysis: describing interaction patterns among human actors

The concept of 'networks' has a long history in the social sciences (e.g. Bott, 1957; Boissevain & Mitchell, 1973; Granovetter, 1973). Typically, the term 'network' suggests the image of a spider web, consisting of points and a pattern of lines between them. In classical forms of network analysis, web-like diagrams are drawn starting from a specific individual (represented as the central point), from which lines are drawn to other individuals with whom the central person has a relationship. When relevant for the analysis (e.g. for studying a person's opinion leadership and/or potential as an entry point for diffusion; see section 8.1.1) the procedure can be repeated with other individuals, so that a network of direct and indirect relationships, from the point of view of the central individual, transpires (see Figure 20.1).

In relation to networks, Mitchell (1969) distinguishes between 'interactional attributes' and 'morphological attributes'. The former express the characteristics of certain relationships, such as the frequency of contact, content, durability, etc. The latter refer to the pattern of relationships; relevant attributes here include, for example, connectedness or density (i.e. the actual number of linkages relative to the possible number of linkages), the centrality of specific individuals and the range of a person's network (i.e. relating to the number of and/or heterogeneity of people that may be

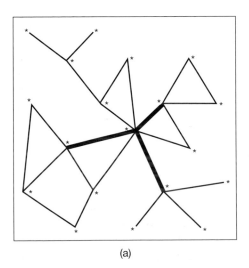 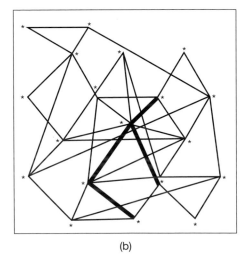

(a)   (b)

Figure 20.1 Two fictitious examples of networks, (a) with a relatively low connectedness and (b) with a relatively high connectedness. Bold lines indicative high intensity contact.

reached). Particularly after the introduction of computer programs for network analysis, this kind of study frequently took the form of mainly quantitative 'sociometric' studies, in which the above mentioned variables regarding communication patterns were calculated (see for overviews Hannerz, 1980; Rogers & Kincaid, 1981). As Rogers and Kincaid (1981) suggest, these kinds of network analysis can improve our insight into issues of adoption and diffusion (see Chapter 8). Moreover, Engel and Salomon (1997) show how similar, even if much less quantitative, forms of network analysis can be informative when looking at organisational and interorganisational communication and co-ordination, which is in fact why several of the RAAKS exercises and tools have been inspired by it (e.g. Table 17.2).

These classical forms of network analysis serve mainly to make a fairly rough summary of network linkages over a particular period, and at a given point in time. Thus, they do not help much to gain in-depth insight into the everyday dynamics of innovation processes, which implies that their value for making process ethnographies is limited merely to providing additional quantitative illustrations and summaries of interaction patterns. Below we introduce some less conventional forms of network analysis that can be of more direct relevance for the purpose of making process ethnographies.

20.4.2   Analysing networks of interrelated events

While the above strand of network analysis focuses on people and their interrelations, it is also possible to look at networks in different ways. One can, for example, look at interrelations between *events* that take place at different points in time or space; i.e. study a network of events. By following, and/or reconstructing in retrospect, a history of different happenings and episodes around a particular innovation process, one may get an in-depth qualitative understanding of how and why it has evolved over time. It can be very illuminating to explore how different actors involved represent and evaluate the 'same' trajectory of events. When looking at interrelated events one can zoom in on one or more particular questions and aspects. One can, for example, focus on who is involved in which events, in what way and to what effect, and how this may be shaped by representations of events at other points in time and place. Or one may look at how, whether and why different kinds of knowledge and ignorance are put forward, made to count, translated, shared, and/or kept hidden during particular events, again in the context of other events. This kind of network analysis, then, is highly compatible with both the idea of process ethnography and the research questions formulated in section 20.2. In methological terms, however, much scope for further development exists, as a clearly outlined procedure is lacking.

An approach that shows some similarity to the idea of looking at interrelated events is what Long calls 'social interface analysis'. Social interfaces are defined by Long (1989:2) as:

> 'critical points of intersection or linkage between different social systems, fields or levels of social order where structural discontinuities, based upon differences of normative value and social interest, are most likely to be found.'

According to Long, our understanding of change processes is likely to be improved most when we focus on 'interface situations'; that is on encounters and events where the tensions and conflicts implied in the notion of social interface are expressed. Thus, one could say that Long proposes, in methodological terms, to focus on (i.e. select) specific types of events and network linkages when performing a network study. However, although the idea of 'interface analysis' fits in nicely with the idea of studying interrelated events, one could also say that it has a wider relevance in that it can be integrated with several other forms of network analysis suggested in the remainder of this chapter.

20.4.3   Looking at the building of networks of human and non-human 'agents'

An interesting approach towards network analysis stems from the field of Science and Technology Studies, and is called 'actor network theory' (Callon et al., 1986; Law, 1986; Latour, 1987; Law & Hassard, 1999). A key element in this approach is that the 'actor networks' constituting individual human agents are not looked on as consisting only of other human agents, but also of man-made artefacts and natural phenomena. This is because the latter (sometimes referred to as 'actants'; see Callon & Latour, 1992) are seen to play a similar role in such networks as human actors. At first, the idea that artefacts and natural phenoma must be looked on as 'actors' may seem strange, but in many ways it makes sense. After all, such entities have an influence on how human beings behave in a given context. A machine may perform certain tasks and hence 'act' in a particular way, and at the same time it structures human behaviour. That is, people must also act in particular ways if they want the technology to work. Similarly, nature may 'respond' in a specific way to human (e.g. agricultural) practices, or 'act' in a more independent way, e.g. by producing climatic conditions. Moreover, human beings tend to feel and act differently in different physical environments (e.g. a modest home, a workplace or a luxury conference centre), which is indicative of the influence such environments have on human identity and action (e.g. Dugdale, 1999). In such ways, artefacts and natural phenomena can indeed be seen to 'act' and *do* something to, with or for us.

But perhaps even more importantly, when interacting with others, people tend to talk about non-human entities in the same sort of way in which they discuss humans. For example, actors frequently make claims not only about what other people (e.g. those that are part of their actor-network) think, have done or will do, but also about how natural phenomena and artefacts function and/or will respond to certain conditions. In a group meeting on biological pest control, for example, a communication worker or researcher may make statements about the way in which nature performs and responds positively or negatively, to human intervention, and also about the support he or she gets from farmers elsewhere and/or from scientists across the world. In essence, then, we see that the communication worker speaks 'on behalf of' (i.e. claims support from) both other people *and* nature – a phenomenon called 'enrolment' in actor network theory (Callon et al., 1986; Law, 1986). Thus, in the vocabulary of actor network theory, actors can be seen to 'enrol' other actors or actants when they acquire the legitimacy or ability to speak on their behalf and/or with their support. And by being enrolled, these other human and non-human agents become part of an actor network. It is significant to note here that non-human agents, of course, cannot be seen to strategically 'enrol' others; hence, the nature of their 'agency' (see section 6.6) remains fundamentally different from that of human beings. There may also be a difference between the way in which man-made artefacts, including technologies, and natural phenomena enter social interaction in a network. Like human actors, both entities can 'act' or be claimed to act in a particular manner. However, it is important to note that man-made artefacts, unlike natural phenomena, are somehow *designed* to act. Drawing an analogy with communication theory (see Chapter 7), we can say that they are encoded with a multi-interpretable 'message' that is transferred through a material communication channel (see section 7.1). Thus, while 'natural phenomena' can be seen to act or respond relatively independently, man-made artefacts are in many ways 'programmed' to act (sometimes even literally, e.g. in the case of computer-based technologies).

Notwithstanding this difference, however, the 'actions' of both types of non-human entities, like those of human agents, remain to be defined through interpretative activity and are thus based in human communication. In all, we can say that the human and non-human are closely intertwined, so that it may well make sense to include the latter in an analysis of networks.

This particular way of looking at networks, then, has been used especially to describe and analyse how scientists and engineers go about mobilising networks in support of certain truth claims and/or prospective policies and innovations. In the analysis of these kinds of social construction processes, much attention has been given to the strategies and methods that people apply to enrol others. Such methods are also referred to as methods for 'translation', as building an actor network essentially revolves around making people accept and draw upon certain interpretations of reality, including also interpretations with regard to the definition and distribution of roles in the network (who should do what), as well as the scenarios and itineraries (e.g. research procedures) that are to be followed (e.g. Callon et al.,1986:glossary). Moreover, translation also involves strategies by which actors present themselves as indispensable to others (i.e. the creation of 'a geography of *obligatory passage points*') in connection with certain tasks, scenarios, etc. According to Callon et al. (1986), the elementary form of translation is '*interessement*' 'which involves one entity attracting a second by coming between that entity and a third'. In the realm of science, interessement often takes the form of '*problematisation*', that is, the translation of a particular problem into one or more scientific problems that need to be further researched. The implied assumption that one needs to solve the scientific problems in order to solve the 'real' problem is in fact an example of an attempt to create one or more obligatory passage points (Callon et al., 1986:glossary).

In many ways, the actor network approach can be understood or used as a specific type and mode of conducting a process ethnography in the context of innovation processes. Rather than describing relationships (e.g. section 20.4.1), the approach focuses specifically on the way in which interrelations are forged and constructed through enrolment and translation of knowledge, and vice versa. Moreover, it proposes a particular conceptual terminology for doing so. Given the significance of knowledge, natural phenomena and technological artefacts in our domain of study, we feel the approach can be of particular use in Communication and Innovation Studies, and especially in connection with issues of knowledge integration and knowledge construction in design processes (see questions 4 and 8 in section 20.2).

20.4.4   Following knowledge constructs and perceptions in networks

An approach that may be particularly relevant to improving our understanding of organisational learning and processes of 'diffusion', may be to study the way in which specific knowledge constructs (see question 4 in section 20.2) and perceptions 'travel' through networks. One can, for example, try to identify what different actors involved in an innovation process 'learn' over time (i.e. how their tacit and/or discursive knowledge changes), and what happens to those 'lessons'. How and to whom are they communicated, or not, and why? And how do recipients interpret and act upon such insights and why (not)? What are the consequences for the development of coherent innovations? When engaging in this kind of network analysis it is important to note that various kinds of perceptions may change (i.e. that different types of lessons can be drawn; see Chapter 9) in relation to a range of social and technical issues and domains (see section 5.1). Also, it must be taken into account that 'lessons' can be expressed or remain implicit in different forms, e.g. as conclusions regarding causal associations, as identified knowledge gaps or questions for research, as guidelines, as problem definitions, as inferences regarding the relevance and importance of the aforementioned knowledge constructs, etc. As has been shown by De Grip (2002), looking at innovation processes in this way can improve one's understanding of how and why certain 'lessons' are selected, made discursive, translated, acted upon or simply forgotten. This, then, may provide a much better insight into the kinds of obstacles to achieving coherence that can exist in innovation networks. At first sight these kinds of network studies may seem to show some similarity to early studies of diffusion in that they focus on how and why ideas 'spread'. An important difference, however, is that the kind of network study discussed in this section takes specific perceptions as the entry point, and tries to understand the micro-interactions through which they 'diffuse', or not. The idea is to 'follow' specific perceptions and how they evolve in a network of interactions through time and space. The result, then, could be called a 'learning process ethnography' regarding specific issues. In contrast, classical diffusion studies (see Chapter 8; Rogers, 1983) were much more interested in specific categories of people (e.g. opinion leaders, innovators, laggards, etc.) and their attributes, and in the global patterns of communication between them (Rogers & Kincaid, 1981).

This approach of following perceptions can also be usefully combined with an, at first sight, radically opposite approach, whereby material products and transformation chains are used as an entry point for analysing networks of social relations. One can, for example, analyse the network of material transformations that is needed for the production of a tractor, mechanical implement or food product (e.g. from raw iron and crude oil to metal parts, plastics and fuels), and can study the interdependent networks of actors that are involved in the process. In the context of

Communication and Innovation Studies, then, it can be of great interest to look at how social learning within and across such networks takes place, or not, and contributes to the emergence of more or less coherent innovations.

20.4.5   Conclusion

We have seen that process ethnographic case-studies can be usefully combined with, or operationalised as, various forms of network analysis. It must be noted that the kinds of network analysis discussed above are not mutually exclusive, and can be combined in several ways. One may, for example, study a network of interrelated events, and focus on the implications of these events for the way in which human and non-human entities are connected in an actor network, paying special attention to processes of knowledge construction and/or learning. Similarly, one can select those events that represent interface situations. The different approaches to network analysis merely propose a specific entry point for making a process ethnography (e.g. people, events, interfaces, perceptions, enrolments) and can at times be associated with a specific research interest (e.g. interaction patterns, network building, social learning, etc.). In many ways, these entry points and interests are to some extent artificial separations and categories produced by researchers, and in everyday social practice are often closely intertwined and inseparable. In that sense, theories and conceptual frameworks always simplify what happens in practice. At the same time such simplification is needed to make it possible to take deliberate action. Thus, we hope that conceptual researchers in the field of Communication and Innovation Studies will be able to use some of our thoughts and proposals for research in order to produce better quality simplifications.

Questions for discussion

(1)   What are your own research interests, and how do they connect with the research areas mentioned in section 20.2?

(2)   Which of the theories and concepts discussed in this book may be relevant to your research? What other theories and concepts would you like to apply as 'sensitising concepts'?

(3)   Given your own research interests, what might be the relevance of a case-study approach? If relevant, what would be suitable boundaries of a case-study?

(4)   What research methods would be consistent with your research interest and the theories/concepts that you find potentially relevant?

References

Aarts, M.N.C. (1998) *Een Kwestie van Natuur: Een Studie naar de Aard en het Verloop van Communicatie over Natuur en Natuurbeleid.* Published doctoral dissertation. Wageningen Agricultural University, Wageningen.

Aarts, M.N.C. & H.F.M. Te Molder (1998) *Natuurontwikkeling: Waarom en Hoe? Een Discourse-Analytische Studie van een Debat.* Rathenau Instituut, Den Haag.

Aarts, M.N.C. & C.M.J. Van Woerkum (1999) Communication and nature policies: the need for an integral approach to policy design. In: C. Leeuwis (Ed.), *Integral Design: Innovation in Agriculture and Resource Management*, pp. 33–47. Mansholt Institute, Wageningen.

Aarts, M.N.C. & C.M.J. Van Woerkum (2002) Dealing with uncertainty in solving complex problems. In: C. Leeuwis & R. Pyburn (Eds), *Wheelbarrows Full of Frogs. Social Learning in Rural Resource Management*, pp. 421–35. Royal Van Gorcum, Assen.

Abbot, J. & I. Guijt (1998) *Changing Views on Change: Participatory Approaches to Monitoring the Environment.* SARL Programme Discussion Paper No. 2. International Institute for Environment and Development (IIED), London.

Abernathy, W.J. & K.B. Clark (1985) Innovation: Mapping the winds of creative destruction. *Research Policy*, **14**, 3–22.

Ackoff, R.L. (1974) *Redesigning the Future.* J. Wiley & Sons, New York.

Ackoff, R.L. (1979) Resurrecting the future of operational research. *Journal of the Operational Research Society*, **30**, 189–99.

Ackoff, R.L. (1981) *Creating the Corporate Future.* J. Wiley & Sons, New York.

Adams, M.E. (1982) *Agricultural Extension in Developing Countries.* Longman, Burnt Mill.

Agritex (1998) *Learning Together Through Participatory Extension. A guide to an approach developed in Zimbabwe.* Agritex/GTZ/IRDEP/ITZ, Harare.

Ajzen, I. & M. Fishbein (1980) *Understanding Attitudes and Predicting Social Behaviour.* Prentice-Hall, Englewood Cliffs.

Ajzen, I. & J.T. Madden (1986) Prediction of goal-directed behavior: attitudes, intentions and perceived behavioral control. *Journal of Experimental Social Psychology*, **22**, 453–74.

Alrøe, H. & E.S. Kristensen (2002), Towards a systemic research methodology in agriculture: Rethinking the role of values in science. *Agriculture and Human Values*, **19**, 3–23.

Amankwah, K. (2000) *Participatory technology development: learning or negotiation process? Case-studies from the Sandema area in Northern Ghana.* MSc thesis, Wageningen Agricultural University.

Arce, A. (1993) *Negotiating Agricultural Development: Entanglements of Bureaucrats and Rural Producers in Western Mexico.* Wageningen Studies in Sociology Nr. 34. Wageningen Agricultural University, Wageningen.

Arce, A. & N. Long (1987) The dynamics of knowledge interfaces between Mexican agricultural bureaucrats and peasants: A case-study from Jalisco. *Boletín de Estudios Latinoamericanos y del Caribe*, **43**, 5–30.

Argyris, C. (1994) Good communication that blocks learning. *Harvard Business Review*, July/August 1994, 77–85.

Argyris, C. & D.A. Schön (1996) *Organizational Learning II. Theory, Method and Practice.* Addison-Wesley Publishing, Reading.

Atsma, G., F. Benedictus, F. Verhoeven & M. Stuiver (2000) *De Sporen van Twee Milieucoöperaties.* Vereniging Eastermars Lânsdouwe & Vereniging Agrarisch Natuurbeheer en Landschap Achtkarspelen (VEL & VANLA), Drachten.

Austin, J.L. (1971) *How to Do Things with Words.* The William James lectures delivered at Harvard University in 1955. Oxford University Press, Oxford.

Backett-Milburn, K. & S. Wilson (2000) Understanding peer-education: insights from a process evaluation. *Health Education Research: Theory and Practice*, **15**, 85–96.

Baland, J.M. & J.P. Platteau (1996) *Halting Degradation of Natural Resources: Is There a Role for Rural Communities?* Oxford University Press, Oxford.

Bandura, A. (1977) Self-efficacy: towards a unifying theory of behavior change. *Psychological Review*, **84**, 191–215.

Bandura, A. (1986) *Social Foundations of Thought and Action: A Social Cognitive Theory.* Prentice-Hall, Englewood Cliffs.

Bartstra, D. (2001) Noord-Holland viert nipte overwinning. De redding van de laatste moerasgronden. *Spil*, **173–4**, 43–4.

Batra, R. & M.L. Ray (1986) Affective responses mediating acceptance of advertising. *Journal of Consumer Research*, **13**, 234–49.

Bawden, R. (1994) Creating learning systems: a metaphor for institutional reform for development. In: I. Scoones & J. Thompson (Eds), *Beyond Farmer First. Rural People's Knowledge, Agricultural Research and Extension Practice*, pp. 258–63. Intermediate Technology Publications, London.

Bawden, R. (1995) On the systems dimensions of FSR. *Journal for Farming Systems Research and Extension*, **5**, 1–9.

Beck, U. (1992) *Risk Society: Towards a New Modernity.* Sage, London.

Benor, D. & M. Baxter (1984) *Training and Visit Extension.* World Bank, Washington.

Benor, D. & J.Q. Harrison (1977) *Agricultural Extension. The Training and Visit System.* World Bank, Washington.

Benvenuti, B. (1982) De technologisch administratieve taakomgeving (TATE) van landbouwbedrijven. *Marquetalia*, **5**, 111–36.

Benvenuti, B. (1991) *Geschriften over Landbouw, Structuur en Technologie.* Wageningen Studies in Sociology Nr. 29. Wageningen Agricultural University, Wageningen.

Berkes, F. & C. Folke (Eds) (1998) *Linking Social and Ecological Systems. Management Practices and Social Mechanisms for Building Resilience.* Cambridge University Press, Cambridge.

Berlo, D.K. (1960) *The Process of Communication. An Introduction to Theory and Practice.* Holt, Rinehart & Winston, New York.

Beynon, J. (1998) *Financing the Future. Options for Agricultural Research and Extension in Sub-Saharan Africa.* Oxford Policy Management, Oxford.

Bierly III, P.E., E.H. Kessler & E.W. Christensen (2000) Organizational learning, knowledge and wisdom. *Journal of Organizational Change Management*, **13**, 595–618.

Biggs, S.D. (1989) *Resource-poor Farmer Participation in Research. A Synthesis of Experiences from Nine Agricultural Research Systems.* On-Farm Client-Oriented Research (OFCOR) Comparative Study Paper No. 3. International Service for National Agricultural Research (ISNAR), The Hague.

Bijker, W., T. Hughes & T. Pinch (Eds) (1987) *The Social Construction of Technological Systems. New Directions in the Sociology and History of Technology.* MIT Press, Cambridge, MA.

Blackburn, D.J. (Ed.) (1994) *Extension Handbook. Processes and Practices*, 2nd edn. Thompson Educational Publishing, Toronto.

Blau, P.M. & W.R. Scott (1962) *Formal Organizations. A Comparative Approach.* Chandler, San Francisco.

Blum, A. (1996) *Teaching and Learning in Agriculture. A Guide for Agricultural Educators.* FAO, Rome.

Boissevain, J. (1974) *Friends of Friends: Networks, Manipulators and Coalitions.* Basil Blackwell, Oxford.

Boissevain, J. & J.C. Mitchell (Eds) (1973) *Network Analysis.* Mouton, The Hague.

Bolding, A., E. Manzungu & P. Van der Zaag (1996) Farmer-initiated irrigation furrows. Observations from the Eastern Highlands. In: E. Manzungu & P. Van der Zaag (Eds), *The Practice of Smallholder Irrigation. Case Studies from Zimbabwe*, pp. 191–218. University of Zimbabwe Publications, Harare.

Bolhuis, E.E. & J.D. Van der Ploeg (1985) *Boerenarbeid en Stijlen van Landbouwbeoefening: Een Socio-Economisch Onderzoek naar de Effekten van Incorporatie en Institutionalisering op Agrarische Ontwikkelingspatronen in Italië en Peru.* Leiden Development Studies, Leiden.

Bos, A.H. (1974) *Oordeelsvorming in Groepen.* Veenman, Wageningen.

Bosman, J., E. Hollander, P. Nelissen, K. Renckstorf, F. Wester & C. Van Woerkum (1989) *Het Omgaan met Kennis – en de Vraag naar Voorlichting: Een Multidisciplinair Theoretisch Referentiekader voor Empirisch Onderzoek naar de Vraag naar Voorlichting.* Katholieke Universiteit Nijmegen, Nijmegen.

Bott, E. (1957) *Family and Social Network.* Tavistock, London.

Bouma, J. (1999) The role of research chains and user interaction in designing multi-functional agricultural production systems. In: C. Leeuwis (Ed.), *Integral Design: Innovation in Agriculture and Resource Management*, pp. 219–35. Mansholt Institute, Wageningen.

Bouman, M. (1999) *The Turtle and the Peacock. Collaboration for Prosocial Change. The Entertainment Education Strategy on Television.* Published doctoral dissertation. Wageningen Agricultural University, Wageningen.

Bourdieu, P. (1990) *The Logic of Practice.* Polity Press, Cambridge.

Brinkerhoff, D.W. & M.D. Ingle (1989) Integrating blueprint and process: A structured flexibility approach to development management. *Public Administration and Development*, **9**, 487–503.

Broerse, J.E.W. & J.F.G. Bunders (1999) Pitfalls in implementation of integral design approaches to innovation: The case of the Dutch Special Programme on Biotechnology. In: C. Leeuwis (Ed.), *Integral Design: Innovation in Agriculture and Resource Management*, pp. 245–65. Mansholt Institute, Wageningen.

Brouwers, J.H.A.M. (1993) *Rural People's Response to Soil Fertility Decline: The Adja Case (Benin).* Published doctoral dissertation. Wageningen Agricultural University, Wageningen.

Brown, D., M. Howes, K. Hussein, C. Longley & K. Swindell (2002) *Participatory methodologies and participatory practices. Assessing PRA use in Gambia.* AgREN Network Paper No. 124. Overseas Development Institute (ODI), London.

Buchanan, D.A. (1992) *The Expertise of the Change Agent. Public Performance and Backstage Activity.* Prentice Hall International, Hemel Hempstead.

Bunker B.B. & B.T. Alban (1997) *Large Group Interventions. Engaging the Whole System for Rapid Change.* Jossey-Bass Publishers, San Franscisco.

Calder, N. (1997) *The Manic Sun. Weather Theories Confounded.* The Pilkington Press, London.

Callon, M. & B. Latour (1992) Don't throw the baby out with the Bath School! A reply to Collins and Yearley. In: A. Pickering (Ed.), *Science as Practice and Culture.* pp. 343–68. University of Chicago Press, Chicago.

Callon, M., J. Law & A. Rip (Eds) (1986) *Mapping the Dynamic of Science and Technology: Sociology of Science in the Real World.* Macmillan, London.

Chambers, R. (1992) *Rural Appraisal: Rapid, Relaxed and Participatory.* IDS discussion paper 311. Institute of Development Studies (IDS), Brighton.

Chambers, R. (1994a) The origins and practice of Participatory Rural Appraisal. *World Development*, **22**, 953–69.

Chambers, R. (1994b) Participatory Rural Appraisal (PRA): Analysis of experience. *World Development*, **22**, 1253–68.

Chambers, R. & B. Ghildyal (1985) Agricultural research for resource poor farmers – the farmer first and last model. *Agricultural Administration*, **10**, 1–30.

Chambers, R., A. Pacey & L.A. Thrupp (Eds) (1989) *Farmer First: Farmer Innovation and Agricultural Research.* Intermediate Technology Publications, London.

Chayanov, A.V. (1966) *The Theory of Peasant Economy*, (Eds D. Thorner, B. Kerblay & R.E.F. Smith). Irwin, Homewood.

Checkland, P.B. (1981) *Systems Thinking, Systems Practice.* J. Wiley & Sons, Chichester.

Checkland, P.B. (1985) From optimizing to learning: a development of systems thinking for the 1990s. *Journal of the Operational Research Society*, **36**, 757–67.

Checkland, P.B. (1988) Soft systems methodology: an overview. *Journal of Applied Systems Analysis*, **15**, 27–30.

Checkland, P.B. & L. Davies (1986) The use of the term 'Weltanschauung' in soft systems methodology. *Journal of Applied Systems Analysis*, **13**, 109–15.

Churchman, C.W. (1971) *The Design of Inquiring Systems.* Basic Books, New York.

Churchman, C.W. (1979) *The Systems Approach and Its Enemies.* Basic Books, New York.

Claar, J.B. & R.P. Bentz (1984), Organizational design and extension administration. In: B.E. Swanson (Ed.) *Agricultural Extension. A Reference Manual*, pp. 161–83. FAO, Rome.

Clegg, S.R., C. Hardy & W.R. Nord (Eds) (1999) *Managing Organizations. Current Issues.* Sage Publications, London.

Cochrane, W.W. (1958) *Farm Prices, Myth and Reality.* University of Minnesota Press, Minneapolis.

Cook, S.D.N. & D. Yanov (1993) Culture and organizational learning. *Journal of Management Inquiry*, **2**, 373–90.

Craig, D. & D. Porter (1997) Framing participation: development projects, professionals and organisations. *Development in Practice*, **7**, 229–36.

Crehan, K. & A. Von Oppen (1988) Understandings of 'development': an arena of struggle. The story of a development project in Zambia. *Sociologia Ruralis*, **28**, 113–45.

Cuilenburg, J.J. (1983) *Overheidsvoorlichting in Overvloed: Over Voorlichtingsonderzoek in het Informatietijdperk.* Vrije Universiteit, Amsterdam.

Dangbégnon, C. (1998) *Platforms for Resource Management. Case Studies of Success or Failure in Benin and Burkina Faso.* Published doctoral dissertation. Wageningen Agricultural University, Wageningen.

Daniels, S.E. & G.B. Walker (1996) Collaborative learning: improving public deliberation in ecosystem-based management. *Environmental Impact Assessment Review*, **16**, 71–102.

Darré, J.P. (1985) *La Parole et la Technique. L'univers de Pensée des Éleveurs du Ternois.* L'Harmattan, Paris.

Dasgupta, S. (1989) *Diffusion of Agricultural Innovations in Village India.* Wiley Eastern, New Delhi.

Dasgupta, P. & I. Serageldin (Eds) (2000) *Social Capital; A Multifaceted Perspective.* World Bank, Washington.

Davis, G.B. & M.H. Olson (1985) *Management Information Systems: Conceptual Foundations, Structure and Development.* McGraw-Hill Book Company, New York.

De Bruin, H. & E. Ten Heuvelhof (1998) Procesmanagement. *Bestuurswetenschappen*, **2**, 120–34.

Defoer, T. & A. Budelman (2000) *Managing Soil Fertility in the Tropics: A Resource Guide for Participatory Learning and Action Research.* Royal Tropical Institute (KIT), Amsterdam.

De Grip, K. (2002) *Over leren, communiceren en innoveren. Een case-study over kennisontwikkeling en kennisuitwisseling met betrekking tot mineralenmanagement in het project Koeien en Kansen.* MSc thesis, Wageningen University.

De Grip, K. & C. Leeuwis (with contributions by L.W.A. Klerkx) (2003) *Lessen over Vraagsturing. Ervaringen met het Steunpunt Mineralen Concept.* Agro Management Tools, Wageningen.

De Jager, H. & A.L. Mok (1974) *Grondbeginselen der Sociologie. Gezichtspunten en Begrippen.* H.E. Stenfert Kroese BV, Leiden.

De Janvry, A. (1981) *The Agrarian Question and Reformism in Latin America.* The John Hopkins University Press, Baltimore.

De Jong, M. & P.J. Schellens (2000) Formatieve evaluatie. In: P.J. Schellens, R. Klaassen & S. De Vries (Eds), *Communicatiekundig Ontwerpen. Methoden, Perspectieven en Toepassingen,* pp. 100–18. Van Gorcum, Assen.

Denzin, N.K. & Y.S. Lincoln (Eds) (1994) *Handbook of Qualitative Research.* Sage Publications, Thousand Oaks, CA.

Dervin, B. (1981) Mass communication: changing conceptions of the audience. In: R.E. Rice & W.J. Paisley (Eds), *Public Communication Campaigns,* pp. 71–88. Sage Publications, Beverly Hills.

Dervin, B. (1983) Information as a user construct: the relevance of perceived information needs to synthesis and interpretation. In: S.A. Ward & L.J. Reed (Eds), *Knowledge, Structure and Use: Implications for Synthesis and Interpretation,* pp. 153–83. Temple University Press, Philadelphia.

Dörner, D. (1996) *The Logic of Failure. Recognizing and Avoiding Error in Complex Situations.* Persues Books, Reading, MA.

Douglas, M. (1992) *Risk and Blame. Essays in Cultural Theory.* Routledge, London.

Dugdale, A. (1999) Materiality: juggling sameness and difference. In: J. Law & J. Hassard, *Actor Network Theory and After* (Eds), pp. 113–35. Blackwell Publishing, Oxford.

Dunn, E.S. (1971) *Economic and Social Development: A Process of Social Learning.* John Hopkins University Press, Baltimore.

Easterby-Smith, M., J. Burgoyne & L. Araujo (1999) *Organizational Learning and the Learning Organization. Developments in Theory and Practice.* Sage, London.

Edmunds, D. & E. Wollenberg (2001) A strategic approach to multi-stakeholder negotiations. *Development and Change,* **32,** 231–53.

Edwards, D. & J. Potter (1992) *Discursive Psychology.* Sage, London.

Elias, N. & J.L. Scotson (1965) *The Established and the Outsiders.* Frank Cass & Co Ltd, London.

Ellis, F. (2000) *Rural Livelihoods and Diversity in Developing Countries.* Oxford University Press, Oxford.

Emery, M. & R.E. Purser (1996) *The Search Conference. A Powerful Method for Planning Organizational Change and Community Action.* Jossey-Bass Publishers, San Francisco.

Engel, P.G.H. (1995) *Facilitating Innovation. An Action-oriented and Participatory Methodology to Improve Innovative Social Practice in Agriculture.* Published doctoral dissertation. Wageningen Agricultural University, Wageningen.

Engel P.G.H. & M. Salomon (1997) *Facilitating Innovation for Development. A RAAKS Resource Box.* Royal Tropical Institute, Amsterdam.

Eshuis, J., M. Stuiver, F. Verhoeven & J.D. Van der Ploeg (2001) *Goede Mest Stinkt Niet: Een Studie over Drijfmest, Ervaringskennis en het Terugdringen van Mineralenverliezen in de Melkveehouderij.* Studies van Landbouw en Platteland Nr. 31. Wageningen University, Wageningen.

Estrella, M. with J. Blauert, D. Campilan, J. Gaventa, J. Gonsalves, I. Guijt, D.A. Johnson & R. Ricafort (Eds) (2000) *Learning From Change: Issues and Experiences in Participatory Monitoring and Evaluation.* Intermediate Technology Publications, London.

Eyben, R. & S. Ladbury (1995) Popular participation in aid-assisted projects: Why more in theory than in practice. In: N. Nelson & S. Wright (Eds), *Power and Participatory Development. Theory and Practice,* pp. 192–200. Intermediate Technology Publications, London.

Fals Borda, O. (1998a) *People's Participation. Challenges Ahead,* pp. 157–73. The Apex Press, New York and Intermediate Technology Publications, London.

Fals Borda, O. (1998b) Theoretical foundations. In: O. Fals Borda (Ed.), *People's Participation. Challenges Ahead*, pp. 157–73. The Apex Press, New York and Intermediate Technology Publications, London.

FAO (1985) *Community Forestry: Participatory Assessment, Monitoring and Evaluation.* Community Forestry Notes. FAO, Rome.

FAO & World Bank (2000) *Agricultural Knowledge and Information Systems for Rural Development (AKIS/RD). Strategic Vision and Guiding Principles.* FAO, Rome and World Bank, Washington.

Farrington, J., D. Carney, C. Ashley & C. Turton (1999) *Sustainable Livelihoods in Practice: Early Applications of Concepts in Rural Areas.* Natural Resources Perspectives No. 42. Overseas Development Institute (ODI), London.

Farrington, J., I. Christoplos & A. Kidd (2002) *Extension, Poverty and Vulnerability. The Scope for Policy Reform.* Final report of the study for the Neuchatel Initiative. Overseas Development Institute (ODI) Working Paper 155. ODI, London.

Fayol, H. (1949) *General and Industrial Management.* Pitman & Sons, London.

Ferguson, J. (1990) *The Anti-politics Machine: Development, Depoliticization and Bureaucratic Power in Lesotho.* University of Minneapolis, Minneapolis.

Festinger, L. (1957) *A Theory of Cognitive Dissonance.* Stanford University Press. Stanford.

Feuerstein, M.T. (1986) *Partners in Evaluation: Evaluating Development and Community Programmes with Participants.* Macmillan Publishers, London.

Feyerabend, P. (1975) *Against Method: Outline of an Anarchistic Theory of Knowledge.* New Left Books, London.

Fishbein, M. & I. Ajzen (1975) *Belief, Attitude, Intention and Behaviour: An Introduction to Theory and Research.* Addison-Wesley, Reading.

Fisher, R. & W. Ury (1981). *Getting to Yes: Negotiating Agreement Without Giving In.* Penguin Books Ltd, Harmondsworth.

Fraser, C. & S. Restrepo-Estrada (1998) *Communicating for Development. Human Change for Survival.* Tauris, London.

Freire, P. (1972) *Pedagogy of the Oppressed.* Penguin Books Ltd, Harmondsworth.

Freud, S. (1905) *Jokes and their Relation to the Unconscious.* Hogarth, London.

Friedman, J. (1992) *Empowerment. The Politics of Alternative Development.* Blackwell, Oxford.

Friedmann, J. (1984) Planning as social learning. In: D.C. Korton & R. Klaus (Eds), *People Centered Development: Contributions Towards Theory and Planning Frameworks*, pp. 189–94. Kumarian Press, West Hartford.

Fuenmayor, R. & H. López-Garay (1991) The scene for interpretative systemology. *Systems Practice*, **4**, 401–18.

Fulk, J., J. Schmitz & C.W. Steinfield (1990) A social influence model of technology use. In: J. Fulk & C.W. Steinfield (Eds) (1990) *Organizations and Communication Technology*, pp. 117–40. Sage, Newbury Park.

Funtowicz, S.O. & J.R. Ravetz (1993) Science for the post-normal age. *Futures*, **25**, 739–55.

Garud, R. & P. Karnoe (2001) *Path Dependence and Creation.* Lawrence Erlbaum, Mahwah, NJ.

Garvin, D.A. (1993) Building a learning organization. Beyond high philosophy and grand themes lie the gritty details of practice. *Harvard Business Review*, July–August, 78–91.

Geels, F. (2002) *Understanding the Dynamics of Technological Transitions. A Co-evolutionary and Socio-technical Analysis.* Twente University Press, Enschede.

Geurtz, J.L. & I. Mayer (1996) *Methods for Participatory Policy Analysis: Towards a Conceptual Model for Research and Development.* Draft paper. Tilburg University, Work and Organisation Research Centre, Tilburg.

Giddens, A. (1976) *New Rules of Sociological Method: A Positive Critique of Interpretative Sociologies*. Hutchinson, London.

Giddens, A. (1979) *Central Problems in Social Theory*. Macmillan, London.

Giddens, A. (1984) *The Constitution of Society: Outline of the Theory of Structuration*. Polity Press, Cambridge.

Gilbreth, F.B. (1911) *Motion Study: A Method of Increasing the Efficiency of the Workman*. D. van Nostrand, Princeton.

Gilbreth, F.B. (1914) *Primer of Scientific Management*. Constable & Company, London.

Gödel, K. (1962) *On Formally Undecidable Propositions*. Basic Books, New York.

Goffman, E. (1959) *The Presentation of Self in Everyday Life*. Doubleday, New York.

Gonzalez, R.M. (2000) *Platforms and Terraces. Bridging Participation and GIS in Joint-learning for Watershed Management with the Ifugaos of the Philippines*. Published doctoral dissertation. ITC, Enschede and Wageningen University, Wageningen.

Gouldner, A.W. & H.P. Gouldner (1963) *Modern Sociology. An Introduction to the Study of Human Interaction*. Harcourt, Brace & World, New York.

Granovetter, M.S. (1973) The strength of weak ties. *American Journal of Sociology*, **78**, 1360–80.

Gray, B. (1997) *Framing and Re-framing of Intractable Environmental Disputes*. Prentice Hall, London.

Griliches, Z. (1957) Hybrid corn: an exploration in the economics of technological change. *Econometrica*, **25**, 501–22.

Gronov, C.J.V. (1995) Shifting power, sharing power: issues from user-group forestry in Nepal. In: N. Nelson & S. Wright, *Power and Participatory Development. Theory and Practice*, pp. 125–32. Intermediate Technology Publications, London.

Groot, A.E., J.P.M. Stuijt & C.A.M. Boon (1995) Changing perspectives on monitoring and evaluation. In: *Agricultural Extension in Africa. Proceedings of an International Workshop Yaoundé, Cameroon*. Vol. I, pp. 121–32. Technical Centre for Agricultural and Rural Co-operation (CTA), Wageningen.

GTZ (1987) ZOPP. *An Introduction to the Method*. Deutsche Gesellschaft für Technische Zusammenarbeit (GTZ), Eschborn.

Guijt, I. (1999) *Participatory Monitoring and Evaluation for Natural Resource Management and Research. Best Practice Guidelines, Socio-economic Methodologies*. Natural Resource Institute, Chatham.

Guijt, I. & A. Cornwall (1995) Critical reflections on the practice of PRA. *PLA Notes*, **24**, 2–7.

Guijt, I. & M.K. Shah (Eds) (1998) *The Myth of Community: Gender Issues in Participatory Development*. Intermediate Technology Publications, London.

Habermas, J. (1970a) On systematically distorted communication. *Inquiry*, **13**, 205–18.

Habermas, J. (1970b) Towards a theory of communicative competence. *Inquiry*, **13**, 360–75.

Habermas, J. (1973) *Theory and Practice*. Beacon Press, Boston.

Habermas, J. (1981) *Theorie des Kommunikativen Handelns. Band 1: Handlungsrationalität und gesellschaftliche Rationalisierung. Band 2: Zur Kritik der funktionalistischen Vernunft*. Suhrkamp Verlag, Frankfurt am Main.

Hamilton, N.A. (1990) *Knowledge systems and the development of computer programs for use by farmers: Towards a new development process. An analysis by case study of Wheat Councellor and Wheatman*. MSc thesis. Wageningen Agricultural University.

Hamilton, N.A. (1995) *Learning to Learn with Farmers. A Case Study of an Adult Learning Extension Project in Queensland, Australia, 1990–1995*. Published doctoral dissertation. Wageningen Agricultural University, Wageningen.

Hannerz, U. (1980) *Exploring the City. Inquiries Towards an Urban Anthropology.* Columbia University Press, New York.

Hannerz, U. (1996) *Transnational Connections. Culture, People, Places.* Routledge, London.

Hanson, J.C. & R.E. Just (2001) The potential for transition to paid extension: Some guiding economic principles. *American Journal of Agricultural Economics*, **83**, 777–84.

Hardin, G. (1968) The tragedy of the commons. *Science*, **162**, 1243–8.

Harrington, J. (1991) *Organizational Structure and Information Technology.* Prentice-Hall, London.

Havelock, R.G. (1973) *Planning for Innovation Through Dissemination and Utilization of Knowledge.* Center for Research in the Utilization of Scientific Knowledge/Institute for Social Research (CRUSK/ISR), Ann Arbor.

Havelock, R.G. (1986) Modelling the knowledge system. In: G.M. Beal, W. Dissanayake & S. Konoshima (Eds), *Knowledge Generation, Exchange and Utilization*, pp. 77–104. Westview Press, Boulder.

Hayami, Y. & V.W. Ruttan (1985) *Agricultural Development: An International Perspective.* John Hopkins, Baltimore.

Hebinck, P. & R. Ruben (1998) *Rural households and livelihood strategies: Straddling farm and non-farm activities.* Proceedings of the 15th International Symposium on Farming Systems Research and Extension 'Going Beyond the Farm Boundary', Pretoria, 29 November to 4 December 1998, pp. 876–85.

Hersoug, B. (1996) Logical framework analysis in an illogical world. *Forum for Development Studies*, **2**, 377–404.

Heymann, F. (1999) Interpersoonlijke communicatie. In: C.M.J. Van Woerkum & R.C.F. Van Meegeren (Eds), *Basisboek Communicatie en Verandering*, pp. 174–94. Boom, Amsterdam.

Hilhorst, D. (2000) *Records and Reputations. Everyday Politics of a Philippine Development NGO.* Published doctoral dissertation. Wageningen Agricultural University, Wageningen.

Hoffmann, V. (2000) *Picture supported communication in Africa. Fundamentals, examples and recommendations for appropriate communication processes in rural development programmes in sub-Saharan Africa.* Margraf Verlag, Weikersheim.

Holling, C.S. (1985) *Adaptive Environmental Assessment and Management.* John Wiley & Sons, Chichester.

Holling, C.S. (1995) What barriers? What bridges? In: L.H. Gunderson, C.S. Holling & S.S. Light (Eds), *Barriers and Bridges to the Renewal of Ecosystems and Institutions*, pp. 3–37. Colombia University Press, New York.

Horton, D., P. Ballantyne, W. Peterson, B. Uribe, D. Gapasin & K. Sheridan (1993) *Monitoring and Evaluating Agricultural Research: a Sourcebook.* CAB International with International Service for National Agricultural Research (ISNAR), Wallingford.

Hounkonnou, D. (2001) *Listen to the Cradle. Building from Local Dynamics for African Renaissance. Case Studies in Rural Areas in Benin, Burkina Faso and Ghana.* Published doctoral dissertation, Wageningen University, Wageningen.

Howell, J. (1982) *Managing Agricultural Extension: The T and V System in Practice.* Discussion paper 8. Agricultural Administration Network. Overseas Development Institute (ODI), London.

Hruschka, E. (1994) Extension problems from a psychological point of view. In: H. Albrecht (Ed.) *Einsicht als Agens des Handelns. Beratung und Angewandte Psychologie*, pp. 25–32. Margraf Verlag, Weikersheim.

Huguenin, P. (1994) *Zakboek voor Onderhandelaars: Vuistregels, Vaardigheden en Valkuilen.* Bohn, Stafleu & Van Lochem, Houten.

Huirne, R., J. Hardaker & A. Dijkhuizen (Eds) (1997) *Risk Management Strategies in Agriculture: State of the Art and Future Perspectives.* Mansholt Institute, Wageningen.

IBM (1988) *Boer en Tuinder op weg naar Managementsystemen.* IBM, Eindhoven.

IFAD (2001) *Rural Poverty Report. The Challenge of Ending Rural Poverty.* Oxford University Press, Oxford.

IFPRI (1995) *A 2020 Vision for Food, Agriculture and the Environment. The Vision, Challenge and Recommended Action.* International Food Policy Research Institute, Washington.

Ison, R.L. & D.B. Russell (2000) *Agricultural Extension and Rural Development: Breaking out of Traditions. A Second-order Systems Perspective.* Cambridge University Press, Cambridge.

Jackson, M.C. (1982) The nature of 'soft' systems thinking: the work of Churchman, Ackoff and Checkland. *Journal of Applied Systems Analysis,* **9**, 17–29.

Jackson, M.C. (1985) Social systems theory and practice: The need for a critical approach. *International Journal of General Systems,* **10**, 135–51.

Janis, I.L. (1972) *Victims of Groupthink.* Houghton-Miffin, Boston.

Jankowski, N.W., C. Leeuwis, P.J. Martin, M.C. Noordhof & J. van Rossum (1999) Teledemocracy in the province: An experiment with Internet-based software and public debate. In: L. d'Haenens (Ed.), *Cyberidentities. Canadian and European Presence in Cyberspace,* pp. 121–32. International Canadian Studies Series, University of Ottawa Press, Ottawa.

Jarvis, P. (1987) *Adult Learning in the Social Context.* Billing and Sons, Worcester.

Jiggins, J. & H. De Zeeuw (1992) Participatory technology development in practice: process and methods. In: C. Reijntjes, B. Haverkort & A. Waters-Bayer (Eds), *Farming for the Future: An Introduction to Low-External-Input and Sustainable Agriculture,* pp. 135–62. Macmillan, London.

Jiggins, J. & D. Gibbon (1998) What does interdisciplinary mean? Experiences from SLU. In: A. Markey, J. Phelan & S. Wilson (Eds), *The Challenges for Extension Education in a Changing Rural World.* Proceedings of the 13th European Seminar on Extension Education, 31 August to 6 September 1997, Dublin, Ireland, pp. 317–25. Department of Agribusiness, Extension and Rural Development. University College Dublin (UCD), Dublin.

Johnson, B.T. & A.H. Eagly (1989) Effects of involvement on persuasion: a meta-analysis. *Psychological Bulletin,* **106**, 290–314.

Jones, G.E. & C. Garforth (1997) The history, development and future of agricultural extension. In: B.E. Swanson, R.P. Bentz & A.J. Sofranko (Eds), *Improving Agricultural Extension. A Reference Manual,* pp. 3–12. FAO, Rome.

Jorna, R.J. (1992) Filosofische achtergronden. In: R.J. Jorna & J.L. Simons (Eds), *Kennis in Organisaties: Toepassingen en Theorie van Kennissystemen,* pp. 29–49. Coutinho, Muiderberg.

Kaimowitz, D. (1990) Moving forces: External pressure and the dynamics of technology systems. *Knowledge in Society: The International Journal of Knowledge Transfer,* **3**, 36–43.

Katz, E. (with contributions by A. Barandun) (2002) *Innovative Approaches to Financing Extension for Agricultural and Natural Resource Management – Conceptual Considerations and Analysis of Experience.* LBL, Swiss Center for Agricultural Extension, Lindau.

Kayanja, V. (2003) *Private-serviced agricultural extension goes to the 'botanical garden'. Multiple realities of institution building and farmers' needs selection during the transition process in Mukono district, Uganda.* MSc thesis, Wageningen University.

Kemp, R., A. Rip & J. Schot (2001) Constructing transition paths through the management of niches. In: R. Garud & P. Karnoe, *Path Dependence and Creation,* pp. 269–99. Lawrence Erlbaum, Mahwah, NJ.

Ketelaars, D. & C. Leeuwis (2002) *Verkenning van de Uitstraling van Mineralenprojecten naar Melkveehouders.* Agro Management Tools Rapport Nr. 6. AMT, Wageningen.

Khamis, S.A. (1998) *Sustainable agriculture in small scale farming; 'facilitating technology development with farmers'. A case study of the IPM programme in Pemba, Zanzibar.* MSc thesis, Wageningen Agricultural University.

Kidd, A.D., J.P.A. Lamers, P.P. Ficarelli & V. Hoffmann (2000) Privatising agricultural extension: caveat emptor. *Journal of Rural Studies*, **16**, 95–102.

King, C. & J. Jiggins (2002) A systemic model and theory for facilitating social learning. In: Leeuwis, C. & R. Pyburn (Eds), *Wheelbarrows Full of Frogs. Social Learning in Rural Resource Management*, pp. 85–104. Royal Van Gorcum, Assen.

Klapper, J.T. (1960) *The Effects of Mass Communication.* Free Press, New York.

Kline, S.J. & N. Rosenberg (1986) An overview of innovation. In: R. Landau & N. Rosenberg (Eds), *The Positive Sum Strategy: Harnessing Technology for Economic Growth*, pp. 275–305. National Academic Press, Washington.

Klink, J.P.M. (1991) *Hoofdlijnen Post-INSP-beleid: Van Informaticastimulering naar Informatiebeleid.* Agritect Advies, Gouda.

Knorr-Cetina, K.D. (1981a) *The Manufacture of Knowledge: An Essay on the Constructivist and Contextual Nature of Science.* Pergamon Press, Oxford.

Knorr-Cetina, K.D. (1981b) The micro-sociological challenge of macro-sociology: Towards a reconstruction of social theory and methodology. In: K.D. Knorr-Cetina & A.V. Cicourel (Eds), *Advances in Social Theory and Methodology. Toward an Integration of Micro- and Macro-sociologies*, pp. 1–47. Routledge & Kegan Paul, London.

Knorr-Cetina, K.D. (1992) The couch, the cathedral, and the laboratory: On the relationship between experiment and laboratory in science. In: A. Pickering (Ed.), *Science as Practice and Culture*, pp. 113–38. University of Chicago Press, Chicago.

Koelen, M.A. (1988) *Tales of logic: A Self-presentational View on Health-related Behaviour.* Published doctoral dissertation. Wageningen Agricultural University, Wageningen.

Koelen, M.A. & N.G. Röling (1994) Sociale dilemmas. In: N.G. Röling, D. Kuiper & R. Janmaat (Eds), *Basisboek Voorlichtingskunde*, pp. 58–74. Boom, Amsterdam.

Kolb, D.A. (1984) *Experiential Learning: Experience as the Source of Learning and Development.* Prentice-Hall, Englewood Cliffs.

Kolb D.M. & J.M. Bartunek (Eds) (1992) *Hidden Conflict in Organizations. Uncovering Behind-the-scenes Disputes.* Sage Publications, Newbury Park.

Koning, N.B.J., P.S. Bindraban & A.J.A. Essers (Eds) (2002) *Wageningen Views on Food Security.* Plant Research International, Wageningen.

Koningsveld, H. & J. Mertens (1986) *Communicatief en Strategisch Handelen. Inleiding tot de Handelingstheorie van Habermas.* Coutinho, Muiderberg.

Kotler, P. (1985) *Marketing for Non-profit Organizations.* Aldine, Chicago.

Kuhn, T.S. (1970) *The Structure of Scientific Revolutions.* University of Chicago Press, Chicago.

Lagerspetz, O. (1998) *Trust: the Tacit Demand.* Kluwer Academic Publishers, Dordrecht.

Langer, I., F. Schulz von Thurn & R. Tausch (1981) *Sich Verständlich Ausdrücken.* Ernst Reinhardt Verlag, München.

Latour, B. (1986) The powers of association. In: J. Law (Ed.), *Power, Action and Belief: A New Sociology of Knowledge?*, pp. 264–80. Routledge, London.

Latour, B. (1987) *Science in Action.* Open University Press, Milton Keynes.

Law, J. (Ed.) (1986) *Power, Action and Belief: A New Sociology of Knowledge?* Routledge, London.

Law, J. & J. Hassard (1999) *Actor Network Theory and After.* Blackwell Publishers, Oxford.

Lee, R.A. (2002) *Interactive Design of Farm Conversion. Linking Agricultural Research and Farmer Learning for Sustainable Small Scale Horticulture Production in Colombia.* Published doctoral dissertation. Wageningen University, Wageningen.

Leeuwis, C. (1989) *Marginalization Misunderstood: Different Patterns of Farm Development in the West of Ireland.* Wageningen Studies in Sociology Nr. 26. Wageningen Agricultural University, Wageningen.

Leeuwis, C. (1993) *Of Computers, Myths and Modelling: The Social Construction of Diversity, Knowledge, Information and Communication Technologies in Dutch Horticulture and Agricultural Extension.* Wageningen Studies in Sociology, Nr. 36. Wageningen Agricultural University.

Leeuwis, C. (1995) The stimulation of development and innovation: Reflections on projects, planning, participation and platforms. *European Journal of Agricultural Education and Extension*, **2**, 15–27.

Leeuwis, C. (Ed.) (1999a) *Integral Design: Innovation in Agriculture and Resource Management.* Mansholt Institute, Wageningen.

Leeuwis, C. (1999b) Integral technology design as a process of learning and negotiation. A social science perspective on interactive prototyping. In: Leeuwis, C. (Ed.), *Integral Design: Innovation in Agriculture and Resource Management*, pp. 123–43. Mansholt Institute, Wageningen.

Leeuwis, C. (1999c) Policy-making and the value of electronic forms of public debate. Under-pinning, assumptions and first experiences. In: d'Haenens, L. (Ed.), *Cyberidentities. Canadian and European Presence in Cyberspace*, pp. 99–109. International Canadian Studies Series. University of Ottawa Press, Ottawa.

Leeuwis, C. (2000a) Re-conceptualizing participation for sustainable rural development. Towards a negotiation approach. *Development and Change*, **31**, 931–59.

Leeuwis, C. (2000b) Learning to be sustainable. Does the Dutch agrarian knowledge market fail? *The Journal of Agricultural Education and Extension*, **7**, 79–92.

Leeuwis, C. (2002a) Making explicit the social dimensions of cognition. In: Leeuwis, C. & R. Pyburn (Eds), *Wheelbarrows Full of Frogs. Social Learning in Rural Resource Management*, pp. 391–406. Royal Van Gorcum, Assen.

Leeuwis, C. (2002b) Hoe effectief spelen de kennisstructuren in op vraag en aanbod? De beperkingen van de 'vraag en aanbod' metafoor bij kennis op het gebied van mineralen-management. In: A. Kuipers (Ed.), *Kloof Tussen Kennisaanbod en Kennisvraag.* Verslag van studiedag 17 April 2002. Agro Management Tools Rapport Nr. 9, pp. 7–10. AMT, Wageningen.

Leeuwis, C., N.W. Jankowski, P.J. Martin, J. van Rossum & M.C. Noordhof (1997) *Besliswijzer Beproefd. Een Onderzoek naar Teledemocratie in de Provincie.* Instituut voor Publiek en Politiek, Amsterdam.

Leeuwis, C., N. Long & M. Villarreal (1990) Equivocations on knowledge systems theory: An actor oriented critique. *Knowledge in Society: The International Journal of Knowledge Transfer*, **3**, 19–27.

Leeuwis, C. & R. Pyburn (Eds) (2002) *Wheelbarrows Full of Frogs. Social Learning in Rural Resource Management.* Royal Van Gorcum, Assen.

Leeuwis, C. & G. Remmers (1999) Accomodating dynamics and diversity in integral design. In: C. Leeuwis (Ed.), *Integral Design: Innovation in Agriculture and Resource Management*, pp. 267–77. Mansholt Institute, Wageningen.

Leeuwis, C., N. Röling & G. Bruin (1998) *Can the farmer field school replace the T&V system of extension in Sub-Saharan Africa? Some answers from Zanzibar.* Proceedings of the 15th International Symposium of the Association for Farming Systems Research-Extension 'Going Beyond the Farm Boundary', Pretoria, 29 November to 4 December 1998, pp. 493–7.

Le Gouis, M. (1991) Alternative financing of agricultural extension. Recent trends and implica-tions for the future. In: W.M. Rivera & D.J. Gustafson (Eds), *Agricultural Extension: Worldwide Institutional Evolution & Forces for Change*, pp. 31–42. Elsevier Science Publishers, Amsterdam.

Lionberger, H. & H.C. Chang (1970) *Farm Information for Modernising Agriculture: The Taiwan System.* Praeger, New York.

Lipton, M. (1989) *New Seeds for Poor People*. Unwin Hyman, London.

Little, S., P. Quintas & T. Ray (2002) *Managing Knowledge. An Essential Reader*. Sage, London.

Litvack, J., J. Ahmad & R. Bird (1998) *Rethinking Decentralisation in Developing Countries*. World Bank, Washington.

LNV (1998) *LNV Kennisbeleid 1999–2003. Sturen op Interactie*. Concept beleidsplan, ongepubliceerd. LNV (Dutch Ministry of Agriculture, Nature Conservation and Fisheries), Den Haag.

Long, N. (Ed) (1989) *Encounters at the Interface: A Perspective on Social Discontinuities in Rural Development*. Wageningen Studies in Sociology Nr. 27. Wageningen Agricultural University, Wageningen.

Long, N. (1990) From paradigm lost to paradigm regained? The case for an actor-oriented sociology of development. *European Review of Latin American and Caribbean Studies*, **49**, 3–32.

Long, N. & A. Long (Eds) (1992) *Battlefields of Knowledge: The Interlocking of Social Theory and Practice in Research and Development*. Routledge, London.

Long, N., & J.D. Van der Ploeg (1989) Demythologising planned intervention. *Sociologia Ruralis*, **29**, 226–49.

Louw, C. & C.J.H. Midden (1991) *Emotiegerichte Milieuvoorlichting*. Rijksuniversiteit Leiden, Leiden.

Luhmann, N. (1982) Autopoiesis, Handlung und kommunikative Verständigung. *Zeitschrift für Soziologie*, **11**, 366–79.

Luhmann, N. (1984) *Soziale Systeme: Grundriss einer allgemeinen Theorie*. Suhrkamp Taschenbuch Wissenschaft 666. Suhrkamp, Frankfurt am Main.

Lyytinen, K. & H. Klein (1985) Critical social theory of Jürgen Habermas (CST) as a basis for the theory of information systems. In: E. Mumford, R. Hirschheim, G. Fitzgerald & A.T. Wood-Harper (Eds), *Research Methods in Information Systems*, pp. 219–36. Elsevier North-Holland, Amsterdam.

Maarleveld, M. & C. Dangbégnon (1999) Managing natural resources: A social learning perspective. *Agriculture and Human Values*, **16**, 267–80.

Maarse, L., W. Wentholt & A. Chibudu (1998) *Making Change Strategies Work: Gender Sensitive, Client Oriented Livestock Extension in Coast Province, Kenya*. Royal Tropical Institute, Amsterdam.

Manzungu, E. & P. Van der Zaag (1996) *The Practice of Smallholder Irrigation. Case Studies from Zimbabwe*. University of Zimbabwe Publications, Harare.

Marglin, S.A. (1991) Alternative agriculture: A systems of knowledge approach. In: H.J. Tillmann, H. Albrecht, M.A. Salas, M. Dhamotharan & E. Gottschalk (Eds), *Proceedings of the International Workshop: Agricultural Knowledge Systems and the Role of Extension*, 21–24 May 1991, pp. 105–26. Institut für Agrarsociologie, landwirtschafliche Beratung und angewandte Psychologie, Hohenheim.

Marion, R. (1999) *The Edge of Organization. Chaos and Complexity Theories of Formal Social Systems*. Sage Publications, Thousand Oaks.

Marsden, D. & P. Oakley (1990) *Evaluating Social Development Projects*. Oxfam Development Guidelines, No. 5. Oxfam, Oxford.

Marsh, S. & D. Pannell (1998) The changing relationship between private and public sector agricultural extension in Australia. *Rural Society*, **8**, 133–49.

Martijn, C. (1995) *Language, Judgement and Attitude Change*. Published doctoral dissertation. University of Amsterdam, Amsterdam.

Martijn, C. & M.A. Koelen (1999) Persuasieve communicatie. In: C.M.J. Van Woerkum & R.C.F. Van Meegeren (Eds), *Basisboek Communicatie en Verandering*, pp. 78–104. Boom, Amsterdam.

Marwick, M. (1974) Is science a form of witchcraft? *New Scientist*, **63**, 578–81. Reprinted as: Witchcraft and the epistemology of science. In: M. Marwick (Ed.), *Witchcraft and Sorcery. Selected readings*, pp. 460–8. Penguin Books Ltd, Harmondsworth.

Marx, K. & F. Engels (1973) *Selected Works*, Vol. 1. Progress Publishers, Moscow.

Mastenbroek, W.F.G. (1997) *Onderhandelen.* Uitgeverij het Spectrum BV, Utrecht.

Maturana, H.R. (1980) Biology of cognition. In: H.R. Maturana & F.J. Varela (Eds), *Autopoiesis and Cognition: The Realization of the Living*, pp. 2–62. Reidel, Dordrecht.

Maturana, H.R. & F.J. Varela (1984) *The Tree of Knowledge: The Biological Roots of Human Understanding.* Shambala, Boston.

Maunder, H. (1973) *Agricultural Extension. A Reference Manual.* FAO, Rome.

Mayo, E. (1933) *The Human Problems of an Industrial Organization.* Macmillan, New York.

McDermott, J.K. (1987) Making extension effective: The role of extension/research linkages. In: W.M. Rivera & S.G. Schram (Eds), *Agricultural Extension World Wide. Issues, Practices and Emerging Priorities*, pp. 89–99. Croom Helm, New York.

McLuhan, M. (1964) *Understanding Media: The Extensions of Man.* The New American Library, New York.

McQuail, D. (1994) *Mass Communication Theory.* Sage, London.

Ménard, C. (1995) Markets as institutions versus organizations as markets? Disentangling some fundamental concepts. *Journal of Economic Behaviour and Organization*, **28**, 161–82.

Merriam, S.B. & R.S. Caffarella (1999) *Learning in Adulthood. A Comprehensive Guide.* Jossey-Bass Publishers, San Francisco.

Merton, R.K. (1957) *Social Theory and Social Structure.* Free Press, Glencoe.

Messick D.M. & M.B. Brewer (1983) Solving social dilemmas. A review. In: L. Wheeler & P. Shaver (Eds), *Review of Personality and Social Psychology*, **4**, 11–44.

Mikkelsen, B. (1995) *Methods for Development Work and Research. A Guide for Practitioners.* Sage Publications, New Delhi.

Mintzberg, H. (1979) *The Structuring of Organizations.* Prentice-Hall, Englewood Cliffs.

Mitchell, J.C. (1969) The concept and use of social networks. In: J.C. Mitchell (Ed.), *Social Networks in Urban Situations.* Manchester University Press, Manchester.

Mohr, L.B. (1992) *Impact Analysis for Program Evaluation.* Sage, Newbury Park, CA.

Mollinga, P. & J. Mooij (1989) *Cracking the Code: Towards a Conceptualization of the Social Content of Technical Artefacts.* Technology Policy Group Occasional Paper No. 18. Open University, Milton Keynes.

Morgan, G. (1998) *Images of Organization. The Executive Edition.* Berret-Koehler Publishers, San Francisco.

Mosse, D. (1995) Local institutions and power. The history and practice of community management of tank irrigation systems in India. In: N. Nelson & S. Wright (Eds), *Power and Participatory Development. Theory and Practice*, pp. 144–56. Intermediate Technology Publications, London.

Mutimukuru, T. (2000) *Facilitation or dictation? A case analysis of Agritex's role in the planning, implementation and management of Gwarada Scheme, in Zimunya Communal Lands, Zimbabwe.* MSc thesis. Wageningen University.

Mutimukuru, T. & C. Leeuwis (2003) Process leadership and natural resource management in Zimunya communal area. In: H.A.J. Moll, C. Leeuwis, E. Manzungu & L.F. Vincent (Eds), *Agrarian Institutions Between Policies and Local Action: Experiences from Zimbabwe.* Weaver Press, Harare.

Nagel, U.J. (1980) Institutionalisation of knowledge flows: an analysis of the extension role of two agricultural universities in India. Special issue of the *Quarterly Journal of International Agriculture*, **30**, DLG Verlag, Frankfurt.

Nelkin, D. (1989) Communicating technological risk: The social construction of risk perception. *Annual Reviews Public Health*, **10**, 95–113.

Nelson, N. & S. Wright (1995) *Power and Participatory Development. Theory and Practice.* Intermediate Technology Publications, London.

Nitsch, U. (1990) Computers and the nature of farm management. *Knowledge in Society: The International Journal of Knowledge Transfer*, **3**, 67–75.

Nonaka, I. & H. Takeuchi (1995) *The Knowledge Creating Company: How Japanese Companies Create the Dynamics of Innovation.* Oxford University Press, Oxford.

Nooij, A.T.J. (1990) *Sociale methodiek: Normatieve en Beschrijvende Methodiek in Grondvormen.* Stenfert Kroese, Leiden.

ODA (1995) *Guidance note on how to do stakeholder analysis of aid projects and programmes.* Overseas Development Administration, Social Development Department, London.

Oerlemans, N., J. Proost & J. Rauwhorst (1997) Farmers' study groups in the Netherlands. In: L. Veldhuizen, A. Waters-Bayer & R. Ramirez (Eds), *Farmers' Research in Practice. Lessons From the Field*, pp. 263–79. Intermediate Technology Publications, London.

Oerlemans, N., E. Van Well & C. Leeuwis (2002) *Diversiteit in Doelgroepen. Naar Gerichte Communicatie over Mineralenmanagement in de Melkveehouderij.* Agro Management Tools Rapport Nr. 11. AMT, Wageningen.

Ofman, D.D. (1992) *Bezieling en Kwaliteit in Organisaties.* Servire, Cothen.

Okali, C., J. Sumberg & J. Farrington (1994) *Farmer Participatory Research. Rhetoric and Reality.* Intermediate Technology Publications, London.

O'Keefe, D.J. (1990) *Persuasion: Theory and Research.* Sage Publications, London.

Oomkes, F.R. (1986) *Communicatieleer. Een Inleiding.* Boom, Meppel.

Ostrom, E. (1990) *Governing the Commons. The Evolution of Institutions for Collective Action.* Cambridge University Press, Cambridge.

Owen, H. (1997) *Open Space Technology: A Users's Guide.* Berrett-Koehler Publishers, San Francisco.

Pacheco, P. (2000) *Participation meets the market. Organisation dynamics and brokerage in the Atlantic zone of Costa Rica.* MSc thesis,Wageningen University.

Parsons, T. (1951) *The Social System.* Free Press, Glencoe.

Paul, S. (1986) *Community Participation in Development Projects. The Worldbank Experience.* Paper presented at the Economic Development Institute workshop on community participation, Washington.

Pepper, G.L. (1995) *Communicating in Organizations. A Cultural Approach.* McGraw-Hill, New York.

Pettigrew, A. (1973) *The Politics of Organizational Decision Making.* Tavistock, London.

Petty, R.E. & J.T. Cacioppo (1986) The elaboration likelihood model of persuasion. In: L. Berkowitz (Ed.), *Advances in Experimental Social Psychology*, **19**, 123–205. Academic Press, New York.

PID (1989) *An Introduction to Participatory Rural Appraisal for Rural Resources Management.* Programme for International Development, Clark University, Worcester MA.

Pijnenburg, B. (2003) *Keeping it vague. Discourses and practices of participation in rural Mozambique.* Doctoral thesis. Wageningen University, Wageningen.

Poate, D. (1993) More 'M' less 'E'. *Rural Extension Bulletin I*, **4**, 9–14.

Polanyi, M. (1958) The stability of scientific theories against experience. In: M. Polanyi (Ed.), *Personal Knowledge*, pp. 286–94. University of Chicago Press, Chicago. Reprinted in: M. Marwick (1974) *Witchcraft and Sorcery*, pp. 452–9. Penguin Books, Harmondsworth.

Potter, J. & M. Wetherell (1987) *Discourse and Social Psychology: Beyond Attitudes and Behaviour.* Sage, London.

Pottier, J. (Ed.) (1993) *Practising Development. Social Science Perspectives.* Routledge, London.

Pretty, J.N. (1994) Alternative systems of inquiry for a sustainable agriculture. *IDS bulletin,* **25**, 39–48.

Pretty, J.N. & R. Chambers (1994) Towards a learning paradigm: new professionalism and institutions for agriculture. In: I. Scoones & J. Thompson (Eds), *Beyond Farmer First. Rural People's Knowledge, Agricultural Research and Extension Practice,* pp. 182–202. Intermediate Technology Publications, London.

Pretty, J.N., I. Guijt, J. Thompson & I. Scoones (1995) *Participatory Learning and Action. A trainer's guide.* International Institute for Environment and Development (IIED), London.

Prigogine, I. & I. Stengers (1990) *Orde uit Chaos: Een Nieuwe Dialoog Tussen de Mens en de Natuur.* Uitgeverij Bert Bakker, Amsterdam.

Pröpper, I.M.A.M. & H.J.M. Ter Braak (1996) Interactie in ontwikkeling. *Bestuurskunde,* **5**, 356–69.

Pruitt, D.G. & P.J. Carnevale (1993) *Negotiation in Social Conflict.* Open University Press, Buckingham.

Purcell, D.R. & J.R. Anderson (1997) *Agricultural Research and Extension: Achievements and Problems in National Systems.* A World Bank Operations Evaluation Study. World Bank, Washington.

Putnam, R.D. (1995) Bowling alone: America's declining social capital. *Journal of Democracy,* **6**, 65–78.

Rabbinge, R., C.A. Van Diepen, J. Dijsselbloem, G.J.H. De Koning, H.C. Van Latesteijn, E. Woltjer & J. Van Zijl (1994) Ground for choices: a scenario study on perspectives for rural areas in the European Community. In: L.O. Fresco, L. Stroosnijder, J. Bouma & H. Van Keulen (Eds), *The Future of the Land: Mobilising and Integrating Knowledge for Land Use Options,* pp. 95–121. John Wiley & Sons, Chichester.

Rafaeli, S. (1988) Interactivity. From new media to communication. In: R.P. Hawkins, J. Wiemann & S. Pingree (Eds), *Advancing Communication Science: Merging Mass and Interpersonal Processes,* pp. 110–34. Sage, Newbury Park.

Rahman, M.A. (1993) *People's Self-development. Perspectives on Participatory Action Research.* Zed Books, London.

Rambaldi, G. & J. Callosa-Tarr (2002) *Participatory 3-dimensional Modelling. Guiding Principles and Applications.* ASEAN Regional Centre for Biodiversity Conservation. Los Baños, Philippines.

Reijntjes, C., M. Minderhoud-Jones & P. Laban (1999) *Leisa in Perspective. 15 years ILEIA.* Special ILEIA Newsletter. Centre for Research and Information on Low External Input and Sustainable Agriculture (ILEIA), Leusden.

Remmers, G.G.A. (1998) *Con Cojones y Maestría: Un Estudio Sociológico-agronómico Acerca del Desarrollo Rural Endógeno y Procesos de Localización en la Sierra de la Contraviesa (España).* Wageningen Studies on Heterogeneity and Relocalization 2. Circle for Rural European Studies. Thela Publishers, Amsterdam.

Richards, P. (1985) *Indigenous Agricultural Revolution.* Hutchinson, London.

Richards, P. (1994) Local knowledge formation and validation: The case of rice production in central Sierra Leone. In: I. Scoones & J. Thompson (Eds), *Beyond Farmer First. Rural People's Knowledge, Agricultural Research and Extension Practice,* pp. 165–70. Intermediate Technology Publications, London.

Rip, A. (1995) Introduction of new technology: Making use of recent insights from sociology and economics of technology. *Technology Analysis & Strategic Management,* **7**, 417–31.

Ritzer, G. (1993) *The McDonaldization of Society: An Investigation into the Changing Character of Contemporary Social Life.* Pine Forge, Thousand Oaks.

Rivera, W.M. (2000) The changing nature of agricultural information and the conflictive global developments shaping extension. *The Journal of Agricultural Education and Extension*, **7**, 31–41.

Rivera, W.M. & D.J. Gustafson (Eds) (1991) *Agricultural Extension: Worldwide Institutional Evolution and Forces for Change*. Elsevier Science Publishers, Amsterdam.

Rivera, W.M. & W. Zijp (Eds) (2002) *Contracting for Agricultural Extension. International Case Studies and Emerging Practices*. CABI Publishing, Wallingford.

Roep, D. (2000) *Vernieuwend Werken. Sporen van Vermogen en Onvermogen*. Circle for Rural European Studies. Wageningen University, Wageningen.

Roep, D., J.D. Van der Ploeg & C. Leeuwis (1991) *Zicht op Duurzaamheid en Kontinuïteit: Bedrijfsstijlen in de Achterhoek*. Wageningen Agricultural University, Wageningen.

Roethlisberger, F.J. & W.J. Dickson (1961) *Management and the Worker*. Harvard University Press, Cambridge.

Rogers, C.R. (1962) The interpersonal relationship: the core of guidance. *Harvard Educational Review*, **32**, 416–529.

Rogers, E.M. (1962) *Diffusion of Innovations*, 1st edn. Free Press, New York.

Rogers, E.M. (1983) *Diffusion of Innovations*, 3rd edn. Free Press, New York.

Rogers, E.M. (1995) *Diffusion of Innovations*, 4th edn. Free Press, New York.

Rogers, E.M. & D.L. Kincaid (1981) *Communication Networks. Toward a New Paradigm for Research*. Free Press, New York.

Röling, N.G. (1988) *Extension Science: Information Systems in Agricultural Development*. Cambridge University Press, Cambridge.

Röling, N.G. (1989) *The Agricultural Research-technology Transfer Interface: A Knowledge Systems Perspective*. International Service for National Agricultural Research (ISNAR), The Hague.

Röling, N.G. (1992) The emergence of knowledge systems thinking: A changing perception of relationships among innovation, knowledge process and configuration. *Knowledge and Policy: The International Journal of Knowledge Transfer and Utilization*, **5**, 42–64.

Röling, N.G. (1994a) Platforms for decision-making about eco-systems. In: L.O. Fresco, L. Stroosnijder, J. Bouma & H. Van Keulen (Eds), *The Future of the Land: Mobilising and Integrating Knowledge for Land Use Options*, pp. 386–93. John Wiley & Sons, Chichester.

Röling, N.G. (1994b) Voorlichting en innovatie. In: N.G. Röling, D. Kuiper & R. Janmaat (Eds), *Basisboek Voorlichtingskunde*, pp. 275–94. Boom, Amsterdam.

Röling, N.G. (1996) Towards and interactive agricultural science. *European Journal of Agricultural Education and Extension*, **2**, 35–48.

Röling, N.G. (1999) Modelling the soft side of land: the potential of multi-agent systems. In: C. Leeuwis (Ed.), *Integral Design: Innovation in Agriculture and Resource Management*, pp. 73–97. Mansholt Institute, Wageningen.

Röling, N.G. (2000) *Gateway to the Global Garden: Beta/Gamma Science for Dealing with Ecological Rationality*. Eighth Annual Hopper Lecture, 24 October 2000. University of Guelph.

Röling, N.G. (2002) Beyond the aggregation of individual preferences. Moving from multiple to distributed cognition in resource dilemmas. In: C. Leeuwis & R. Pyburn (Eds), *Wheelbarrows Full of Frogs. Social Learning in Rural Resources Management*, pp. 25–47. Royal Van Gorcum, Assen.

Röling, N.G. & P.G.H. Engel (1990) IT from a knowledge systems perspective: Concepts and issues. *Knowledge in Society: The International Journal of Knowledge Transfer*, **3**, 6–18.

Röling, N.G. & A. Groot (1999) Het onmaakbare van innovatie. In: C.M.J. Van Woerkum & R.C.F. Van Meegeren (Eds), *Basisboek Communicatie en Verandering*, pp. 30–58. Boom, Amsterdam.

Röling, N.G. & J. Jiggins (1998) The ecological knowledge system. In: N.G. Röling & M.A.E. Wagemakers (Eds), *Facilitating Sustainable Agriculture. Participatory Learning and Adaptive Management in Times of Environmental Uncertainty*, pp. 283–307. Cambridge University Press, Cambridge.

Röling, N.G., J. Jiggins & C. Leeuwis (1998) *Treadmill Success and Failure: The Challenge for FSR/E*. Proceedings of the 15th International Symposium of the Association for Farming Systems Research-Extension, 'Going Beyond the Farm Boundary', Pretoria, 29 November to 4 December 1998, pp. 860–6.

Röling, N.G. & D. Kuiper (1994) Wat is voorlichting? In: N.G. Röling, D. Kuiper & R. Janmaat (Eds), *Basisboek Voorlichtingskunde*, pp. 17–36. Boom, Amsterdam.

Röling, N.G., D. Kuiper & R. Janmaat (Eds) (1994) *Basisboek Voorlichtingskunde*. Boom, Amsterdam.

Röling, N.G. & C. Leeuwis (2001) Strange bedfellows: how knowledge systems came Longer and why they will never be Long. In: P. Hebinck & G. Verschoor (Eds), *Resonances and Dissonances in Development. Actors, Networks and Cultural Repertoires*, pp. 47–64. Royal van Gorcum, Assen.

Röling, N.G. & E. Van de Fliert (1994) Transforming extension for sustainable agriculture: the case of integrated pest management in rice in Indonesia. *Agriculture and Human Values*, **11**, 96–108.

Röling, N.G. & M.A.E. Wagemakers (Eds) (1998) *Facilitating Sustainable Agriculture. Participatory Learning and Adaptive Management in Times of Environmental Uncertainty*. Cambridge University Press, Cambridge.

Rossi, P.H. & H.E. Freeman (1993) *Evaluation: A Systematic Approach*, 5th edn. Sage, Newbury Park.

Rossing, W.A.H., J.E. Jansma, F.J. De Ruijter & J. Schans (1997) Operationalizing sustainability: exploring options for environmentally friendlier flower bulb production systems. *European Journal of Plant Pathology*, **103**, 217–34.

Rossing, W.H.A., M.K. Van Ittersum, H.F.M. Ten Berge & C. Leeuwis (1999) Designing land use options and policies. Fostering co-operation between Kasparov and Deep Blue? In: C. Leeuwis (Ed.), *Integral Design: Innovation in Agriculture and Resource Management*, pp. 49–72. Mansholt Institute, Wageningen.

Rotmans, J., R. Kemp & M.B.A. Van Asselt (2001) More evolution than revolution: transition management in public policy. *Foresight*, **3**, 15–31.

Rousseau, J.J. (1968, originally 1762) *The Social Contract*, (Trans. M. Cranston). Penguin Books Ltd, Harmondsworth.

Ryan, B. & N.C. Gross (1943) The diffusion of hybrid seed corn in two Iowa communities. *Rural Sociology*, **8**, 15–24.

Sadomba, W.Z. (1996) *Use of proverbs and taboos as oral archives of indigenous knowledge: A participatory method for studying indigenous knowledge systems*. Unpublished paper.

Sadomba, W.Z. (1999) *The impact of settler colonialism on indigenous agricultural knowledge in Zimbabwe: fusion, confusion or negation?* MSc thesis, Wageningen Agricultural University, Wageningen.

Scarborough, V., S. Killough, D.A. Johnson & J. Farrington (Eds) (1997) *Farmer-led Extension. Concepts and Practices*. Intermediate Technology Publications, London.

Schein, E.H. (1992) *Organizational Culture and Leadership*, 2nd edn. Jossey-Bass Publishers, San Francisco.

Schutz, A. & T. Luckmann (1974) *The Structures of the Life-world*. Heinemann Educational Books, London.

Scoones, I. & J. Thompson (Eds) (1994) *Beyond Farmer First. Rural People's Knowledge, Agricultural Research and Extension Practice*. Intermediate Technology Publications, London.

Scott, J.C. (1998) *Seeing Like a State: How Certain Schemes to Improve the Human Condition Have Failed.* Yale University Press, New Haven.

Searle, J.R. (1969) *Speech Acts: An Essay in the Philosophy of Language.* Cambridge University Press, Cambridge.

Senge, P.M. (1993) *The Fifth Discipline. The Art and Practice of the Learning Organization.* Century Business, London.

Senge, P.M. (1996) The leader's new work. Building learning organizations. In: K. Starkey (Ed.), *How Organizations Learn?* pp. 288–315. International Thomson Business Press, London.

Shambare, E. (2000) *Collective management of resources: A case study of Chamakareva dam, irrigation and tourism development in Dora Ward of Mutare District.* Doctoral thesis, Wageningen University.

Shapiro, C. & H.R. Varian (1999) *Information Rules. A Strategic Guide to the Network Economy.* Harvard Business School Press, Cambridge.

Sherif, C.W., M. Sherif & R.E. Nebergall (1965) *Attitude and Attitude Change. The Social Judgement-involvement Approach.* W.B. Saunders, Philadelphia.

Shields, D. (1993) What is the logical framework? *Rural Extension Bulletin 1,* **4**, 15–20.

Simon, H.A. (1976) *Administrative Behaviour: A Study of Decision-making Processes in Administrative Organization*, 3rd edn. Free Press, New York.

Singhal, A. & E.M. Rogers (1999) *Entertainment Education.* Lawrence Erlbaum Associates, Mahwah.

Slovic, P. (1987) Perception of risk. *Science,* **236**, 280–5.

Smits, R. (2000) *Innovatie in de Universiteit.* Inaugurele rede. Universiteit Utrecht, Utrecht.

Sorrentino, R.M. & C.J.R. Roney (2000) *The Uncertain Mind: Individual Differences in Facing the Unknown.* University of Cardiff, Miles Hewstone.

Stappers, J.G., A.D. Reijnders & W.A.J. Möller (1997) *De Werking van Massamedia. Een Overzicht van Inzichten.* Uitgeverij de Arbeiderspers, Amsterdam.

Stiglitz, J. (2002) *Globalization and its Discontents.* W.W. Norton & Company, New York.

Stolzenbach, A. (1994) Learning by improvization: farmers' experimentation in Mali. In: I. Scoones & J. Thompson (Eds), *Beyond Farmer First. Rural People's Knowledge, Agricultural Research and Extension Practice*, pp. 155–9. Intermediate Technology Publications, London.

Stolzenbach, A.F.V. & C. Leeuwis (1996) *Leren van de Mineralenbalans.* Onderzoeksverslag 146, DOBI-rapport 4. Landbouw-Economisch Instituut, Den Haag.

Strauss, A.L. & J.M. Corbin (1990) *Basics of Qualitative Research. Grounded Theory Procedures and Techniques.* Sage Publications, Newbury Park.

Stuiver, M., C. Leeuwis & J.D. Van der Ploeg (2003) The force of experience: farmers' knowledge and sustainable innovations in agriculture. In: J.S.C. Wiskerke & J.D. Van der Ploeg (Eds), *Seeds of Transition. Essays on Novelties, Niches and Regimes in Agriculture.* Royal Van Gorcum, Assen.

Sulaiman, V.R. & A. Hall (2002) Beyond technology dissemination: reinventing agricultural extension. *Outlook on Agriculture,* **31**, 225–33.

Supe, S.V. (1983) *An Introduction to Extension Education.* Oxford and IBH Publishing, New Delhi.

Susskind, L. & J. Cruikshank (1987) *Breaking the Impasse: Consensual Approaches to Resolving Public Disputes.* Basic Books, New York.

Swanson, B.E., R.P. Bentz & A.J. Sofranko (Eds) (1997) *Improving Agricultural Extension. A Reference Manual.* FAO, Rome.

Swanson, B.E. & J.B. Claar (1984) The history and development of agricultural extension. In: B.E. Swanson (Ed.), *Agricultural Extension. A Reference Manual*, pp. 1–19. FAO, Rome.

Tatenhove, J. & P. Leroy (1995) Beleidsnetwerken: een kritische analyse. *Beleidswetenschap*, **8**, 128–45.

Taylor, F.W. (1947) *Scientific Management.* Harper & Row, New York.

Te Molder, H. (1995) *Discourse of Dilemmas: An Analysis of Government Communicators' Talk.* Published doctoral dissertation. Wageningen Agricultural University, Wageningen.

Te Molder, H.F.M. & C. Leeuwis (1998) Overheid en nieuwe media: de belofte van inter-activiteit. In: V. Frissen & H.F.M. Te Molder (Eds), *Van Forum tot Supermarkt? Consumenten en Burgers in de Informatiesamenleving*, pp. 71–82. Acco (Academisch Coöperatief c.v.), Leuven.

Ten Berge, H.F.M. & J.J.M. Riethoven (1997) Applications of a simple rice-nitrogen model. In: T. Ando et al. (Eds), *Plant Nutrition – For Sustainable Food Production and Environment*, pp. 793–8. Kluwer Academic Publishers, Dordrecht.

Te Velde, H., M.N.C. Aarts & C.M.J. Van Woerkum (2002) Dealing with ambivalence: farmers' and consumers' perceptions of animal welfare in livestock breeding. *Journal of Agricultural and Environmental Ethics*, **15**, 203–19.

Thomas, A.R. & M. Lockett (1979) Marxism and systems research: values in practical action. In: R.F. Ericson (Ed.), *Improving the Human Condition: Quality and Stability in Social Systems.* Proceedings of the silver anniversary international meeting of the Society for General Systems Research (SGSR), 1979 August 20–24, London. pp. 284–93. SGSR, London.

Thompson, M., R. Ellis & A. Wildavsky (1990) *Cultural Theory.* Westview Press, Boulder.

Thrupp, L.A., B. Cabarle & A. Zazueta (1994) Participatory methods and political processes: linking grassroots actions and policy-making for sustainable development in Latin America. In: I. Scoones & J. Thompson (Eds), *Beyond Farmer First. Rural People's Knowledge, Agricultural Research and Extension Practice*, pp. 170–7. Intermediate Technology Publications, London.

Torres, R.T., H.S. Preskill & M.E. Piontek (1996) *Evaluation Strategies for Communicating and Reporting. Enhancing Learning in Organizations.* Sage, Thousand Oaks.

Turkle, S. (1995) *Life on Screen: Identity in the Age of the Internet.* Simon & Schuster, New York.

Turner, G. & J. Shepherd (1999) A method in search of a theory: peer education and health promotion. *Health Education Research: Theory and Practice*, **14**, 235–47.

Uitdewilligen, E.A.W., P. Van Meegeren & C. Martijn (1993) *De Perceptie, Acceptatie en Communicatie van Milieurisico's.* Ministerie van Volkshuisvesting, Ruimtelijke Ordening en Milieubeheer, Den Haag.

Ulrich, W. (1988) Systems thinking, systems practice, and practical philosophy: A program of research. *Systems Practice*, **1**, 137–63.

Umali, D.L. & L. Schwarz (1994) *Public and Private Agricultural Extension: Beyond Traditional Boundaries.* World Bank Technical Paper 247. World Bank, Washington.

Uphoff, N. (1989) *Participatory Evaluations of Participatory Development: A Scheme for Measuring and Monitoring Local Capacity.* Cornell University, New York.

Uphoff, N. (2000) Understanding social capital: Learning from the analysis and experience of participation. In: P. Dasgupta & I. Serageldin (Eds) (2000) *Social Capital; A Multifaceted Perspective*, pp. 215–49. World Bank, Washington.

Van Arkel, M. & A. Versteeg (1997) *Participation in local development: Experiences from a participatory rural appraisal regarding livestock production in Nianeni area, Machakos, Kenya.* MSc thesis, Wageningen Agricultural University.

Van Beek, P.G.H. (1991) The Queensland dairy AKIS: A systems approach to the management of research and extension. In: D. Kuiper & N.G. Röling (Eds), *The Edited Proceedings of the European Seminar on Knowledge Management and Information Technology*, pp. 30–44. Wageningen Agricultural University, Wageningen.

Van de Fliert, E. (1993) *Integrated Pest Management: Farmer Field Schools Generate Sustainable Practices. A Case Study in Central Java Evaluating IPM Training.* Wageningen Agricultural University papers 93–3, Wageningen.

Van den Ban, A.W. (1963) *Boer en Landbouwvoorlichting: De Communicatie van Nieuwe Landbouwmethoden.* Pudoc, Wageningen.

Van den Ban, A.W. (1974) *Inleiding tot de Voorlichtingskunde.* Boom, Meppel.

Van den Ban, A.W. (1997) Successful agricultural extension agencies are learning organizations. In: R.K. Samanta & S.K. Arora (Eds), *Management of Agricultural Extension in Global Perspective*, pp. 47–77. B.R. Publishing Corporation, Delhi.

Van den Ban, A.W. (2000) *Different Ways of Financing Agricultural Extension.* AgREN: Agricultural Research and Extension Network Paper 106b. Overseas Development Institute (ODI), London.

Van den Ban, A.W. (2002) Poverty alleviation among farmers. The role of knowledge. In: C. Leeuwis & R. Pyburn (Eds), *Wheelbarrows Full of Frogs. Social Learning in Rural Resource Management*, pp. 183–96. Royal Van Gorcum, Assen.

Van den Ban, A.W. & H.S. Hawkins (1988) *Agricultural Extension.* Longman Scientific & Technical, Burnt Mill.

Van den Ban, A.W. & H.S. Hawkins (1996) *Agricultural Extension*, 2nd edn. Blackwell Science, Oxford.

Van den Ban, A.W., H.S. Hawkins, J.H.A.M. Brouwers & C.A.M. Boon (1994) *Vulgarisation Rurale en Afrique.* Karthala, Paris.

Van den Hamsvoort, C., H. Hillebrand & L. de Savornin Lohman (1999) Vermarkting van natuur en landschap. In: H. Korevaar, A. Van der Werf & M.J.M. Oomes (Eds), *Meervoudig duurzaam Landgebruik: Van Visie naar Realisatie, Verslag van een Symposium*, pp. 139–48. AB-DLO, Wageningen.

Van der Ploeg, J.D. (1987) *De Verwetenschappelijking van de Landbouwbeoefening.* Wageningen Studies in Sociology Nr. 21. Wageningen Agricultural University, Wageningen.

Van der Ploeg, J.D. (1990) *Labor, Markets, and Agricultural Production.* Westview Press, Boulder.

Van der Ploeg, J.D. (1991) *Landbouw als Mensenwerk: Arbeid en Technologie in de Agrarische Ontwikkeling.* Coutinho, Muiderberg.

Van der Ploeg, J.D. (1994) Styles of farming: an introductory note on concepts and methodology. In: J.D. Van der Ploeg & A. Long (Eds), *Born from Within: Practices and Perspectives of Endogenous Development*, pp. 7–30. Van Gorcum, Assen.

Van der Ploeg, J.D. (1999) *De Virtuele Boer.* Van Gorcum, Assen.

Van der Veen, J. & P. Glasbergen (1992) De consensusbenadering; Verkenning van een innovatieve werkvorm om regionale milieuconflicten te doorbreken. *Bestuurskunde*, **1**, 228–37.

Van der Werf, W., C. Leeuwis & W.A.H. Rossing (1999) Quality of modelling for integrated crop management: issues for discussion. In: P.S. Wagemakers, W. Van der Werf & Ph. Blaise (Eds), *Proceedings of the Fifth International Symposium on Computer Modelling in Fruit Research and Orchard Management*, Wageningen, 28–31 July, 1998. Published as: *IOBC/WPRS Bulletin, Bulletin OILB/SROP*, **22**/*Acta Horticulturae*, **499**, 151–9.

Van Deursen, S. (2000) *Wie betaalt, bepaalt. Een kwalitatief onderzoek naar de gevolgen van invoering van marktwerking in het kennisnetwerk van de Nederlandse glastuinbouwsector.* MSc thesis, Wageningen University, Wageningen.

Van Dusseldorp, D. (1990) Planned development via projects: its necessity, limitations and possible improvements. *Sociologia Ruralis*, **30**, 336–52.

Van Gent, B. & J. Katus (Eds) (1980) *Voorlichting, Theorieën, Werkwijzen en Terreinen.* Alphen aan den Rijn, Samsom.

Van Ginniken, J. (1999) *Breinbevingen. Snelle Omslagen in Opinie en Communicatie.* Boom, Amsterdam.

Van Meegeren, R.C.F. (1997) *Communicatie en Maatschappelijke Acceptatie van Milieubeleid. Een Onderzoek naar de Houding ten Aanzien van de 'Dure Afvalzak' in Barendrecht.* Published doctoral dissertation. Wageningen Agricultural University, Wageningen.

Van Meegeren, R.C.F (1999) De planning van instrumentele communicatie. In: C.M.J. Van Woerkum & R.C.F. Van Meegeren (Eds), *Basisboek Communicatie en Verandering,* pp. 226–47. Boom, Amsterdam.

Van Meegeren, R.C.F. & C. Leeuwis (1999) Towards an interactive design methodology: guidelines for communication. In: C. Leeuwis (Ed.), *Integral Design: Innovation in Agriculture and Resource Management,* pp. 205–17. Mansholt Institute, Wageningen.

Van Riel, C.B.M. (1992) *Identiteit en Imago: Een Inleiding in de Corporate Communication.* Academic Service, Schoonhoven.

Van Schoubroeck, F.H.J. (1999) *Learning to Fight a Fly: Developing Citrus IPM in Bhutan.* Published doctoral dissertation. Wageningen University, Wageningen.

Van Schoubroeck, F. & C. Leeuwis (1999) Enhancing social cognition for combating the Chinese citrus fly in Bhutan. In: C. Leeuwis (Ed.), *Integral Design: Innovation in Agriculture and Resource Management,* pp. 145–71. Mansholt Institute, Wageningen.

Van Twist, M.J.W. & L. Schaap (1991) Introduction to autopoiesis theory and autopoietic steering. In: R.J. in't Veld, L. Schaap, C.J.A.M. Termeer & M.J.W. van Twist (Eds), *Autopoiesis and Configuration Theory: New Approaches to Societal Steering,* pp. 31–44. Kluwer Academic Publishers, Dordrecht.

Van Veldhuizen, L., A. Waters-Bayer & H. De Zeeuw (1997) *Developing Technology with Farmers. A Trainer's Guide for Participatory Learning.* Zed Books, London.

Van Velsen, J. (1967) The extended-case method and situational analysis. In: A.L. Epstein (Ed.), *The Craft of Social Anthropology,* pp. 129–49. Tavistock Publications – Social Science Paperbacks, London.

Van Woerkum, C.M.J. (1982) *Voorlichtingskunde en massacommunicatie: Het Werkplan van de Massamediale Voorlichting.* Published doctoral dissertation. Wageningen Agricultural University, Wageningen.

Van Woerkum, C.M.J. (1990a) Het instrumentele nut van voorlichting in beleidsprocessen. *Massacommunicatie,* **18,** 263–78.

Van Woerkum, C.M.J. (1990b) Het besturen van kennissystemen: Waar zit het stuur? *Massacommunicatie,* **18,** 99–116.

Van Woerkum, C.M.J. (1991) De emotionele benadering in de voorlichting. *Massacommunicatie,* **19,** 265–77.

Van Woerkum, C.M.J. (1994) Massamediale voorlichting. In: N.G. Röling, D. Kuiper & R. Janmaat (Eds), *Basisboek Voorlichtingskunde,* pp. 101–26. Boom, Amsterdam.

Van Woerkum, C.M.J. (1997) *Communicatie en Interactieve Beleidsvorming.* Bohn Stafleu Van Loghum, Houten.

Van Woerkum, C.M.J. (1999) Massamediale communicatie. In: C.M.J. Van Woerkum & R.C.F. Van Meegeren (Eds), *Basisboek Communicatie en Verandering,* pp. 146–57. Boom, Amsterdam.

Van Woerkum, C.M.J. (2002) Orality in environmental planning. *European Environment,* **12,** 160–72.

Van Woerkum, C.M.J. & M.N.C. Aarts (2002) *Wat Maakt het Verschil. Over de Waarde van Pluriformiteit in Interactieve Beleidsprocessen.* Innovatie Netwerk Groene Ruimte en Agrocluster, Den Haag.

Van Woerkum, C.M.J., D. Kuiper & E. Bos (1999) *Communicatie en Innovatie: Een Inleiding.* Samsom, Alphen aan den Rijn.

Van Woerkum, C.M.J. & R.C.F. Van Meegeren (Eds) (1999) *Basisboek Communicatie en Verandering.* Boom, Amsterdam.

Veldboer, L. (1996) *De Inspraak Voorbij; Ervaringen van Burgers en Lokale Bestuurders met Nieuwe Vormen van Overleg.* Instituut voor Publiek en Politiek, Amsterdam.

Vereijken, P. (1997) A methodical way of prototyping integrated and ecological arable farming systems (I/EAFS) in interaction with pilot farms. *European Journal of Agronomy,* 7, 235–50.

Verkaik, A.P. & N.A. Dijkveld-Stol (1989) *Commercialisering van Kennis en het Functioneren van het Landbouwkennissysteem.* NRLO-rapport nr. 89/32. National Council for Agricultural Research (NRLO), Den Haag.

Verkaik, A.P., J.M. Rutten & N.A. Dijkveld Stol (1997) *Vitaliteit van Agrosector en Landbouwkennissysteem.* NRLO-Rapport nr. 97/1. National Council for Agricultural Research (NRLO), Den Haag.

Vermeulen, W.J., J.F.M. Van der Waals, H. Ernste & P. Glasbergen (1997) *Duurzaamheid als Uitdaging. De Afweging van Ecologische en Maatschappelijke Risico's in Confrontatie en Dialoog.* Wetenschappelijke Raad voor het Regeringsbeleid, Voorstudies en achtergronden (V101). Sdu Uitgevers, Den Haag.

Verplanken, B. (1989) *Persuasive Communication of Technological Risks: A Test of the Elaboration Likelihood Model.* Published doctoral dissertation. Rijksuniversiteit Leiden, Leiden.

Vickers, G. (1983) *Human Systems are Different.* Harper & Row, London.

Vijverberg, A.J. (1997) *Glastuinbouw in Ontwikkeling. Beschouwingen over de Sector en de Beïnvloeding er van door de Wetenschap.* Eburon, Delft.

Vonk, R. (1990) *Prototyping: The Effective Use of CASE Technology.* Prentice-Hall, London.

Wagemans, M.C.H. (1987) *Voor de Verandering: Een op Ervaringen Gebaseerde Studie naar de Spanning tussen de Theorie en Praktijk van het Besturen.* Published doctoral dissertation. Wageningen Agricultural University, Wageningen.

Wagemans, M.C.H. (1998) *Geregeld Mis. Gedachten over Zingeving en Reductie Binnen het Publieke Domein.* Eburon, Delft.

Wagemans, M.C.H. (2002) Institutional conditions for transformations. A plea for policy making from the perspective of constructivism. In: C. Leeuwis & R. Pyburn (Eds), *Wheelbarrows Full of Frogs. Social Learning in Rural Resource Management,* pp. 245–58. Royal Van Gorcum, Assen.

Wagemans, M.C.H. & Boerma, J. (1998) The implementation of nature policy in the Netherlands: platforms designed to fail. In: N.G. Röling & M.A.E. Wagemakers (Eds), *Facilitating Sustainable Agriculture. Participatory Learning and Adaptive Management in Times of Environmental Uncertainty,* pp. 250–71. Cambridge University Press, Cambridge.

Wallerstein, I.M. (1974) *The Modern World-system: Capitalist Agriculture and the Origins of the European World-economy in the Sixteenth Century.* Academic Press, New York.

Warren, D.M. (1991) The role of indiginous knowledge in facilitating a participatory approach to agricultural extension. In: H.J. Tillmann, H. Albrecht, M.A. Salas, M. Dhamotharan & E. Gottschalk (Eds), *Proceedings of the International Workshop: Agricultural Knowledge Systems and the Role of Extension,* 1991 May 21–24, pp. 161–77. Institut für Agrarsociologie, landwirtschafliche Beratung und angewandte Psychologie, Höhenheim.

Watzlawick, P., J. Helminck Beavin & D.D. Jackson (1967) *Pragmatics of Human Communication.* W.W. Norton & Company, New York.

Weber, M. (1947) *The Theory of Social and Economic Organization.* The Free Press, New York.

Weber, M. (1968) *Economy and Society: An Outline of Interpretative Sociology.* Bedminster Press, New York.

Webler, T. & O. Renn (1995) A brief primer on participation: philosophy and practice. In: O. Renn, T. Webler & P. Wiedemann (Eds), *Fairness and Competence in Citizen Participation. Evaluating Models for Environmental Discourse*, pp. 17–33. Kluwer Academic Publishers, Dordrecht.

Weggeman, M. (2000) *Kennismanagement: de Praktijk.* Scriptum Management, Schiedam.

Weick, K.E. & F. Westley (1999) Organizational Learning. Affirming an oxymoron. In: S.R. Clegg, C. Hardy & W.R. Nord (Eds), *Managing Organizations. Current Issues*, pp. 190–208. Sage Publications, London.

Weisbord, M.R. & S. Janoff (1995) *Future Search. An Action Guide to Finding Common Ground in Organizations and Communities.* Berrett-Koehler Publishers, San Francisco.

Wetherell, M. (1996) *Identities, Groups and Social Issues.* Sage Publications, London.

Williamson, O.E. (1998) *The Economic Institutions of Capitalism.* The Free Press, New York.

Wilson, M. (1991) Reducing the costs of public extension services. In: W.M. Rivera & D.J. Gustafson (Eds), *Agricultural Extension: Worldwide Institutional Evolution and Forces for Change*, pp. 13–21. Elsevier Science Publishers, Amsterdam.

Windahl, S., B.H. Signitzer & J.T. Olson (1992) *Using Communication Theory: An Introduction to Planned Communication.* Sage, London.

Winograd, T. & C.F. Flores (1986) *Understanding Computers and Cognition: A New Foundation for Design.* Ablex Publishing Corporation, Norwood.

Wipfler, C., M. Spittel, S. Maalim Hamad & A. van Huis (1998) *IPM in development.* Unpublished paper. Plant Protection Semrice (PPS), Pemba.

Wiskerke, J.S.C. & J.D. Van der Ploeg (Eds) (2003) *Seeds of Transition. Essays on Novelties, Niches and Regimes in Agriculture.* Royal Van Gorcum, Assen.

Wittgenstein, L. (1969) *The Blue and Brown Books: Preliminary Studies for the 'Philosophical Investigations'.* Basil Blackwell, Oxford.

Woodhill, J. (2002) Sustainability, social learning and the democratic imperative. Lessons from the Australian Landcare movement. In: C. Leeuwis & R. Pyburn (Eds), *Wheelbarrows Full of Frogs. Social Learning in Rural Resource Management*, pp. 317–31. Royal Van Gorcum, Assen.

World Bank (1997) *World Development Report 1997.* Oxford University Press, New York.

World Bank (1998) *World Development Report 1998/1999. Knowledge for Development.* Oxford University Press, New York.

World Bank (2000) *World Development Report 2000/2001. Attacking Poverty.* Oxford University Press, New York.

WRR (1992) *Ground for Choices. Four Perspectives for the Rural Areas of the European Community.* Reports to the Government Nr. 42. Netherlands Scientific Council for Government Policy (WRR), The Hague.

Wynne, B. (1996) May the sheep safely graze? A reflexive view of the expert-lay knowledge divide. In: S. Lash, B. Szerszynshi & B. Wynne (Eds), *Risk, Environment and Modernity. Towards a New Ecology*, pp. 44–83. Sage Publications, London.

Yin, R.K. (1994) *Case Study Research: Design and Methods*, 2nd edn. Sage, Thousand Oaks.

Zijp, W. (1998) *Unleashing the Potential. Changing the Way the World Bank Thinks about and Supports Agricultural Extension.* Discussion paper. World Bank, Washington.

Zimbardo, P.G. & M.R. Leippe (1991) *The Social Psychology of Attitude Change and Social Influence.* McGraw Hill, Boston.

Zuñiga Valerin, A.G. (1998) *Participatory methods in practice. An analysis and comparison of two development platforms (Basic Agricultural Centres) in Costa Rica.* MSc thesis. Wageningen Agricultural University, Wageningen.

Zuurbier, P.J.P. (1984) *De Besturing van de Landbouwvoorlichtingsdienst.* Published doctoral dissertation. Wageningen Agricultural University, Wageningen.

Index